Institute for Mathematics and its Applications
IMA

The **Institute for Mathematics and its Applications** was established by a grant from the National Science Foundation to the University of Minnesota in 1982. The IMA seeks to encourage the development and study of fresh mathematical concepts and questions of concern to the other sciences by bringing together mathematicians and scientists from diverse fields in an atmosphere that will stimulate discussion and collaboration.

The IMA Volumes are intended to involve the broader scientific community in this process.

Avner Friedman, Director
Willard Miller, Jr., Associate Director

* * * * * * * * * *

IMA PROGRAMS

1982-1983	**Statistical and Continuum Approaches to Phase Transition**
1983-1984	**Mathematical Models for the Economics of Decentralized Resource Allocation**
1984-1985	**Continuum Physics and Partial Differential Equations**
1985-1986	**Stochastic Differential Equations and Their Applications**
1986-1987	**Scientific Computation**
1987-1988	**Applied Combinatorics**
1988-1989	**Nonlinear Waves**
1989-1990	**Dynamical Systems and Their Applications**

* * * * * * * * * *

SPRINGER LECTURE NOTES FROM THE IMA:

The Mathematics and Physics of Disordered Media
Editors: Barry Hughes and Barry Ninham
(Lecture Notes in Math., Volume 1035, 1983)

Orienting Polymers
Editor: J.L. Ericksen
(Lecture Notes in Math., Volume 1063, 1984)

New Perspectives in Thermodynamics
Editor: James Serrin
(Springer-Verlag, 1986)

Models of Economic Dynamics
Editor: Hugo Sonnenschein
(Lecture Notes in Econ., Volume 264, 1986)

The IMA Volumes in Mathematics and Its Applications

Volume 19

Series Editors
Avner Friedman Willard Miller, Jr.

Dennis Stanton
Editor

Invariant Theory and Tableaux

Springer-Verlag
New York Heidelberg Berlin
London Paris Tokyo Hong Kong

Dennis Stanton
School of Mathematics
University of Minnesota
Minneapolis, MN 55455, USA

Invariant theory and tableaux / Dennis Stanton.
p. cm. — (IMA volumes in mathematics and its applications : 19)
Proceedings of the Workshop on Invariant Theory and Tableux held at the IMA on March 21–25, 1988.
ISBN 0-387-97170-X
1. Invariants—Congresses. 2. Representations of groups—Congresses. 3. Young tableaux—Congresses. I. Stanton, Dennis. II. Workshop on Invariant Theory and Tableaux (1988 : University of Minnesota) III. Series.
QA201.I58 1989
512.5—dc20 89-28842

Mathematical Subject Classification: 15A72, 20C30.

Printed on acid-free paper.

Camera-ready text prepared by the Institute for Mathematics and Its Applications.
Printed and bound by Edwards Brothers, Incorporated, Ann Arbor, Michigan.
Printed in the United States of America.

9 8 7 6 5 4 3 2 1

ISBN 0-387-97170-X Springer-Verlag New York Berlin Heidelberg
ISBN 3-540-97170-X Springer-Verlag Berlin Heidelberg New York

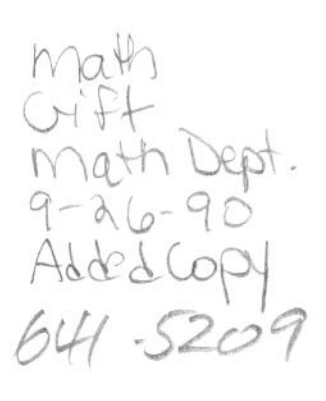

The IMA Volumes
in Mathematics and its Applications

Current Volumes:

Volume 1: Homogenization and Effective Moduli of Materials and Media
Editors: Jerry Ericksen, David Kinderlehrer, Robert Kohn, J.-L. Lions

Volume 2: Oscillation Theory, Computation, and Methods of Compensated Compactness
Editors: Constantine Dafermos, Jerry Ericksen, David Kinderlehrer, Marshall Slemrod

Volume 3: Metastability and Incompletely Posed Problems
Editors: Stuart Antman, Jerry Ericksen, David Kinderlehrer, Ingo Muller

Volume 4: Dynamical Problems in Continuum Physics
Editors: Jerry Bona, Constantine Dafermos, Jerry Ericksen, David Kinderlehrer

Volume 5: Theory and Applications of Liquid Crystals
Editors: Jerry Ericksen and David Kinderlehrer

Volume 6: Amorphous Polymers and Non-Newtonian Fluids
Editors: Constantine Dafermos, Jerry Ericksen, David Kinderlehrer

Volume 7: Random Media
Editor: George Papanicolaou

Volume 8: Percolation Theory and Ergodic Theory of Infinite Particle Systems
Editor: Harry Kesten

Volume 9: Hydrodynamic Behavior and Interacting Particle Systems
Editor: George Papanicolaou

Volume 10: Stochastic Differential Systems, Stochastic Control Theory and Applications
Editors: Wendell Fleming and Pierre-Louis Lions

Volume 11: Numerical Simulation in Oil Recovery
Editor: Mary Fanett Wheeler

Volume 12: Computational Fluid Dynamics and Reacting Gas Flows
Editors: Bjorn Engquist, M. Luskin, Andrew Majda

Volume 13: Numerical Algorithms for Parallel Computer Architectures
Editor: Martin H. Schultz

Volume 14: Mathematical Aspects of Scientific Software
Editor: J.R. Rice

Volume 15: Mathematical Frontiers in Computational Chemical Physics
Editor: D. Truhlar

Volume 16: Mathematics in Industrial Problems
by Avner Friedman

Volume 17: Applications of Combinatorics and Graph Theory to the Biological and Social Sciences
Editor: Fred Roberts

Volume 18: q-Series and Partitions
Editor: Dennis Stanton

Volume 19: Invariant Theory and Tableaux
Editor: Dennis Stanton

Volume 24: Mathematics in Industrial Problems, Part 2
by Avner Friedman

Forthcoming Volumes:

1987-1988: *Applied Combinatorics*
Coding Theory and Applications
Design Theory and Applications

Summer Program 1988: *Signal Processing*
Signal Processing (Part 1)
Signal Processing (Part 2)

1988-1989: *Nonlinear Waves*
Solitons in Physics and Mathematics
Solitons in Nonlinear Optics and Plasma Physics
Two Phase Waves in Fluidized Beds, Sedimenation, and Granular Flows
Nonlinear Evolution Equations that Change Type
Computer Aided Proofs in Analysis
Multidimensional Hyperbolic Problems and Computations (2 Volumes)
Microlocal Analysis and Nonlinear Waves

CONTENTS

FOREWORD

This IMA Volume in Mathematics and its Applications

Invariant Theory and Tableaux

is based on the proceedings of a workshop which was an integral part of the 1987-88 IMA program on APPLIED COMBINATORICS. We are grateful to the Scientific Committee: Victor Klee (Chairman), Daniel Kleitman, Dijen Ray-Chaudhuri and Dennis Stanton for planning and implementing an exciting and stimulating year-long program. We especially thank the Workshop Organizer, Dennis Stanton, for organizing a workshop which brought together many of the major figures in a variety of research fields in which invariant theory and tableaux are used.

Avner Friedman

Willard Miller, Jr.

PREFACE

This volume contains the Proceedings of the Workshop on Invariant Theory and Tableaux held at the IMA on March 21-25, 1988. Also included by invitation is a paper by Sheila Sundaram.

The papers by Rota and Sturmfels and Sagan are introductory papers on Invariant Theory and Tableaux (respectively), and can be read with a minimum of preparation. Kung, Olver, and Sundaram wrote survey papers which are also easily accessible to non-experts. The papers have been roughly grouped into four parts: Invariant Theory, Coxeter groups, enumeration, and tableaux.

I would like to thank the staff of the IMA, and its directors, Avner Friedman and Willard Miller, Jr., for providing a wonderful environment for the Workshop. Patricia Brick and Kaye Smith prepared the manuscripts.

Dennis Stanton

INTRODUCTION TO INVARIANT THEORY IN SUPERALGEBRAS

GIAN-CARLO ROTA* AND BERND STURMFELS†

Table of Contents

1. Introduction.

This paper is based on a series of five lectures presented by G.-C. Rota during a workshop on *"Invariant Theory and Tableaux"*, held in March 1988 at the Institute for Mathematics and its Applications, University of Minnesota, Minneapolis. These lectures provided an elementary introduction to the invariant theory of symmetric and skew-symmetric tensors, and they showed how the classical *symbolic method* ties in with some recent developments in combinatorics and invariant theory. Notes were taken and worked out in detail by B. Sturmfels.

The present exposition is essentially self-contained and requires no previous knowledge other than basic abstract algebra. It is intended to introduce the non-specialist to some fundamental problems and important applications of invariant theory, both classical and modern. We hope that this introduction will be useful to a large group of researchers and students, in particular those interested in combinatorics, computational algebra, geometry, and applications thereof. It will serve

*Department of Mathematics, Massachusetts Institute of Technology, Cambridge, Mass. 02139, U.S.A.
†Department of Mathematics, Cornell University, Ithaca, New York 14853

readers with a further interest in invariant theory as a guide to original literature, mainly to the monograph on superalgebras by Grosshans, Rota & Stein [8].

This text is organized as follows. Chapter 2 deals with an important special case of symmetric tensors, namely homogeneous polynomials in two variables or *binary forms*. We will illustrate all fundamental ideas and prove the main general results for this special case.

The basic question in the invariant theory of binary forms can be phrased as follows: Characterize all algebraic and geometric properties of binary forms which are invariant under linear changes of variables!

This problem can be solved, at least in principle, by the so-called *symbolic method*. The symbolic method works as follows. The desired invariants form a subalgebra of the polynomial algebra freely generated by the coefficients of the given binary forms. This subalgebra is constructed as the image of the more familiar *bracket algebra* under a vector space homomorphism, called the *umbral operator*. An important tool for working in the bracket algebra is the *straightening algorithm*. This is a combinatorial normal form algorithm which rewrites any bracket expression as a linear combination of certain special bracket monomials, represented by rectangular *standard Young tableaux*.

In Chapter 3, which is the heart of this paper, these concepts will be generalized to the setting of arbitrary symmetric and skew-symmetric tensors. While complete proofs were given for binary forms, we will now restrict ourselves to stating and explaining the main theorems of the general theory.

A variety of examples is given throughout the exposition. Most of these examples are non-trivial, and they are chosen to include familiar invariants such as the resultant, the discriminant, the Hessian, and the Pfaffian, as well as important geometric applications such as planar quadrics, line complexes in projective 3-space, and Grassmann-Plücker relations.

The authors wish to thank the I.M.A. for their support and the very stimulating atmosphere during the workshop on *Invariant Theory and Tableaux* and the 1987/88 year on *Applied Combinatorics* respectively.

2. Invariant theory of binary forms.

The first chapter on binary forms is elementary and serves as an introduction to some fundamental ideas, problems and methods of classical invariant theory. It is based on the comprehensive exposition of J.P.S. Kung and G.C. Rota [11].

A *binary form of degree* n is a homogeneous polynomial in two variables

$$(1) \qquad f(x_1, x_2) \quad = \quad \sum_{k=0}^{n} \binom{n}{k} a_k \, x_1^k \, x_2^{n-k}$$

where the coefficients a_k are in a field K which is generally assumed to be infinite. Depending on the point of view, binary forms can be seen as special cases of symmetric tensors or as arrangements of points in projective 1-space. Let us start out with the classical interpretation of binary forms as homogenized univariate polynomials.

Given a univariate polynomial

$$(2) \qquad p(t) \quad = \quad a_n t^n \; + \; \binom{n}{n-1} a_{n-1} t^{n-1} \; + \; \ldots \; + \; \binom{n}{1} a_1 t \; + \; a_0,$$

we are interested in properties of p which remain invariant under the following three types of changes of variables

$$(3) \qquad t \mapsto t + c, \quad t \mapsto c \cdot t, \quad \text{and} \quad t \mapsto \frac{1}{t}, \qquad \text{where } c \in K \text{ is a constant.}$$

These transformations can be handled better if one uses homogeneous coordinates: Replacing t by x_1/x_2 in $p(t)$ and clearing denominators, we obtain the binary form $f(x_1, x_2)$ in (1). If we assume that K has characteristic zero, then the form (1) can be rewritten over the algebraic closure $\overline{K}$ of K as

$$(4) \qquad f(x_1, x_2) \quad = \quad (\mu_1 x_1 - \nu_1 x_2) \cdot (\mu_2 x_1 - \nu_2 x_2) \cdot \ldots \cdot (\mu_n x_1 - \nu_n x_2).$$

The vectors $(\mu_i, -\nu_i) \in \overline{K}^2$ are the *homogenized roots* of f.

We will be looking at three classes of invariant properties of univariate polynomials or binary forms.

- *General properties of the roots* (e.g. multiplicity of roots)
- *Special position properties of the roots.* (e.g. "the cross ratio of four roots equals 1")
- *Canonical forms* (e.g. can p be written as n-th power of a linear polynomial: $p(t) = (ct + d)^n$? Or $p(t) = (ct + d)^n + (et + f)^n$? What is the minimum number of terms required for such a sum of powers representation ?)

The group of linear fractional transformations generated by the transformations in (3) corresponds to the canonical action of the group $GL(K^2)$ of invertible 2×2-matrices on the set of binary forms. Every linear transformation $(c_{ij}) \in GL(K^2)$ induces a change of variables

$$(5) \qquad x_1 = c_{11}\bar{x}_1 + c_{12}\bar{x}_2, \qquad x_2 = c_{21}\bar{x}_1 + c_{22}\bar{x}_2$$

where $\det\begin{pmatrix} c_{11} & c_{12} \\ c_{21} & c_{22} \end{pmatrix} \neq 0$. Under this transformation the binary form $f(x_1, x_2)$ in (1) is mapped into

$$\begin{aligned} \bar{f}(\bar{x}_1, \bar{x}_2) &= \sum_{k=0}^{n} \binom{n}{k} a_k \, (c_{11}\bar{x}_1 + c_{12}\bar{x}_2)^k \, (c_{21}\bar{x}_1 + c_{22}\bar{x}_2)^{n-k} \\ &= \sum_{l=0}^{n} \binom{n}{l} \bar{a}_l \, \bar{x}_1^l \, \bar{x}_2^{n-l}. \end{aligned} \tag{6}$$

This equation expresses the new coefficients $\bar{a}_l$ as polynomial functions in the a_k and c_{ij}.

A polynomial $I(a_0, a_1, \ldots, a_n) = \mathcal{I}(f)$ is said to be an *invariant* of f provided

$$I(\bar{a}_0, \bar{a}_1, \ldots, \bar{a}_n) \quad = \quad \det(c_{ij})^g \cdot I(a_0, a_1, \ldots, a_n) \tag{7}$$

for every linear transformation (c_{ij}). If I is an invariant of f, then the equation $I(a_0, a_1, \ldots, a_n) = 0$ describes a *projective geometric property*. This means that its solution set $\{ f : \mathcal{I}(f) = 0 \}$ is invariant under the action of $GL(K^2)$.

However, not all projective properties arise from invariants. We need to slightly generalize the above definition. A polynomial $C(a_0, a_1, \ldots, a_n, x_1, x_2)$ is a *covariant* if

$$C(\bar{a}_0, \bar{a}_1, \ldots, \bar{a}_n, \bar{x}_1, \bar{x}_2) \quad = \quad \det(c_{ij})^g \cdot C(a_0, a_1, \ldots, a_n, x_1, x_2). \tag{8}$$

for every linear transformation (c_{ij}). *Gram's principle* states that an algebraic set is projectively invariant if and only if it can be expressed by the vanishing and non-vanishing of a finite set of covariants. See Dieudonné and Carrell [6, Section 2.8] for a precise formulation of Gram's theorem and its proof.

We summarize three basic questions in the invariant theory of binary forms.

- Determine as explicitly as possible all covariants of f, and, in particular, determine all invariants of f.
- Find their geometric significance.
- Conversely, given a geometric property, find the corresponding covariant.

The concept of invariants and covariants can be generalized in a straightforward manner to *joint invariants* and *joint covariants* of a set of binary forms. A joint invariant of k forms

(9)

$$f_1 = \sum_{l=0}^{n_1} \binom{n_1}{l} a_l \, x_1^l x_2^{n_1-l}, \quad f_2 = \sum_{l=0}^{n_2} \binom{n_2}{l} b_l \, x_1^l x_2^{n_2-l}, \quad \ldots, \quad f_k = \sum_{l=0}^{n_k} \binom{n_k}{l} c_l \, x_1^l x_2^{n_k-l}$$

is a polynomial

$$\mathfrak{I}(f_1, f_2, \ldots, f_k) \quad = \quad I(a_1, \ldots, a_{n_1}, b_1, \ldots, b_{n_2}, \ldots, c_1, \ldots, c_{n_k})$$

satisfying the condition (7) simultaneously for all forms.

EXAMPLE 2.1. *(The resultant of a 3-form and a 2-form)*

A joint invariant of the binary 3-form

$$f_1(x_1,x_2) \quad = \quad a_0\, x_2^3 + 3\, a_1\, x_1 x_2^2 + 3\, a_2 x_1^2 x_2 + a_3 x_1^3$$

and the binary 2-form

$$f_2(x_1,x_2) \quad = \quad b_0\, x_2^2 + 2\, b_1\, x_1 x_2 + b_2\, x_1^2$$

is given by the expression

$$\begin{aligned}\mathcal{R}(f_1,f_2) \;=\; R(a_0,a_1,a_2,a_3,b_0,b_1,b_2) \;=\; & \\ a_3^2 b_0^3 - 8a_0a_3b_1^3 + a_0^2b_2^3 + 6a_0a_3b_0b_1b_2 - 18a_1a_2b_0b_1b_2 + 9a_1^2b_0b_2^2 + 9a_2^2b_0^2b_2 & \\ - 6a_1a_3b_0^2b_2 - 6a_0a_2b_0b_2^2 - 6a_0a_1b_1b_2^2 - 6a_2a_3b_0^2b_1 + 12a_1a_3b_0b_1^2 + 12a_0a_2b_1^2b_2 &\end{aligned}$$

The invariant $\mathcal{R}(f_1,f_2)$ is the *resultant* of the f_1 and f_2. We have $\mathcal{R}(f_1,f_2) = 0$ if and only if f_1 and f_2 have a common root over $\overline{K}$. □

2.1. Joint invariants of linear forms.

In this section we shall work out in detail the simplest possible case. A complete description will be given for all joint invariants and all joint covariants of k linear binary forms

$$\begin{aligned} f_1(x_1,x_2) &= \alpha_1 x_1 + \alpha_2 x_2 \\ f_2(x_1,x_2) &= \beta_1 x_1 + \beta_2 x_2 \\ &\cdots \\ f_k(x_1,x_2) &= \gamma_1 x_1 + \gamma_2 x_2. \end{aligned} \tag{10}$$

Consider the coefficient vectors $\alpha := (\alpha_1,\alpha_2)$, $\beta := (\beta_1,\beta_2), \ldots, \gamma := (\gamma_1,\gamma_2)$. The *bracket* of a pair of vectors, which is the determinant of their coordinates, will be denoted as follows $[\alpha\,\beta] := \alpha_1\beta_2 - \alpha_2\beta_1$, $[\alpha\,\gamma] := \alpha_1\gamma_2 - \alpha_2\gamma_1, \ldots$. Using the standard rules of matrix multiplication, we find

$$[\bar{\alpha}\,\bar{\beta}] \quad = \quad \det(c_{ij}) \cdot [\alpha\,\beta]. \tag{11}$$

Therefore every bracket is a joint invariant of the k linear forms. Since the sum and the product of invariants are again invariants, any polynomial in the brackets $[\alpha\,\beta]$, $[\alpha\,\gamma]$, ..., $[\beta\,\gamma]$ is invariant, provided it is homogeneous.

The first fundamental theorem of invariant theory states that, conversely, every invariant is a polynomial function in the brackets.

FIRST FUNDAMENTAL THEOREM (FOR LINEAR BINARY FORMS).
A polynomial $I(\alpha_1, \alpha_2, \beta_1, \beta_2, \ldots, \gamma_1, \gamma_2)$ is a joint invariant of k linear forms if and only if I can be written as a polynomial in the brackets $[\alpha\,\beta]$, $[\alpha\,\gamma]$, $\ldots$, $[\beta\,\gamma]$.

Proof. The proof of this theorem is based on the so-called *straightening algorithm*. We fix a total order, say $\alpha < \beta < \ldots < \gamma$, on the alphabet $A = \{\alpha, \beta, \ldots, \gamma\}$. Bracket monomials over A will be represented by rectangular *tableaux*

$$T \quad = \quad \begin{bmatrix} \alpha & \beta \\ \gamma & \delta \\ \vdots & \vdots \\ \sigma & \tau \end{bmatrix} \quad := \quad [\alpha\,\beta][\gamma\,\delta]\ldots[\sigma\,\tau]. \tag{11}$$

A tableau T is *standard* if the rows of T are increasing and the columns of T are non-decreasing with respect to the specified order on A.

We will see that every tableaux can be written as a unique K-linear combination of standard tableaux. This normal form computation, called *straightening*, can be done using the following *syzygy* among the brackets

$$[\alpha\,\beta][\gamma\,\delta] \;+\; [\gamma\,\alpha][\beta\,\delta] \;+\; [\beta\,\gamma][\alpha\,\delta] \quad = \quad 0 \tag{12}$$

This identity, which is also known as the *Grassmann-Plücker relation*, follows easily from Cramer's rule. Consider the system of linear equations $\alpha_i = x\gamma_i + y\beta_i$, $i = 1, 2$. Solving for (x, y) and plugging in the solution yields

$$\alpha_i \quad = \quad \frac{[\beta\,\alpha]}{[\beta\,\gamma]}\gamma_i \;+\; \frac{[\alpha\,\gamma]}{[\beta\,\gamma]}\beta_i\,, \qquad i = 1, 2.$$

Clearing denominators we obtain the vector identity

$$[\alpha\,\beta]\,\gamma \;+\; [\gamma\,\alpha]\,\beta \;+\; [\beta\,\gamma]\,\alpha \quad = \quad 0.$$

Application of the linear functional $[\,\cdot\;\delta\,]$ implies (12).

The straightening algorithm works as follows. Given any non-zero tableau T, we first rewrite T in such a way that the rows of T are increasing and the first column of T is non-decreasing. Suppose that the resulting T is non-standard and that the first violation (from the top) occurs in the rows i and $i+1$. We denote these rows by $[\alpha\,\delta]$ and $[\beta\,\gamma]$ where $\alpha < \beta < \gamma < \delta$. We multiply the syzygy

$$[\alpha\,\delta][\beta\,\gamma] \;+\; [\alpha\,\beta][\gamma\,\delta] \;-\; [\alpha\gamma\,][\beta\,\delta]$$

by the brackets of T not corresponding to the rows i and $i+1$, and we subtract the resulting bracket polynomial from T. This process rewrites T as a linear combination of two tableaux both of which are lexicographically earlier than T. Iterative application of this procedure eventually terminates, and hence the standard tableaux span the bracket algebra as a K-vector space.

The fact that the standard tableaux are linearly independent can be shown by induction on the number of rows [11, Lemma 3.3]. Hence every bracket polynomial can be written uniquely as $\sum_i c_i T_i$ where $c_i \in K$ and the T_i are standard tableaux.

Now let $I(\alpha_1, \alpha_2, \beta_1, \ldots, \gamma_1, \gamma_2)$ be a joint invariant of (10). Using the abbreviations $e^{(1)} := (0,1)$ and $e^{(2)} := (-1,0)$, we get $\alpha_1 = [\alpha\, e^{(1)}]$ and $\alpha_2 = [\alpha\, e^{(2)}]$. With this we can rewrite

$$I = I\big([\alpha e^{(1)}], [\alpha e^{(2)}], [\beta e^{(1)}], \ldots, [\gamma e^{(1)}], [\gamma e^{(2)}]\big). \tag{13}$$

Consider a linear change of variables given by the transformation $\alpha_i \mapsto d_{i1}\alpha_1 + d_{i2}\alpha_2$, $i = 1, 2$. This is equivalent to the transformation $[\alpha\, e^{(1)}] \mapsto [\alpha\, u^{(1)}]$, $[\alpha\, e^{(2)}] \mapsto [\alpha\, u^{(2)}]$ where $u^{(1)} := (-d_{12}, d_{11})$ and $u^{(2)} := (-d_{22}, d_{21})$. Conversely, every linear change of variables can be expressed in this way for suitable $u^{(1)}$, $u^{(2)}$.

I being an invariant means that

$$\begin{aligned} (14) \qquad & I\big([\alpha u^{(1)}], [\alpha u^{(2)}], [\beta u^{(1)}], \ldots, [\gamma u^{(1)}], [\gamma u^{(2)}]\big) \\ = \; & [u^{(1)} u^{(2)}]^g \cdot I\big([\alpha e^{(1)}], [\alpha e^{(2)}], [\beta e^{(1)}], \ldots, [\gamma e^{(1)}], [\gamma e^{(2)}]\big) \end{aligned}$$

Now consider another change of variables $[\alpha\, e^{(1)}] \mapsto [\alpha\, v^{(1)}]$, $[\alpha\, e^{(2)}] \mapsto [\alpha\, v^{(2)}]$. Then (14) implies

$$\begin{aligned} (15) \qquad & [v^{(1)} v^{(2)}]^g \cdot I\big([\alpha u^{(1)}], [\alpha u^{(2)}], [\beta u^{(1)}], \ldots, [\gamma u^{(1)}], [\gamma u^{(2)}]\big) \\ = \; & [u^{(1)} u^{(2)}]^g \cdot I\big([\alpha v^{(1)}], [\alpha v^{(2)}], [\beta v^{(1)}], \ldots, [\gamma v^{(1)}], [\gamma v^{(2)}]\big) \end{aligned}$$

We choose the order $u^{(1)} < u^{(2)} < \alpha < \beta < \ldots < \gamma < v^{(1)} < v^{(2)}$ on the letters, and we apply the straightening algorithm to both sides of the equation (15). The left hand side straightens to an expression $\sum_i c_i \cdot T_i \cdot [v^{(1)} v^{(2)}]^g$ where the T_i are standard tableaux in the letters $u^{(1)} < u^{(2)} < \alpha < \beta < \ldots < \gamma$. The right hand side straightens to $\sum_j d_j \cdot [u^{(1)} u^{(2)}]^g \cdot T_j'$ where the T_j' are standard tableaux in the letters $\alpha < \beta < \ldots < \gamma < v^{(1)} < v^{(2)}$. Since every bracket expression is a <u>unique</u> linear combination of standard tableaux, the standard tableaux as well as their coefficients are the same in both expansions. Hence (15) equals

$$\sum_i c_i \cdot T_i'' \cdot [u^{(1)} u^{(2)}]^g \cdot [v^{(1)} v^{(2)}]^g \tag{16}$$

where the T_i'' are standard tableaux in $\alpha < \beta < \ldots < \gamma$. This implies

$$I\big([\alpha v^{(1)}], [\alpha v^{(2)}], [\beta v^{(1)}], \ldots, [\gamma v^{(1)}], [\gamma v^{(2)}]\big) = \sum_i c_i \cdot T_i'' \cdot [v^{(1)} v^{(2)}]^g. \tag{17}$$

Specializing $v^{(1)} := e^{(1)}$, $v^{(2)} := e^{(2)}$, we obtain

$$I(\alpha_1, \alpha_2, \beta_1, \ldots, \gamma_1, \gamma_2) = \pm \sum_i c_i \cdot T_i''. \tag{18}$$

Since the T_i'' are tableaux only containing the original letters $\alpha, \beta, \ldots, \gamma$, this completes the proof of the first fundamental theorem for k linear binary forms. □

2.2. The symbolic method.

The *First Fundamental Theorem* for linear forms, which was proved in the previous section, states that every joint invariant of k linear forms is a polynomial in the brackets $[\alpha\,\beta]$, $[\alpha\,\gamma]$, ... This result fulfills the program of invariant theory because the only geometric conditions on the projective line are of the form

$$\begin{array}{rcll} [\alpha\,\beta] & = & 0 & \text{(“two points agree”)} \\ \dfrac{[\alpha\,\beta][\gamma\,\delta]}{[\alpha\,\gamma][\beta\,\delta]} & = & ? & \text{(conditions on the cross ratios)} \end{array}$$

In this section we extend that result to arbitrary forms of degree n. To this end we develop the so-called *symbolic method.* This method has been criticized very much during the 19th century and early 20th century because of its lack of rigor. Developed by the Germans and some British mathematicians, this method was rejected by the French school, and, after the death of Alfred Young it was essentially abandoned. It is the objective of the present exposition to describe the symbolic method in mathematically rigorous terms.

Suppose we are given a fixed set of k binary forms $f_1(x_1, x_2), \ldots, f_k(x_1, x_2)$ of arbitrary degree. With each of these forms we associate an infinite alphabet A_i, and we set $A := A_1 \cup A_2 \cup \cdots \cup A_k$. Moreover, we assume that A is linearly ordered and that $A_i \cap A_j = \varnothing$ whenever $i \neq j$. A *symbol* $\alpha \in A$ is said to *belong to* f_i if $\alpha \in A_i$. Two symbols are called *equivalent* if they belong to the same form.

With every symbol α we associate a pair of independent transcendentals α_1, α_2, and we think of α as the vector $\alpha = (\alpha_1, \alpha_2)$. The polynomial ring $K[\{\alpha_1, \alpha_2 : \alpha \in A\}]$ freely generated by all these (infinitely many) transcendentals is called the *umbral space.*

DEFINITION. We define a linear functional U on the umbral space, called the *umbral operator*, as follows

(1) Suppose $\alpha \in A_i$ belongs to $f_i(x_1, x_2) = \sum_{k=0}^{n} \binom{n}{k} a_k x_1^k x_2^{n-k}$. Then we set

$$\langle\, U \mid \alpha_1^k \alpha_2^{n-k} \,\rangle \quad := \quad a_k$$

Note that for any equivalent $\beta \in A_i$ we also have $\langle\, U \mid \beta_1^k \beta_2^{n-k} \,\rangle = a_k$.

(2) Given $\alpha \in A_i$, we set

$$\langle\, U \mid \alpha_1^r \alpha_2^s \,\rangle \quad := \quad 0 \qquad \text{if} \quad r + s \neq n = \deg f_i$$

(3) Given any monomial m in the umbral space $K[\{\alpha_1, \alpha_2 : \alpha \in A\}]$, we can write

$$m \quad = \quad \prod_{\alpha \in A} m(\alpha)$$

where $m(\alpha) = \alpha_1^i \alpha_2^j$, and we define

$$\langle U \mid m \rangle \quad := \quad \prod_{\alpha \in A} \langle U \mid m(\alpha) \rangle$$

In short, the umbral operator U is multiplicative with respect to distinct symbols. This is well defined since the A_i are assumed to be mutually disjoint.

(4) Finally, we extend the umbral operator U to the entire umbral space by K-linearity.

(5) Later we will adjoin other variables x_i which will be algebraically independent transcendentals over the umbral space. For these extra variables we set

$$\langle U \mid x_i \rangle \quad := \quad x_i.$$

This completes the definition of the umbral operator U.

REMARK. *Every polynomial $\mathcal{P}(f_1, \ldots, f_k)$ in the coefficients of the forms f_i is in the image of the umbral operator. That is, we can write*

$$\mathcal{P}(f_1, \ldots, f_k) \quad = \quad \langle U \mid Q \rangle$$

for some polynomial Q in the umbral space.

Instead of giving a complete proof for this remark, we show in several examples how an *umbral representation* can be found.

EXAMPLE 2.2.

(i) How can we represent umbrally a_k^2, the square of a coefficient of the form f_i ? We need two distinct but equivalent letters α, β, and we obtain

$$\begin{aligned} a_k^2 \quad &= \quad \langle U \mid \alpha_1^k \alpha_2^{n-k} \beta_1^k \beta_2^{n-k} \rangle \\ &= \quad \langle U \mid \alpha_1^k \alpha_2^{n-k} \rangle \langle U \mid \beta_1^k \beta_2^{n-k} \rangle \end{aligned}$$

(ii) Observe that the form f can be viewed as a covariant of itself, i.e., it is an umbral image. We have the umbral representation

$$\begin{aligned} f(x_1, x_2) \quad &= \quad \langle U \mid (\alpha_1 x_1 + \alpha_2 x_2)^n \rangle \\ &= \quad \sum_{k=0}^{n} \binom{n}{k} \langle U \mid \alpha_1^k \alpha_2^{n-k} \rangle \, x_1^k \, x_2^{n-k} \\ &= \quad \sum_{k=0}^{n} \binom{n}{k} a_k \, x_1^k \, x_2^{n-k}. \end{aligned}$$

Suppose we have $\mathcal{P}(f_i) = \langle U \mid Q \rangle$ where $Q = Q(\alpha_1, \alpha_2, \ldots, \gamma_2)$ is a polynomial in the umbral space. What is the effect of a linear change of variables $f_i \mapsto \bar{f}_i$ on this umbral representation ?

$$\mathcal{P}(\bar{f}_i) \quad = \quad \langle U \mid ?? \rangle$$

A linear change of variables for f_i induces also a linear change of variables in the umbral space because U is a K-linear map. Hence this transformation can be described as in the previous section by

$$\alpha_1 \mapsto [\alpha\, u^{(1)}] \qquad \alpha_2 \mapsto [\alpha\, u^{(2)}] \qquad\qquad \text{where} \quad [u^{(1)}\, u^{(2)}] \neq 0.$$

We obtain

$$\mathcal{P}(\bar{f}_i) \quad = \quad \langle\, U \mid Q(\, [\alpha\, u^{(1)}], [\alpha\, u^{(2)}], \ldots, [\gamma\, u^{(2)}]\,)\,\rangle$$

This shows that the umbral evaluation of every bracket polynomial gives rise to an invariant. We shall see below that all invariants can be obtained in this way.

EXAMPLE 2.3. *(The discriminant of a quadratic binary form)*
Suppose we have a quadratic form $f_i(x_1, x_2) = a_0 x_1^2 + 2a_1 x_1 x_2 + a_2 x_2^2$. Taking two equivalent symbols $\alpha, \beta \in A_i$, we compute

$$\begin{aligned}
\langle\, U \mid [\alpha\, \beta]^2\,\rangle \quad &= \quad \langle U \mid (\alpha_1\beta_2 - \alpha_2\beta_1)^2\,\rangle \\
&= \quad \langle U \mid \alpha_1^2\beta_2^2 - 2\alpha_1\alpha_2\beta_1\beta_2 + \alpha_2^2\beta_1^2\,\rangle \\
&= \quad \langle U \mid \alpha_1^2\,\rangle\langle U \mid \beta_2^2\rangle \;-\; 2\langle U \mid \alpha_1\alpha_2\,\rangle\langle U \mid \beta_1\beta_2\,\rangle \;+\; \langle U \mid \alpha_2^2\,\rangle\langle U \mid \beta_1^2\,\rangle \\
&= \quad a_0 a_2 \;-\; 2a_1^2 \;+\; a_0 a_2 \\
&= \quad 2\,(a_0\, a_2 \;-\; a_1^2\,)
\end{aligned}$$

This invariant is the *discriminant* of the quadratic form f_i. It vanishes if and only if f_i has a double root. □

We can now state the

FIRST FUNDAMENTAL THEOREM (FOR BINARY FORMS).
Every invariant $\mathfrak{I}(f_i)$ *of a form* f_i *of degree* n *can be written as*

$$\mathfrak{I}(f_i) \quad = \quad \langle\, U \mid \textit{some bracket polynomial in umbral space} \,\rangle.$$

This theorem yields a straightforward yet impractical algorithm for computing a finite set of algebra generators for the ring of invariants. By the straightening algorithm as described in Section 2.1, we need only enumerate all standard tableaux with letters from A_i (e.g. lexicographically). The set of umbral evaluations of all standard tableaux is a generating set for the K-vector space of invariants. By Hilbert's famous basis theorem [9] we may terminate this process after a finite number of steps: the set of invariants is finitely generated as a K-algebra. For degree bounds in Hilbert's basis theorem see Dieudonné & Carrell [6] and Kempf [10].

The proof of the first fundamental theorem consists of two steps. First one shows the validity of the statement for multilinear invariants, and in a second step the general case will be reduced to the multilinear case. Here a joint invariant $I = \mathfrak{I}(f_1, \ldots, f_k)$ of k binary forms is called *multilinear* if I is a linear function in the coefficients of f_i for each $i = 1, 2, \ldots, k$.

LEMMA. *Suppose* $\mathfrak{I}(f_1, \ldots, f_k)$ *is a multilinear joint invariant of the* f_i*'s. Then* $\mathfrak{I}$ *is the umbral evaluation of some bracket polynomial in the umbral space.*

Proof of the lemma. By multilinearity, we can write

$$\mathfrak{I}(f_1, f_2, \ldots, f_k) \quad = \quad \langle\, U \,|\, Q(\alpha, \beta, \ldots, \gamma) \,\rangle$$

where α belongs to f_1, β belongs to f_2, ... and γ belongs to f_k. In other words, all occurring letters are pairwise non-equivalent.

We will first prove that the umbral operator is injective when restricted to the subspace of umbral polynomials with pairwise non-equivalent letters. To see this, we consider k forms which are perfect powers

$$p_1 = (\alpha_1 x_1 + \alpha_2 x_2)^{n_1} \;,\; p_2 = (\beta_1 x_1 + \beta_2 x_2)^{n_2} \;,\; \ldots \;,\; p_k = (\gamma_1 x_1 + \gamma_2 x_2)^{n_k}.$$

With respect to these particular forms the umbral operator acts like the identity map; e.g., we have

$$\langle\, U \,|\, \alpha_1^k \alpha_2^{n_1 - k} \,\rangle \quad = \quad \text{the } k\text{-th coefficient of the form } p_1 \quad = \quad \alpha_1^k \alpha_2^{n_1 - k}.$$

This implies that the invariant $\mathfrak{I}$ evaluated at the forms $p_1, p_2, \ldots, p_k$ is equal to its umbral representation Q. We have the identity

$$Q(\alpha, \beta, \ldots, \gamma) \quad = \quad \mathfrak{I}(p_1, p_2, \ldots, p_k).$$

This shows that

$$\mathcal{I}(f_1, f_2, \ldots, f_k) = \langle U \,|\, Q(\alpha, \beta, \ldots, \gamma) \rangle = 0 \iff Q(\alpha, \beta, \ldots, \gamma) = 0.$$

This injectivity property generalizes to umbral representations of covariants which involve only one letter per form.

Now the proof of the lemma is derived as follows. $\mathcal{I}$ being an invariant means

$$\mathcal{I}(\bar{f}_1, \ldots, \bar{f}_\ell) = [u^{(1)} u^{(2)}]^g \cdot \mathcal{I}(f_1, \ldots, f_k)$$

and

$$\mathcal{I}(\bar{\bar{f}}_1, \ldots, \bar{\bar{f}}_k) = [v^{(1)} v^{(2)}]^g \cdot \mathcal{I}(f_1, \ldots, f_k)$$

for linear transformations given by $u^{(i)}$ and $v^{(i)}$ respectively. By crossmultiplication we obtain

$$[v^{(1)} v^{(2)}]^g \cdot \mathcal{I}(\bar{f}_1, \ldots, \bar{f}_\ell) = [u^{(1)} u^{(2)}]^g \cdot \mathcal{I}(\bar{\bar{f}}_1, \ldots, \bar{\bar{f}}_k).$$

This identity can be rewritten using the umbral operator:

$$\begin{aligned} & [v^{(1)} v^{(2)}]^g \cdot \langle\, U \mid Q(\,[\alpha u^{(1)}], [\alpha u^{(2)}], \ldots, [\gamma u^{(2)}]\,)\,\rangle \\ = \; & [u^{(1)} u^{(2)}]^g \cdot \langle\, U \mid Q(\,[\alpha v^{(1)}], [\alpha v^{(2)}], \ldots, [\gamma v^{(2)}]\,)\,\rangle \end{aligned}$$

Notice that both sides of this equation contain only one symbol per form. Since $\langle\, U \,|\, \cdot\, \rangle$ is injective on the corresponding subspace in the umbral space, we obtain

$$\begin{aligned} & [v^{(1)} v^{(2)}]^g \cdot Q(\,[\alpha u^{(1)}], [\alpha u^{(2)}], \ldots, [\gamma u^{(2)}]\,) \\ = \; & [u^{(1)} u^{(2)}]^g \cdot Q(\,[\alpha v^{(1)}], [\alpha v^{(2)}], \ldots, [\gamma v^{(2)}]\,) \end{aligned}$$

By the arguments of the previous section, we conclude that Q can be written as a bracket polynomial. This completes the proof of the lemma. □

Now we are ready to consider the general case. Let $\mathcal{I}(f) = I(a_0, \ldots, a_n)$ be an invariant of degree k of one form f. Using a process called *restitution*, we will replace the degree k invariant $\mathcal{I}(f)$ of one form f by a multilinear invariant $Q_{11\ldots1}(f_1, \ldots, f_k)$ of as many as k forms $f_1, \ldots, f_k$.

To this end we consider k forms $f_1, \ldots, f_k$ all of degree k, and we introduce k (transcendental) "slack variables" $\lambda_1, \ldots, \lambda_k$. Then $\lambda_1 f_1 + \ldots + \lambda_k f_k$ is a form of degree k for which we can evaluate the given invariant $\mathcal{I}(\quad)$. We get

$$\mathfrak{I}(f_1, \ldots, f_k) := \mathcal{I}(\,\lambda_1 f_1 + \ldots + \lambda_k f_k\,).$$

Next we view that expression as a polynomial function in the slack variables λ_i and we compute its Taylor expansion

$$\mathfrak{I}(f_1, \ldots, f_k) = \sum_{i_1, \ldots, i_k} \frac{\lambda_1^{i_1} \ldots \lambda_k^{i_k}}{i_1! \ldots i_k!} Q_{i_1 \ldots i_k}(f_1, \ldots, f_k).$$

The specific coefficient $Q_{11\dots1}(f_1, \dots, f_k)$ in that Taylor expansion is a multilinear polynomial in the coefficients of the f_i which is invariant if and only if the original expression $\mathcal{I}(f)$ is invariant. By the previous lemma, $Q_{11\dots1}(f_1, \dots, f_k)$ has an umbral representation which is a bracket polynomial.

Finally observe that the specialization $f_i \mapsto f$ in that expression yields the original invariant up to a scalar factor

$$Q_{11\dots1}(f, f, \dots, f) \quad = \quad const \cdot \mathcal{I}(f).$$

Consequently also $\mathcal{I}(f)$ has an umbral representation which is a bracket polynomial. This completes the proof. □

The statement and the proof of the fundamental theorem generalize immediately to covariants. Let us close this section with an interesting example, showing how the restitution process can be used for efficiently computing an umbral representation of a non-multilinear covariant.

EXAMPLE 2.4. *(The Hessian of a cubic binary form)*

The ***Hessian*** of a cubic binary form

$$f(x_1, x_2) \quad = \quad a_0\, x_2^3 \,+\, 3a_1\, x_1 x_2^2 \,+\, 3a_2\, x_1^2 x_2 \,+\, a_3\, x_1^3$$

is a covariant of degree 2, given by the expressions

$$\begin{aligned} \mathcal{H}(f) \quad &:= \quad \det \begin{pmatrix} f_{xx} & f_{xy} \\ f_{xy} & f_{yy} \end{pmatrix} \quad = \\ H(a_0, a_1, a_2, a_3, x_1, x_2) \quad &= \quad 36\,[\,(a_1 a_3 - a_2^2)\, x_1^2 \,+\, (a_0 a_3 - a_1 a_2)\, x_1 x_2 \,+\, (a_0 a_2 - a_1^2)\, x_2^2\,]. \end{aligned}$$

The condition $\mathcal{H}(f) \equiv 0$ is equivalent to f being a perfect cube. This means geometrically that f has a triple root.

In order to compute an umbral representation for the Hessian, we consider the specific binary cubic

$$g(x_1, x_2) \quad := \quad \lambda_1 [\alpha\, x]^3 \,+\, \lambda_2 [\beta\, x]^3 \quad = \quad \lambda_1 (\alpha_1 x_1 - \alpha_2 x_2)^3 \,+\, \lambda_2 (\beta_1 x_1 - \beta_2 x_2)^3.$$

Expanding its Hessian in terms of the "slack variables" λ_1 and λ_2, we get

$$\mathcal{H}(g) \quad = \quad Q_{20}(\alpha_i, \beta_j, x_1, x_2)\lambda_1^2 \,+\, Q_{11}(\alpha_i, \beta_j, x_1, x_2)\lambda_1\lambda_2 \,+\, Q_{02}(\alpha_i, \beta_j, x_1, x_2)\lambda_2^2,$$

where $Q_{20} = Q_{02} = 0$, and

$$Q_{11}(a_i, b_j, x_1, x_2) \quad = \quad 36\,(\beta_2\alpha_1 - \alpha_2\beta_1)^2 (\alpha_1 x_1 - \alpha_2 x_2)(\beta_1 x_1 - \beta_2 x_2) \quad = \quad 36\,[\alpha\beta]^2 [\alpha x][\beta x].$$

Interpreting both α and β as symbols belonging to the original generic form f, this is (up to a scalar factor) the desired umbral representation of the Hessian, because the above computation combines the restitution process with a subsequent application of the "multilinear lemma". As is easily verified,

$$\langle\, U \mid 18\,[\alpha\beta]^2[\alpha x][\beta x]\,\rangle \quad = \quad \mathcal{H}(f). \qquad \square$$

2.3. Apolarity and canonical representations.

We saw in the previous section that for a given binary form

$$f(x_1, x_2) \quad = \quad \sum_{k=0}^{n} \binom{n}{k}\, a_k\, x_1^k\, x_2^{n-k},$$

or, more generally for a finite set of binary forms, every covariant can be written as

$$\mathcal{C}(F) \;=\; c(a_0, a_1, \ldots, a_n, x_1, x_2) \;=\; \langle U \mid \text{bracket polynomial in } [\alpha\,\beta], [\alpha\,\gamma], \ldots, [\alpha\, x], \ldots \rangle$$

In this section the previously developed invariant-theoretic concepts will be applied to the computation of canonical representations for binary forms. To this end we introduce a certain bilinear form and the notion of apolarity on the vector space of binary forms. Throughout Section 2.3 we assume that K has characteristic zero.

Given two forms f, g of the same degree n, we define their *apolar covariant* $\{f, g\}_{n,n}$ in terms of its umbral representation as

$$\{f, g\}_{n,n} \quad := \quad \langle\, U \mid [\alpha\,\beta]^n\,\rangle$$

where α belongs to f and β belongs to g

$$= \quad \langle U \mid (\alpha_1\,\beta_2 - \alpha_2\,\beta_1)^n \rangle$$

$$= \quad \sum_{k=0}^{n} \binom{n}{k} (-1)^k\, a_k\, b_{n-k}$$

What does the vanishing of the apolar covariant mean geometrically ? Suppose first that g is a perfect n-th power

$$g(x_1, x_2) \quad = \quad (c_1 x_1 + c_2 x_2)^n.$$

Then the equation

$$\{f, g\}_{n,n} \quad = \quad \sum_{k=0}^{n} \binom{n}{k} (-1)^k\, a_k\, c_1^{n-k}\, (-c_2)^k = 0$$

can be rewritten as $f(-c_2, c_1) = 0$. Or, equivalently, the linear form $(c_1 x_1 + c_2 x_2)$ is a factor of $f(x_1, x_2)$. Conversely, taking any linear factor of f and raising it to the n-th power we get a form

$$g \quad = \quad (c_1 x_1 + c_2 x_2)^n$$

which is apolar to f. Hence the linear subspace

$$f^\perp \quad := \quad \{\, g \,:\, \{f,g\}_{n,n} \,=\, 0\,\}$$

contains the n-th powers of linear factors of f. By a dimension argument [11, Prop. 5.1] one can see that the space $f^\perp$ is actually spanned by the n-th powers of linear factors of f.

Let us extend the notion of apolarity to forms of different degree. Assume that g is of degree $m \leq n = \deg f$. In that case we define

$$\{f,g\}_{n,m} \quad := \quad \{f,h\}_{n,n} \qquad \text{where} \quad h(x_1,x_2) \quad := \quad g(x_1,x_2)\cdot(y_1x_1+y_2x_2)^{n-m},$$

and where y_1, y_2 are new transcendentals.

This means that in general $\{f,g\}_{n,m}$ is not an invariant but only a covariant, depending on y_1 and y_2. We have the umbral representation

$$\{f,g\}_{n,m} \quad = \quad \langle\, U \mid [\alpha\,\beta]^m[\alpha\,y]^{n-m}\,\rangle.$$

Given a fixed form g of degree m and a fixed integer $n \geq m$, we would like to determine the K-dimension of the vector subspace $g^\perp$ of n-forms f apolar to g. As before, if a linear form $(c_1x_1 + c_2x_2)$ divides $g(x_1,x_2)$, then $(c_1x_1 + c_2x_2)^n$ is apolar to g and vice versa. This implies $\dim(g^\perp) = m$. A word of caution: this and some other results to be discussed in this section are only valid *generically*, i.e., they hold only for a Zariski-dense subset in the K-vector space of degree m forms. It is generally sufficient to assume that all occurring forms have distinct roots.

Given $m \leq n$ as before, we now fix a form f of degree n. What is the K-dimension of the subspace $f^\perp$ of m-forms g apolar to f ? This question is not so easy and requires a more detailed analysis. The covariant equation $\{f,g\}_{n,m} = 0$ translates into the system

$$\sum_{i=0}^{m}(-1)^{m-i}\binom{m}{i}\, a_{i+j}\, b_{m-i} \quad = \quad 0 \qquad \text{for} \quad j = 0,1,\ldots,n-m$$

of $n-m+1$ homogeneous linear equations in $m+1$ unknowns $b_0, b_1, \ldots, b_m$. The desired magnitude $\dim(f^\perp)$ equals the dimension of the solution space of the linear system, and hence $\dim(f^\perp) \geq m+1-(n-m+1) = 2m-n$. These two dimension counts constitute the main theorem on apolarity of binary forms.

Let us see how these results can be applied to canonical forms. We consider the special case $n = 2r+1$ and $m = r+1$. Given any fixed n-form f, we then have $\dim(f^\perp) \geq 2m - n = 1$. Hence there exists an m-form

$$g(x_1,x_2) \quad = \quad \sum_{k=0}^{m}\binom{m}{k}\, b_k\, x_1^k\, x_2^{m-k} \quad = \quad \prod_{i=1}^{m}(c_{i1}x_1 + c_{i2}x_2)$$

apolar to f. Moreover, we can compute g by solving the above linear system.

Changing our point of view, we have also that f is contained in the m-dimensional K-vector space $g^{\perp}$ which is generated by the perfect n-th powers $(c_{i1}x_1 + c_{i2}x_2)^n$. We obtain

$$f(x_1, x_2) \quad = \quad \lambda_1\,(c_{11}x_1{+}c_{12}x_2)^n + \lambda_2\,(c_{21}x_1{+}c_{22}x_2)^n + \ldots + \lambda_m\,(c_{m1}x_1{+}c_{m2}x_2)^n$$

for suitable $\lambda_i \in K$. This implies that the form f can be written as a sum of at most m n-th powers with respect to the algebraic closure $\overline{K}$ of K.

We have proved

SYLVESTER'S THEOREM. *Generically, every binary form f of odd degree $2r+1$ can be written as the sum of $r+1$ perfect powers.*

Let us specialize the situation even further. Let $n = 3, m = 2$, and fix a form f of degree 3. We solve the system of linear equations $\{f, g\}_{3,2} = 0$ for the coefficients of the binary 2-form $g = (c_1x_1 + c_2x_2)(d_1x_1 + d_2x_2)$ as above. Observe that (up to a scalar) $g(x_1, x_2)$ equals the Hessian $\mathcal{H}(f)$ of f. By the above discussion, we have a representation

$$f(x_1, x_2) \quad = \quad \lambda\,(c_1x_1 + c_2x_2)^3 \; + \; \mu\,(d_1x_1 + d_2x_2)^3 = 0.$$

Taking cube roots in the resulting identity provides a solution in terms of radicals for the general cubic equation. This method is not only efficient, but it can be remembered easily.

Let us finally consider a fixed form of degree of $n = 5$. Again, find g such that

$$\{f, g\}_{5,3} \quad = \quad 0,$$

by computing the coefficients of g from a system of linear equations. We obtain the representation

$$f(x_1, x_2) \quad = \quad (c_1x_1 + c_2x_2)^5 \; + \; (d_1x_1 + d_2x_2)^5 \; + \; (e_1x_1 + e_2x_2)^5,$$

which "suggests" that the general quintic equation is not solvable by radicals. □

It is possible and important to generalize the theory of apolarity to forms in more than 2 variables. Such a general theory of apolarity mimics the existence of roots even though there are no roots in higher degree.

3. Invariant theory of symmetric and skew-symmetric tensors.

This chapter generalizes the invariant theory of binary forms to higher-dimensional symmetric tensors and skew-symmetric tensors. The crucial problem in this generalization is to find a correct definition for the umbral space and the umbral operator. The *superalgebra of letterplaces* which will be defined in Section 3.3 is the key to the solution of that problem.

A detailed exposition of this subject and complete proofs are found in the monograph by Grosshans, Rota and Stein [8]. Here we restrict ourselves to describing the fundamental ideas and results and to supplying a variety of geometric examples.

3.1. Invariants of symmetric tensors and the bracket algebra.

A *symmetric tensor of degree n and step d* is a homogeneous polynomial

$$f(x_1, x_2, \ldots, x_d) \quad := \quad \sum \binom{n}{i_1 i_2 \ldots i_d} a_{i_1 i_2 \ldots i_d} \, x_1^{i_1} \, x_2^{i_2} \cdots x_d^{i_d}$$

of degree n in d variables. Sometimes f will also be referred to as an *n-form* in d variables. The general linear group $GL(K^d)$ acts by linear substitution on the $\binom{n+d-1}{n}$-dimensional K-vector space of n-forms in d variables.

A *covariant* of that action is a polynomial $\mathcal{C}(f)$ in the coefficients $a_{i_1 i_2 \ldots i_d}$ and the indeterminates x_j which is invariant (up to scalar multiples) under the action of $GL(K^d)$. If $\mathcal{C}(f)$ does not depend on the indeterminates x_j, then it is an *invariant.*

EXAMPLE 3.1. *(Decomposability of ternary 2-forms)*

The symmetric tensor of degree 2 and step 3

$$f(x_1, x_2, x_3) \quad = \quad a_{200} \, x_1^2 + a_{020} \, x_2^2 + a_{002} \, x_3^2 + 2 \, a_{110} \, x_1 x_2 + 2 \, a_{101} \, x_1 x_3 + 2 \, a_{011} \, x_2 x_3.$$

describes geometrically a quadric in projective 2-space. Consider its invariant

$$\begin{aligned} \mathcal{I}(f) \quad &:= \quad I(a_{200}, a_{020}, a_{002}, a_{110}, a_{101}, a_{011}) \\ &:= \quad a_{200} a_{020} a_{002} \; - \; a_{200} a_{011}^2 \; - \; a_{020} a_{101}^2 \; - \; a_{002} a_{110}^2 \; + \; 2 \, a_{110} a_{101} a_{011}. \end{aligned}$$

The geometric interpretation of that invariant is the following: $\mathcal{I}(f)$ vanishes if and only if the curve $f(x_1, x_2, x_3) = 0$ is the union of two lines in the projective plane. Algebraically, we have the decomposability criterion

$$\mathcal{I}(f) = 0 \quad \text{if and only if} \quad f(x_1, x_2, x_3) \; = \; (b_1 x_1 + b_2 x_2 + b_3 x_3)(c_1 x_1 + c_2 x_2 + c_3 x_3)$$

for some $b_1, b_2, b_3, c_1, c_2, c_3 \in \overline{K}$. □

As in the previous section it is our objective to describe as explicitly as possible all invariants and covariants of an arbitrary n-forms in d variables. In order to define the umbral space, we associate an infinite alphabet A with the form f. For every $\alpha \in A$ we introduce d independent transcendentals

$$(\alpha \,|\, 1), \; (\alpha \,|\, 2), \; \ldots \;, \; (\alpha \,|\, d).$$

The *umbral space* is defined to be the K-polynomial ring freely generated by all the variables $(\alpha \,|\, i)$, $\alpha \in A$, $i \in \{1, 2, \ldots, d\}$.

The *umbral operator* is a K-linear map from the umbral space onto the polynomial ring generated by the coefficients (and indeterminates) of the form f. It is defined by the formula

$$\langle\, U \mid (\alpha \,|\, 1)^{i_1} (\alpha \,|\, 2)^{i_2} \ldots (\alpha \,|\, d)^{i_d} \,\rangle \quad := \quad a_{i_1 i_2 \ldots i_d}$$

provided $i_1 + i_2 + \cdots + i_d = n$; otherwise the value of the above expression is 0. For arbitrary monomials the umbral operator is multiplicative with respect to different symbols:

$$\begin{aligned}
&\langle\, U \mid (\alpha \,|\, 1)^{i_1} (\alpha \,|\, 2)^{i_2} \ldots (\alpha \,|\, d)^{i_d} \, (\beta \,|\, 1)^{j_1} (\beta \,|\, 2)^{j_2} \ldots (\beta \,|\, d)^{j_d} \,\rangle \quad := \\
&\qquad \langle\, U \mid (\alpha \,|\, 1)^{i_1} (\alpha \,|\, 2)^{i_2} \ldots (\alpha \,|\, d)^{i_d} \,\rangle \langle\, U \mid (\beta \,|\, 1)^{j_1} (\beta \,|\, 2)^{j_2} \ldots (\beta \,|\, d)^{j_d} \,\rangle.
\end{aligned}$$

The umbral operator is extended to the entire umbral space by K-linearity.

Given $\alpha_1, \alpha_2, \ldots, \alpha_d \in A$, we define the *bracket*

$$\begin{aligned}
[\alpha_1, \alpha_2, \ldots, \alpha_d] \quad &:= \quad (-1)^d \cdot \det\big((\alpha_i \,|\, j) \big) \\
&= \quad (-1)^d \sum_\sigma (\operatorname{sign} \sigma) (\alpha_1 \,|\, \sigma_1) \ldots (\alpha_d \,|\, \sigma_d) \\
&= \quad (-1)^d \sum_\sigma (\operatorname{sign} \sigma) (\alpha_{\sigma_1} \,|\, 1) \ldots (\alpha_{\sigma_d} \,|\, d) \\
&= \quad (\, \alpha_1, \alpha_2, \ldots, \alpha_d \mid 1,\, 2,\, \ldots,\, d\,).
\end{aligned}$$

Finally, we adjoin algebraically independent transcendentals $(x \,|\, i)$ to the umbral space in order to describe covariants. For these additional variables the umbral operator is defined by the formula $\langle\, U \,|\, (x \,|\, i)^j \,\rangle := x_i{}^j$. The first fundamental theorem discussed earlier for binary forms generalizes to this setting

FIRST FUNDAMENTAL THEOREM (FOR SYMMETRIC TENSORS). *Every invariant $\mathcal{I}(f)$, and more generally, every covariant $\mathcal{C}(f)$ of a symmetric tensor f is the umbral evaluation of some bracket polynomial in the umbral space*

$$\mathcal{C}(f) \quad = \quad \langle\, U \mid \textit{some bracket polynomial in} \ldots [\alpha_1, \alpha_2, \ldots, \alpha_i, x_1, \ldots, x_{d-i}] \ldots \,\rangle$$

Conversely, the umbral evaluation of any bracket polynomial yields a covariant.

EXAMPLE 3.1. (CONTINUED).

We describe an umbral representation of the invariant $\mathcal{I}(f)$ which characterizes the decomposability of symmetric tensors of degree 2 and step 3. Given the alphabet $A = \{\alpha, \beta, \gamma, \ldots\}$, then the umbral space is the polynomial K-algebra freely generated by the indeterminates

$$(\alpha|1),\, (\alpha|2),\, (\alpha|3),\, (\beta|1),\, (\beta|2),\, (\beta|3),\, (\gamma|1),\, (\gamma|2),\, (\gamma|3) \,\ldots\,.$$

Consider the element $[\alpha,\beta,\gamma]^2$ in the subring of bracket polynomials in the umbral space. We obtain

$$[\alpha,\beta,\gamma]^2 = \left\{\det\begin{pmatrix}(\alpha|1) & (\beta|1) & (\gamma|1)\\ (\alpha|2) & (\beta|2) & (\gamma|2)\\ (\alpha|3) & (\beta|3) & (\gamma|3)\end{pmatrix}\right\}^2 =$$

$$\begin{aligned}
&(\alpha|1)^2(\beta|2)^2(\gamma|3)^2 + (\alpha|1)^2(\beta|3)^2(\gamma|2)^2 + (\alpha|2)^2(\beta|1)^2(\gamma|3)^2\\
&+ (\alpha|2)^2(\beta|3)^2(\gamma|1)^2 + (\alpha|3)^2(\beta|1)^2(\gamma|2)^2 + (\alpha|3)^2(\beta|2)^2(\gamma|1)^2\\
&- 2(\alpha|1)^2(\beta|2)(\beta|3)(\gamma|2)(\gamma|3) - 2(\alpha|2)(\alpha|3)(\beta|1)^2(\gamma|2)(\gamma|3) - 2(\alpha|2)(\alpha|3)(\beta|2)(\beta|3)(\gamma|1)^2\\
&- 2(\alpha|2)^2(\beta|1)(\beta|3)(\gamma|1)(\gamma|3) - 2(\alpha|1)(\alpha|2)(\beta|3)^2(\gamma|1)(\gamma|2) - 2(\alpha|1)(\alpha|3)(\beta|1)(\beta|3)(\gamma|2)^2\\
&- 2(\alpha|3)^2(\beta|1)(\beta|2)(\gamma|1)(\gamma|2) - 2(\alpha|1)(\alpha|3)(\beta|2)^2(\gamma|1)(\gamma|3) - 2(\alpha|1)(\alpha|2)(\beta|1)(\beta|2)(\gamma|3)^2\\
&+ 2(\alpha|1)(\alpha|2)(\beta|2)(\beta|3)(\gamma|1)(\gamma|3) + 2(\alpha|1)(\alpha|2)(\beta|1)(\beta|3)(\gamma|2)(\gamma|3)\\
&+ 2(\alpha|1)(\alpha|3)(\beta|1)(\beta|2)(\gamma|2)(\gamma|3) + 2(\alpha|1)(\alpha|3)(\beta|2)(\beta|3)(\gamma|1)(\gamma|2)\\
&+ 2(\alpha|2)(\alpha|3)(\beta|1)(\beta|2)(\gamma|1)(\gamma|3) + 2(\alpha|2)(\alpha|3)(\beta|1)(\beta|3)(\gamma|1)(\gamma|2).
\end{aligned}$$

From this expansion it can be read off easily that

$$\langle\, U \mid [\alpha,\beta,\gamma]^2\,\rangle = 6\,a_{200}a_{020}a_{002} - 6\,a_{200}a_{011}^2 - 6\,a_{020}a_{101}^2 - 6\,a_{002}a_{110}^2 + 12\,a_{110}a_{101}a_{011}.$$

This proves that $\frac{1}{6}[\alpha,\beta,\gamma]^2$ is an umbral representation of the invariant $\mathfrak{I}(f)$. □

For the subalgebra generated by the brackets in the umbral space there exists a normal form algorithm due to Young, Hodge and Littlewood, called the *straightening algorithm*. The straightening algorithm normalizes umbral representations of covariants, and, analogously to Section 2.2, it is used to prove the first fundamental theorem.

For a recent exposition of the straightening algorithm and its interpretation in terms of contemporary computational algebra we refer to the paper on *Gröbner bases and invariant theory* by Sturmfels and White [14]. Here we restrict ourselves to describing the possible normal forms resulting from the straightening algorithm.

We fix a total order on the alphabet $A = \{\alpha_i, \beta_j, \gamma_k, \dots\}$. As before, bracket monomials are represented by rectangular *tableaux*

$$\begin{bmatrix}\alpha_1 & \alpha_2 & \dots & \alpha_d\\ \beta_1 & \beta_2 & \dots & \beta_d\\ \vdots & \vdots & \dots & \vdots\\ & & \ddots & \\ \vdots & \vdots & \dots & \vdots\\ \gamma_1 & \gamma_2 & \dots & \gamma_d\end{bmatrix} := [\alpha_1,\alpha_2,\dots,\alpha_d][\beta_1,\beta_2,\dots,\beta_d]\cdots\cdots[\gamma_1,\gamma_2,\dots,\gamma_d]$$

A tableau is *standard* if its rows are increasing and its columns are nondecreasing, i.e. if $\alpha_i \le \beta_i \le \dots \le \gamma_i$. The result of the straightening algorithm is the following

THEOREM. *The standard tableaux form a $\mathbb{Z}$-basis for the space of bracket polynomials, that is, every tableaux can be written as a unique linear combination of standard tableaux. Moreover, all coefficients occurring in that expansion are integers.*

EXAMPLE 3.2. Consider the total order $\alpha_1 < \alpha_2 < \alpha_3 < \alpha_4 < \alpha_5 < \alpha_6 < \dots$ on the alphabet $A := \{\alpha_1, \alpha_2, \alpha_3, \dots\}$. Then the tableau

$$[\alpha_1, \alpha_4, \alpha_5][\alpha_2, \alpha_3, \alpha_6][\alpha_3, \alpha_4, \alpha_5] \quad = \quad \begin{bmatrix} \alpha_1 & \alpha_4 & \alpha_5 \\ \alpha_2 & \alpha_3 & \alpha_6 \\ \alpha_3 & \alpha_4 & \alpha_5 \end{bmatrix}$$

is not standard because its second and third column are not non-decreasing. Its unique expansion in terms of standard tableaux equals

$$\begin{bmatrix} \alpha_1 & \alpha_2 & \alpha_4 \\ \alpha_3 & \alpha_4 & \alpha_5 \\ \alpha_3 & \alpha_5 & \alpha_6 \end{bmatrix} - \begin{bmatrix} \alpha_1 & \alpha_2 & \alpha_5 \\ \alpha_3 & \alpha_4 & \alpha_5 \\ \alpha_3 & \alpha_4 & \alpha_6 \end{bmatrix} - \begin{bmatrix} \alpha_1 & \alpha_3 & \alpha_4 \\ \alpha_2 & \alpha_4 & \alpha_5 \\ \alpha_3 & \alpha_5 & \alpha_6 \end{bmatrix} + \begin{bmatrix} \alpha_1 & \alpha_3 & \alpha_5 \\ \alpha_2 & \alpha_4 & \alpha_5 \\ \alpha_3 & \alpha_4 & \alpha_6 \end{bmatrix}.$$

In other words, these two expressions will be equal when all brackets $[\alpha_i, \alpha_j, \alpha_k]$ are expanded as 3×3-determinants $\det \begin{pmatrix} (\alpha_i|1) & (\alpha_i|2) & (\alpha_i|3) \\ (\alpha_j|1) & (\alpha_j|2) & (\alpha_j|3) \\ (\alpha_k|1) & (\alpha_k|2) & (\alpha_k|3) \end{pmatrix}$. □

3.2. Invariants of skew-symmetric tensors and projective geometry.

So far we have dealt exclusively with homogeneous polynomials in the commutative polynomial ring $K[x_1, x_2, \dots, x_d]$ or, equivalently, with symmetric tensors. Indeed, this polynomial ring is naturally isomorphic to the symmetric algebra $\mathrm{Sym}_K(\{x_1, \dots, x_d\})$ freely generated by the variables x_i.

In this section we focus our attention on a non-commutative algebra which is of considerable importance in geometry. The *exterior algebra*

$$\mathrm{Ext}_K(\{x_1, \dots, x_d\})$$

generated by the variables x_i is the free associative K-algebra generated by the set $\{x_1, \dots, x_d\}$ subject to the relations

$$\text{for all } i, j \qquad \begin{cases} x_i x_j &= -x_j x_i \\ x_i^2 &= 0. \end{cases}$$

The homogeneous elements of the exterior algebra are called *skew-symmetric tensors*. Using the above relations, every skew-symmetric tensor can be written uniquely as a skew-symmetric homogeneous polynomial

$$t(x_1, x_2, \dots, x_d) \quad = \sum_{1 \le i_1 < i_2 < \dots < i_n \le d} A_{i_1 i_2 \dots i_n} \, x_{i_1} x_{i_2} \dots x_{i_n}.$$

The degree n of that polynomial is also called the *step* of the tensor t.

This polynomial representation implies that the K-vector space $\mathrm{Ext}_K^n(\{x_1,\ldots,x_d\})$ of skew-symmetric tensors of step n in d variables has dimension $\binom{d}{n}$. Hence the exterior algebra

$$\mathrm{Ext}_K(\{x_1,\ldots,x_d\}) \quad = \quad \bigoplus_{n=0}^{d} \mathrm{Ext}_K^n(\{x_1,\ldots,x_d\})$$

is a K-vector space of (finite) dimension 2^d.

The group $GL(K^d)$ of invertible $d \times d$-matrices acts on the K-vector space $\mathrm{Ext}_K^n(\{x_1,\ldots,x_d\})$ by linear substitution. As before, this action gives rise to *invariants* $\mathfrak{I}(t) = I(\ldots, A_{i_1 i_2 \ldots i_n}, \ldots)$ and *covariants* $\mathfrak{C}(t) = C(\ldots, A_{i_1 i_2 \ldots i_n}, \ldots, x_1, \ldots, x_d)$, and we have an umbral representation for the invariants and covariants.

First Fundamental Theorem (for skew-symmetric tensors).
Every invariant $\mathfrak{I}(t)$ of a skew-symmetric tensor t is the umbral evaluation

$$\mathfrak{I}(t) \quad = \quad \langle\, U \mid \textit{polynomial in } \ldots [\alpha_1, \alpha_2, \ldots, \alpha_d]\, \rangle$$

of some bracket polynomial and conversely. This generalizes to covariants.

However, there is a major distinction between this theorem and its analogues discussed earlier because both the umbral operator U and the umbral space are defined differently. In particular, the umbral space will be an algebra whose multiplication is (in general) not commutative any longer. As a consequence also the brackets occurring in umbral representations of skew-symmetric tensors behave quite differently from the brackets we introduced in the previous sections.

At this point we will not give the "correct" definitions because those will follow as a special case from the definition of the *superalgebra* in Section 3.3. Indeed, umbral representations for invariants of skew-symmetric tensors are one of the main motivations for studying superalgebras. In order to fully appreciate that theory it is therefore indispensable to get a better feeling for skew-symmetric tensors, their invariants and their geometric interpretation and application.

A skew-symmetric tensor t is *decomposable* if it factors into linear terms

$$t \quad = \quad (a_1x_1 + a_2x_2 + \cdots + a_dx_d)(b_1x_1 + b_2x_2 + \cdots + b_dx_d) \ldots (c_1x_1 + c_2x_2 + \cdots + c_dx_d).$$

As is well known, the coordinates of a decomposable skew-symmetric tensor are the *Plücker coordinates* of some d-dimensional linear subspace of K^n. The invariant theory of skew-symmetric tensors deals with the geometric properties of these subspaces, and consequently it is, in a sense, easier than the invariant theory of symmetric tensors.

In the remainder of this section we will study in some detail the smallest nontrivial case, namely invariants of skew-symmetric 2-forms in 4 variables. If decomposable, these forms will be interpreted as lines in projective 3-space.

EXAMPLE 3.4. *(Skew-symmetric tensors and lines in projective 3-space)* Consider a skew-symmetric tensor

$$t_1(x_1,x_2,x_3,x_4) \quad = \quad A_{12}x_1x_2 + A_{13}x_1x_3 + A_{14}x_1x_4 + A_{23}x_2x_3 + A_{24}x_2x_4 + A_{34}x_3x_4$$

in

$$\mathrm{Ext}_K^2(\{x_1,\ldots,x_4\}) \quad = \quad \mathrm{span}\,\{x_1x_2\,,\, x_1x_3\,,\, x_1x_4\,,\, x_2x_3\,,\, x_2x_4\,,\, x_3x_4\}.$$

The group $GL(K^4)$ of invertible 4×4-matrices acts on the 6-dimensional K-vector space $\mathrm{Ext}_K^2(\{x_1,\ldots,x_4\})$ by linear substitutions. Given a linear change of variables

$$\bar{x}_i \quad = \quad c_{i1}\,x_1 \,+\, c_{i2}\,x_2 \,+\, c_{i3}\,x_3 \,+\, c_{i4}\,x_4\,, \qquad i=1,\ldots,4,$$

where $(c_{ij}) \in GL(K^4)$, we obtain

$$\begin{aligned} &t_1(\bar{x}_1,\bar{x}_2,\bar{x}_3,\bar{x}_4) \quad = \\ &= A_{12}\cdot(c_{11}x_1+c_{12}x_2+c_{13}x_3+c_{14}x_4)\cdot(c_{21}x_1+c_{22}x_2+c_{23}x_3+c_{24}x_4) + \ldots\ldots \\ &= A_{12}\cdot\big[(c_{11}c_{22}-c_{12}c_{21})x_1x_2 + (c_{11}c_{23}-c_{13}c_{21})x_1x_3 + \ldots + (c_{13}c_{24}-c_{14}c_{23})x_3x_4)\big] + \ldots \\ &= \bar{A}_{12}\,x_1x_2 \,+\, \bar{A}_{13}\,x_1x_3 \,+\, \bar{A}_{14}\,x_1x_4 \,+\, \bar{A}_{23}\,x_2x_3 \,+\, \bar{A}_{24}\,x_2x_4 \,+\, \bar{A}_{34}\,x_3x_4 \\ &=: \bar{t}_1(x_1,x_2,x_3,x_4) \end{aligned}$$

where

$$\bar{A}_{kl} \quad := \quad \sum_{1\le i<j\le 4} (\,c_{ik}\,c_{jl} \,-\, c_{il}\,c_{jk}\,)\,A_{ij}.$$

The polynomial

$$\mathcal{I}(t_1) \quad = \quad I(A_{12},A_{13},A_{14},A_{23},A_{24},A_{34}) \quad = \quad A_{12}A_{34} \,-\, A_{13}A_{24} \,+\, A_{14}A_{23}.$$

is an invariant of the skew-symmetric tensor t_1 because

$$\mathcal{I}(\bar{t}_1) \quad = \quad \bar{A}_{12}\bar{A}_{34} \,-\, \bar{A}_{13}\bar{A}_{24} \,+\, \bar{A}_{14}\bar{A}_{23} \quad = \quad [\det(c_{ij})]\cdot I(t_1).$$

The *Grassmann condition* $\mathcal{I}(\bar{t}_1) = 0$ is equivalent to t_1 being decomposable into two linear factors

$$t_1(x_1,x_2,x_3,x_4) \quad = \quad (a_{11}x_1+a_{12}x_2+a_{13}x_3+a_{14}x_4)(a_{21}x_1+a_{22}x_2+a_{23}x_3+a_{24}x_4)$$

where $a_{ij}\in K$. In that case we have $A_{ij} = \det\begin{pmatrix} a_{1i} & a_{1j} \\ a_{2i} & a_{2j}\end{pmatrix}$ for all i,j, and t_1 is the Plücker coordinate vector of the row space of the 2×4-matrix (a_{ij}). We will interpret that 2-dimensional linear subspace in terms of homogeneous coordinates as a 1-dimensional (projective) subspace L_1 of projective 3-space over K.

Let

$$t_2(x_1,x_2,x_3,x_4) \;=\; B_{12}x_1x_2 \;+\; B_{13}x_1x_3 \;+\; B_{14}x_1x_4 \;+\; B_{23}x_2x_3 \;+\; B_{24}x_2x_4 \;+\; B_{34}x_3x_4$$

$$\ldots \qquad \ldots \qquad \ldots \qquad \ldots \qquad \ldots$$

$$t_6(x_1,x_2,x_3,x_4) \;=\; F_{12}x_1x_2 \;+\; F_{13}x_1x_3 \;+\; F_{14}x_1x_4 \;+\; F_{23}x_2x_3 \;+\; F_{24}x_2x_4 \;+\; F_{34}x_3x_4.$$

be five additional tensors. We will consider the following two polynomials in the coefficients of the t_i:

$$\mathcal{J}(t_1,t_2) \;=\; A_{12}B_{34} \;-\; A_{13}B_{24} \;+\; A_{14}B_{23} \;+\; A_{23}B_{14} \;-\; A_{24}B_{13} \;+\; A_{34}B_{12}$$

and

$$\mathcal{B}(t_1,t_2,t_3,t_4,t_5,t_6) \;=\; \det\begin{pmatrix} A_{12} & B_{12} & C_{12} & D_{12} & E_{12} & F_{12} \\ A_{13} & B_{13} & C_{13} & D_{13} & E_{13} & F_{13} \\ A_{14} & B_{14} & C_{14} & D_{14} & E_{14} & F_{14} \\ A_{23} & B_{23} & C_{23} & D_{23} & E_{23} & F_{23} \\ A_{24} & B_{24} & C_{24} & D_{24} & E_{24} & F_{24} \\ A_{34} & B_{34} & C_{34} & D_{34} & E_{34} & F_{34} \end{pmatrix}.$$

It can be seen that $\mathcal{J}(\bar{t}_1,\bar{t}_2) = \det(c_{ij})\cdot\mathcal{J}(t_1,t_2)$ and

$$\mathcal{B}(\bar{t}_1,\ldots,\bar{t}_6) \;=\; [\det(c_{ij})]^3\cdot\mathcal{B}(t_1,\ldots,t_6),$$

i.e., $\mathcal{J}$ is a joint invariant of two tensors and $\mathcal{B}$ is a joint invariant of six tensors.

In order to geometrically interpret these invariants we assume that all tensors are decomposable:

$$\mathcal{I}(t_1) = \mathcal{I}(t_2) = \ldots = \mathcal{I}(t_6) \;=\; 0.$$

As above we write

$$t_2(x_1,x_2,x_3,x_4) \;=\; (b_{11}x_1+b_{12}x_2+b_{13}x_3+b_{14}x_4)(b_{21}x_1+b_{22}x_2+b_{23}x_3+b_{24}x_4)$$

$$\ldots \qquad \ldots \qquad \ldots \qquad \ldots \qquad \ldots$$

$$t_6(x_1,x_2,x_3,x_4) \;=\; (f_{11}x_1+f_{12}x_2+f_{13}x_3+f_{14}x_4)(f_{21}x_1+f_{22}x_2+f_{23}x_3+f_{24}x_4),$$

and we let $L_1, L_2, \ldots, L_6$ denote the corresponding lines in projective 3-space.

Then the lines L_1 and L_2 are incident if and only if the joint invariant $\mathcal{J}(t_1,t_2)$ vanishes. If K is a subfield of the real numbers, and if $\sum A_{ij}^2 = \sum B_{ij}^2 = 1$, then $\mathcal{J}(t_1,t_2)$ is proportional to the oriented distance between the lines L_1 and L_2.

We say that the six lines $\{L_1,L_2,L_3,L_4,L_5,L_6\}$ form a *line complex* if and only if their joint invariant $\mathcal{B}(t_1,t_2,t_3,t_4,t_5,t_6)$ vanishes. This is equivalent to the following non-trivial geometric condition: Either there exists a line L such that $L_1, L_2, \ldots, L_6$ all intersect L, or all six lines are reciprocal to a common screw. The study of line complexes in 3-space is one of the central subjects of classical projective geometry. See [2] for details, and see [18] for applications of these concepts to rigidity theory.

Let us close this section by giving an expansion of the invariant $\mathcal{B}(t_1,t_2,t_3,t_4,t_5,t_6)$ in terms of the 48 indeterminates $a_{ij}, b_{ij}, \ldots f_{ij}$. This expansion has been obtained by T. McMillan, and it improves an earlier result of N. White [16]. We abbreviate the 4×4-determinants spanned by vectors from $\{a_1, a_2, b_1, \ldots, f_2\}$ by brackets, e.g.

$$[a_1, c_2, e_2, f_1] \quad := \quad \det \begin{pmatrix} a_{11} & a_{12} & a_{13} & a_{14} \\ c_{21} & c_{22} & c_{23} & c_{24} \\ e_{21} & e_{22} & e_{23} & e_{24} \\ f_{11} & f_{12} & f_{13} & f_{14} \end{pmatrix}.$$

Using that bracket notation, the McMillan–White expansion is given by

$$\begin{aligned} &\mathcal{B}(t_1,t_2,t_3,t_4,t_5,t_6) \quad = \\ &\sum_{\sigma,\tau,\pi} \operatorname{sign}(\sigma)\operatorname{sign}(\tau)\operatorname{sign}(\pi)\,[a_1,a_2,d_{\sigma_1},e_{\tau_1}][b_1,b_2,d_{\sigma_2},f_{\pi_1}][c_1,c_2,e_{\tau_2},f_{\pi_2}] \quad - \\ &\sum_{\sigma,\tau,\pi} \operatorname{sign}(\sigma)\operatorname{sign}(\tau)\operatorname{sign}(\pi)\,[d_1,d_2,a_{\sigma_1},b_{\tau_1}][e_1,e_2,a_{\sigma_2},c_{\pi_1}][f_1,f_2,b_{\tau_2},c_{\pi_2}] \end{aligned}$$

where both sums are ranging over all permutations σ, τ, π of $\{1,2\}$. Hence the above expansion consists of 16 bracket monomials as summands.

3.3. Superalgebra and joint invariants.

In this section we come to the main goal of the present paper. A symbolic method will be introduced for representing joint invariants of symmetric and skew-symmetric tensors. As a special case thereof we shall obtain invariants of skew-symmetric tensors alone. In the end of this section we will derive "super-algebraic" umbral representations for the invariants in Example 3.4 related to lines in projective 3-space.

The umbral space for joint invariants is a certain non-commutative algebra whose structure depends on the types of tensors involved. For the invariant theoretic applications to be discussed in this paper we may restrict ourselves to the specific case of superalgebras generated by *letterplaces*. We refer to the monograph of Grosshans, Rota and Stein [8] for a discussion of superalgebras in full generality.

In the following we shall study joint invariants of two symmetric tensors

$$\begin{aligned} f_1(x_1,\ldots,x_d) &= \sum \binom{n}{i_1 i_2 \ldots i_d} a_{i_1 i_2 \ldots i_d}\, x_1^{i_1} x_2^{i_2} \cdots x_d^{i_d} \\ f_2(x_1,\ldots,x_d) &= \sum \binom{n}{i_1 i_2 \ldots i_d} b_{i_1 i_2 \ldots i_d}\, x_1^{i_1} x_2^{i_2} \cdots x_d^{i_d} \end{aligned}$$

and two skew-symmetric tensors

$$t_1(x_1,\dots,x_d) = \sum_{1\le i_1<i_2<\cdots<i_n\le d} A_{i_1 i_2\dots i_n}\, x_{i_1}\, x_{i_2}\,\dots\, x_{i_n}$$
$$t_2(x_1,\dots,x_d) = \sum_{1\le i_1<i_2<\cdots<i_n\le d} B_{i_1 i_2\dots i_n}\, x_{i_1}\, x_{i_2}\,\dots\, x_{i_n}$$

all of degree n in d variables. The definitions and results about joint invariants of f_1, f_2, t_1 and t_2 will generalize in a straightforward manner to covariants and to larger collections of symmetric and skew-symmetric tensors, also of different degrees.

For the skew-symmetric tensors t_1 and t_2 we introduce alphabets L_1^+ and L_2^+ respectively, both containing an infinite number of *positively signed letters*. Positive letters $\alpha \in L_1^+$ are said to *belong* to t_1, and positive letters $\beta \in L_2^+$ *belong* to t_2. For the two symmetric tensors f_1 and f_2 we introduce alphabets L_3^- and L_4^- respectively, both containing an infinite number of *negatively signed letters.* Negative letters $\alpha \in L_3^-$ *belong* to f_1, and negative letters $\beta \in L_4^-$ *belong* to f_2. We write

$$L \;:=\; L_1^+ \cup L_2^+ \cup L_3^- \cup L_4^-$$

for the alphabet of all letters, both negative and positive. We suppose that the alphabets L_1^+, L_2^+, L_3^- and L_4^- are pairwise disjoint and that the set L is totally ordered.

As in Section 3.1, the letters or symbols $\alpha \in L$ will be viewed as vectors

$$\alpha \;=\; \big(\, (\alpha\,|\,1),\, (\alpha\,|\,2),\, \dots\, ,\, (\alpha\,|\,d)\,\big)$$

whose coordinates $(\alpha\,|\,i)$ are algebraically independent transcendentals, called *letterplaces.* Thus a letterplace $(\alpha\,|\,i)$ is a tuple consisting of a *letter* $\alpha \in L$ and a *place* $i \in P := \{1,2,\dots,n\}$. We write $[L\,|\,P]$ for the set of letterplaces.

The *superalgebra* $\mathrm{Super}[L\,|\,P]$ is defined to be the free associative algebra generated by the infinite set $[L\,|\,P]$ subject to the following commutation rules

$$(\alpha\,|\,i)(\beta\,|\,j) \;=\; \begin{cases} (\beta\,|\,j)(\alpha\,|\,i) & \text{if } \alpha \text{ \underline{or} } \beta \text{ is a \textit{negative} letter,} \\ -(\beta\,|\,j)(\alpha\,|\,i) & \text{if both } \alpha \text{ \underline{and} } \beta \text{ are \textit{positive} letters.} \end{cases}$$

This superalgebra can also be described abstractly as a K-algebra tensor product of a symmetric algebra with an exterior algebra

$$\mathrm{Super}[L\,|\,P] \;=\; \mathrm{Sym}\big(\{(\alpha\,|\,i) : \alpha \in L_3^- \cup L_4^-\}\big) \otimes \mathrm{Ext}\big(\{(\alpha\,|\,i) : \alpha \in L_1^+ \cup L_2^+\}\big).$$

If the underlying field K has finite characteristic, then that definition is not quite accurate as we will see below. For the time being let us assume that K has characteristic 0.

We next introduce certain elements in the algebra $\text{Super}[L\,|\,P]$ which generalize the basic letterplaces $(\alpha\,|\,i)$ to strings of letters and strings of places. Let $1 \le k \le d$. Given k letters $\alpha_1, \ldots, \alpha_k \in L$ and k places $i_1, i_2, \ldots, i_k \in P$, we define

$$(\alpha_1\alpha_2\ldots\alpha_k \mid i_1\, i_2 \ldots i_k) \quad := \quad \sum_{\sigma\in S_k} (\text{sign}\,??)(\alpha_{\sigma_1}\,|\,i_1)(\alpha_{\sigma_2}\,|\,i_2)\,\ldots\,(\alpha_{\sigma_k}\,|\,i_k),$$

where the sign factors $(\text{sign}\,??) \in \{-1,+1\}$ are determined by the following rule: Compare the two words

$$\begin{aligned} w_\sigma &= \alpha_{\sigma_1}\, i_1\, \alpha_{\sigma_2}\, i_2\, \ldots\, \alpha_{\sigma_k}\, i_k \qquad \text{and} \\ w &= \alpha_1\, \alpha_2\, \ldots\, \alpha_k\, i_1\, i_2\, \ldots\, i_k. \end{aligned}$$

Then the sign "(sign ??)" of the summand corresponding to w_σ is the parity of the number of transpositions not involving positive letters needed to transform the word w into the word w_σ. In other words: we consider only transpositions of two places, transpositions of a place with a negative letter, or transpositions of two negative letters. With respect to this crucial sign rule the places can be thought of as negatively signed symbols.

In the formula defining $(\alpha_1\alpha_2\ldots\alpha_k \mid i_1\, i_2 \ldots i_k)$ we sum over all permutations of the k letters. It follows from the defining commutation rules that we may also sum over all permutations of the places. Doing so, we obtain the identity

$$(\alpha_1\alpha_2\ldots\alpha_k \mid i_1\, i_2 \ldots i_k) \quad = \quad \sum_{\sigma\in S_k} (\text{sign}\,??)(\alpha_1\,|\,i_{\sigma_1})(\alpha_2\,|\,i_{\sigma_2})\,\ldots\,(\alpha_k\,|\,i_{\sigma_k}),$$

where (sign ??) equals the parity of the number of transpositions not involving positive letters needed to transform the word

$$\alpha_1\, i_{\sigma_1}\, \alpha_2\, i_{\sigma_2}\, \ldots\, \alpha_k\, i_{\sigma_k}$$

into the word

$$\alpha_1\, \alpha_2\, \ldots\, \alpha_k\, i_1\, i_2\, \ldots\, i_k.$$

EXAMPLE 3.5. Let $\alpha, \beta \in L_1^+$. Then

$$\begin{aligned} (\alpha\,\beta\,|\,1\,2) &= (\alpha\,|\,1)(\beta\,|\,2) + (\beta\,|\,1)(\alpha\,|\,2) \\ &= (\alpha\,|\,1)(\beta\,|\,2) - (\alpha\,|\,2)(\beta\,|\,1). \end{aligned}$$

This implies that

$$(\alpha\,\beta\,|\,1\,2) \quad = \quad (\beta\,\alpha\,|\,1\,2)$$

for positive letters α and β.

Let us now consider the special case $k = d$. Given d letters $\alpha_1, \ldots, \alpha_d \in L$, we abbreviate

$$\begin{aligned} [\alpha_1, \alpha_2, \ldots, \alpha_d] \quad &:= \quad (\alpha_1\alpha_2\ldots\alpha_d \mid 1\; 2 \ldots d) \\ &= \sum_\sigma (\text{sign}\,??)(\alpha_{\sigma_1}\,|\,1)(\alpha_{\sigma_2}\,|\,2)\,\ldots\,(\alpha_{\sigma_d}\,|\,d) \\ &= \sum_\sigma (\text{sign}\,??)(\alpha_1\,|\,\sigma_1)(\alpha_2\,|\,\sigma_2)\,\ldots\,(\alpha_d\,|\,\sigma_d) \end{aligned}$$

The expression $[\alpha_1, \alpha_2, \ldots, \alpha_d]$ is called the *superbracket* or the *superdeterminant* of the letters $\alpha_1, \ldots, \alpha_d$.

If $d = 2$, then Example 3.5 shows that $[\alpha, \beta] = [\beta, \alpha]$ for positive letters α and β (belonging to a skew-symmetric tensor). If α and β are negative letters (belonging to a symmetric tensor), then the bracket is anti-symmetric: $[\alpha, \beta] = -[\beta, \alpha]$, as in the previous sections.

In general, we have the following sign rule for the superbracket. Permuting two letters α_i and α_j in a bracket

$$[\alpha_1, \ldots, \alpha_i, \ldots, \alpha_j, \ldots, \alpha_d] \quad = \quad \pm [\alpha_1, \ldots, \alpha_j, \ldots, \alpha_i, \ldots, \alpha_d]$$

gives a global minus sign if and only if the two letters are negatively signed.

Similarly, the above definition implies a unique sign rule for commuting entire brackets:

$$[\alpha_1, \ldots, \alpha_d][\beta_1, \ldots, \beta_d] \quad = \quad \pm [\beta_1, \ldots, \beta_d][\alpha_1, \ldots, \alpha_d].$$

If all letters are negative, then the superbracket reduces to the ordinary bracket, obeying the usual rules for determinants.

The *Laplace expansion* for ordinary determinants generalizes to superbrackets. Using a suitable combinatorial sign rule, we can derive, for example, an expansion

$$(\, \alpha_1 \ldots \alpha_d \mid 1 2 \ldots d\,) \quad = \quad \sum_{\substack{\sigma_1 < \sigma_2 \\ \sigma_3 < \cdots < \sigma_d}} (\text{sign}\, ??)\, (\alpha_1 \alpha_2 \,|\, \sigma_1 \sigma_2)\, (\alpha_3 \ldots \alpha_d \,|\, \sigma_3 \ldots \sigma_d).$$

This formula can be thought of as the Laplace expansion with respect to the first two columns of a $d \times d$-matrix

$$\begin{array}{c} \\ \alpha_1 \\ \alpha_2 \\ \alpha_3 \\ \vdots \\ \alpha_d \end{array} \begin{array}{c} 1 \quad 2 \quad 3 \quad \ldots \quad d \\ \left(\begin{array}{ccccc} & & & & \\ & & & & \\ & & & & \\ & & & & \\ & & & & \end{array} \right) \end{array}$$

whose rows are indexed by letters and whose columns are indexed by places.

All we have said so far about the superbracket is entirely characteristic-free as long as the letters $\alpha_1, \ldots, \alpha_d$ are distinct. What happens if some letters are equal ? Clearly, we have

$$(\alpha\alpha \,|\, 1 2) \quad = \quad 0 \qquad\qquad \text{if } \alpha \text{ is negative,}$$

because in this case we are dealing with ordinary determinants. However, if α is positive the we get

$$(\alpha\alpha \,|\, 1 2) \quad = \quad 2!\, (\alpha \,|\, 1)\, (\alpha \,|\, 2).$$

In general we obtain

$$(\alpha\alpha\ldots\alpha\,|\,12\ldots k) \quad = \quad k!\,(\alpha\,|\,1)(\alpha\,|\,2)\ldots(\alpha\,|\,k).$$

for positive letters α.

In order to make the theory characteristic-free, one has to get rid of these factorials. This can be achieved by using the so-called *divided powers algebra.* The basic idea is to introduce new "compound letters" $\alpha^{(k)}$ which satisfy the factorial-free relations

$$(\alpha^{(k)}\mid 12\ldots k) \quad = \quad (\alpha\,|\,1)(\alpha\,|\,2)\ldots(\alpha\,|\,k).$$

In order to keep the present exposition as introductory as possible, we will circumvent the divided powers algebra. Instead we continue to assume that K has characteristic 0, and we simply define

$$(\alpha^{(k)}\mid 12\ldots k) \quad := \quad \frac{1}{k!}(\alpha\alpha\ldots\alpha\,|\,12\ldots k)$$

for all positive letters α and $1 \leq k \leq d$. Moreover, we generalize this definition to expressions containing several letters:

$$(\cdots \quad \alpha^{(l)} \quad \cdots\mid 12\ldots k) \quad := \quad \frac{1}{l!}(\cdots \quad \underbrace{\alpha\,\alpha\ldots\alpha}_{l\ \text{times}} \quad \cdots\mid 12\ldots k)$$

Finally, we extend this divided powers notation in the obvious way to brackets

$$[\cdots \quad \alpha^{(l)} \quad \cdots] \quad := \quad (\cdots \quad \alpha^{(l)} \quad \cdots\mid 12\ldots d).$$

Studying the general characteristic-free definition for the symbols $\alpha^{(l)}$ given in [8] and [13], the reader will notice that the above relations are valid identities in characteristic zero.

Here are two examples of Laplace expansions in the superalgebra which involve the divided powers notation.

EXAMPLE 3.6.

(a) Let α be a positive letter, β a negative letter, and $d = 3$. Then we have

$$\begin{aligned}[\overset{+}{\alpha}{}^{(2)}\,\overset{-}{\beta}] &= (\alpha^{(2)}\beta\,|\,123)\\ &= (\alpha^{(2)}\,|\,12)(\beta\,|\,3) - (\alpha^{(2)}\,|\,13)(\beta\,|\,2) + (\alpha^{(2)}\,|\,23)(\beta\,|\,1)\\ &= (\alpha\,|\,1)(\alpha\,|\,2)(\beta\,|\,3) - (\alpha\,|\,1)(\alpha\,|\,3)(\beta\,|\,2) + (\alpha\,|\,2)(\alpha\,|\,3)(\beta\,|\,1)\end{aligned}$$

(b) Let $d = 4$, and suppose that both α and β are positive letters. Then we have

$$\begin{aligned}&[\alpha^{(2)}\beta^{(2)}] \quad = \quad (\alpha^{(2)}\beta^{(2)}\,|\,1234) \quad = \\ = \quad &(\alpha^{(2)}|12)(\beta^{(2)}|34) - (\alpha^{(2)}|13)(\beta^{(2)}|24) + (\alpha^{(2)}|14)(\beta^{(2)}|23)\\ &+(\alpha^{(2)}|23)(\beta^{(2)}|14) - (\alpha^{(2)}|24)(\beta^{(2)}|13) + (\alpha^{(2)}|34)(\beta^{(2)}|12)\\ = \quad &(\alpha|1)(\alpha|2)(\beta|3)(\beta|4) - (\alpha|1)(\alpha|3)(\beta|2)(\beta|4) + (\alpha|1)(\alpha|4)(\beta|2)(\beta|3)\\ &+(\alpha|2)(\alpha|3)(\beta|1)(\beta|4) - (\alpha|2)(\alpha|4)(\beta|1)(\beta|3) + (\alpha|3)(\alpha|4)(\beta|1)(\beta|2)\end{aligned}$$

These definitions give us the tools to extend the main theorems of invariant theory to skew-symmetric tensors and symmetric tensors jointly. The superalgebra $\text{Super}[L \mid P]$ plays the role of the umbral space. The *umbral operator* U is the unique K-linear map from $\text{Super}[L \mid P]$ onto the polynomial ring $K[\ldots, a_{i_1 i_2 \ldots i_d}, b_{i_1 i_2 \ldots i_d}, A_{i_1 i_2 \ldots i_d}, B_{i_1 i_2 \ldots i_d}, \ldots]$ generated by the coefficients of the symmetric tensors f_1, f_2 and the skew-symmetric tensors t_1, t_2, which is defined by the following rules

1. Suppose $\alpha \in L$ is positive and belongs to the symmetric tensor

$$f_1(x_1, \ldots, x_d) \quad = \quad \sum \binom{n}{i_1 i_2 \ldots i_d} a_{i_1 i_2 \ldots i_d} \, x_1^{i_1} \, x_2^{i_2} \cdots x_d^{i_d}.$$

Then we set

$$\langle \, U \mid (\alpha \,|\, 1)^{i_1} (\alpha \,|\, 2)^{i_2} \ldots (\alpha \,|\, d)^{i_d} \, \rangle \quad := \quad a_{i_1 i_2 \ldots i_d} \qquad \text{if} \quad i_1 + i_2 + \cdots + i_d = n.$$

and

$$\langle \, U \mid (\alpha \,|\, 1)^{i_1} (\alpha \,|\, 2)^{i_2} \ldots (\alpha \,|\, d)^{i_d} \, \rangle \quad := \quad 0 \qquad \text{if} \quad i_1 + i_2 + \cdots + i_d \neq n,$$

and similarly for letters belonging to the symmetric tensor f_2. This is just a repetition of the definition in Section 3.1.

2. Suppose $\alpha \in L$ is negative and belongs to a skew-symmetric tensor

$$t_1 \quad = \quad \sum A_{i_1 i_2 \ldots i_n} \, x_{i_1} \, x_{i_2} \, \ldots \, x_{i_n}$$

Then we set

$$\langle \, U \mid (\alpha \,|\, i_1)(\alpha \,|\, i_2) \ldots (\alpha \,|\, i_n) \rangle \quad := \quad A_{i_1 i_2 \ldots i_n},$$

and

$$\langle \, U \mid (\alpha \,|\, i_1) \ldots (\alpha \,|\, i_k) \, \rangle \quad := \quad 0 \qquad \text{if} \quad k \neq n.$$

3. Using the total order on the alphabet L and allowing a possible sign change, we can write every monomial in $\text{Super}[L \mid P]$ as a product of letterplace powers with strictly increasing letters. The umbral operator U is multiplicative with respect to that decomposition. For example, if α and β are equivalent letters belonging to the skew-symmetric tensor t, and if $\alpha < \beta$, then we get

$$\langle \, U \mid (\alpha \,|\, i_1) \ldots (\alpha \,|\, i_n) (\beta \,|\, j_1) \ldots (\beta \,|\, j_n) \, \rangle \quad := \quad A_{i_1 \ldots i_n} \, A_{j_1 \ldots j_n}.$$

4. Finally, we extend the umbral operator U by K-linearity to the entire K-vector space $\text{Super}[L \mid P]$.

The following result is the main theorem on joint invariants of symmetric and skew-symmetric tensors.

FIRST FUNDAMENTAL THEOREM. *Every joint invariant of symmetric and skew-symmetric tensors is expressible in the form $\langle\, U \mid$ superbracket polynomial $\rangle$.*

EXAMPLE 3.7. *(Skew-symmetric tensors and lines in projective 3-space)*

Let $n = 2, d = 4$ and consider the skew-symmetric tensors t_1 and t_2 as discussed in Example 3.4.

Suppose that α and β are positive letters both belonging to t_1. Then

$$\langle\, U \mid [\alpha^{(2)}\,\beta^{(2)}\,]\,\rangle \quad = \quad 2\cdot(A_{12}A_{34} \;-\; A_{13}A_{24} \;+\; A_{14}A_{23})$$

as is easily seen from the expansion in Example 3.6 (b). Hence $\frac{1}{2}[\,\alpha^{(2)}\,\beta^{(2)}\,]$ is a "super-algebraic" umbral representation for the Grassmann invariant $\mathfrak{I}(t_1)$ which characterizes the decomposability of a skew-symmetric tensor degree 2 in 4 variables. In other words, $\langle\, U \mid [\,\alpha^{(2)}\,\beta^{(2)}\,]\,\rangle \;=\; 0$ if and only if the tensor t_1 corresponds to a line in projective 3-space.

Now suppose that α and β are positive letters belonging to distinct decomposable tensors t_1 and t_2 respectively. Then the expansion in Example 3.6 (b) implies

$$\langle\, U \mid [\,\alpha^{(2)}\,\beta^{(2)}\,]\,\rangle \quad = \quad A_{12}B_{34} - A_{13}B_{24} + A_{14}B_{23} + A_{23}B_{14} - A_{24}B_{13} + A_{34}B_{12},$$

and hence $[\,\alpha^{(2)}\,\beta^{(2)}\,]$ is an umbral representation for the invariant $\mathfrak{Z}(t_1,t_2)$ which is non-zero if and only if the lines corresponding to t_1 and t_2 are *skew*, i.e., if they do not intersect in the ambient projective 3-space. □

An umbral representation for a joint invariant of symmetric and skew-symmetric tensors, or equivalently, a polynomial in the subalgebra of Super$[L \mid P]$ generated by the superbrackets can be reduced to a standard form using a generalization of the straightening algorithm. That normal form procedure will be referred to as the *super-straightening algorithm.*

Recall that the alphabet L was assumed to be totally ordered. Generalizing the tableaux in Section 2.1 and Section 3.1, superbracket monomials are also represented by rectangular tableaux. In this generalized tableaux representation the compound letters $\alpha^{(l)}$ are decomposed into strings of letters $\underbrace{\alpha\,\alpha\,\ldots\,\alpha}_{l\text{ times}}$. Such a tableau T is *superstandard* if

(1) both its rows and its columns are nondecreasing,

(2) if α,β are negative and successive in the same row then $\alpha < \beta$

(3) if γ,δ are positive and successive in the same column then $\gamma < \delta$.

For example, let $d = 5$, and let $\alpha,\beta,\gamma,\delta,\epsilon$ be positive letters. Then the tableau

$$\begin{bmatrix} \alpha & \alpha & \beta & \beta & \gamma \\ \beta & \gamma & \gamma & \delta & \delta \\ \delta & \delta & \delta & \epsilon & \epsilon \end{bmatrix} \quad := \quad [\,\alpha^{(2)}\,\beta^{(2)}\,\gamma\,][\,\beta\,\gamma^{(2)}\,\delta^{(2)}\,][\,\delta^{(3)}\,\epsilon^{(2)}\,]$$

is superstandard. However, the tableau

$$\begin{bmatrix} \alpha & \alpha & \beta & \beta & \gamma \\ \beta & \beta & \beta & \delta & \delta \\ \delta & \delta & \delta & \epsilon & \epsilon \end{bmatrix} \quad := \quad [\alpha^{(2)}\,\beta^{(2)}\,\gamma][\beta^{(3)}\,\delta^{(2)}][\delta^{(3)}\,\epsilon^{(2)}]$$

is not superstandard because its third column is not strictly increasing.

The ***super-straightening algorithm*** which is described in detail in [8] expresses any tableau as a unique K-linear combination of superstandard tableaux. Hence the superstandard tableaux form a K-vector space basis for the subalgebra generated by the superbrackets in Super$[L \mid P]$. Moreover, all tableaux occurring in that expansion will have integer coefficients. Hence the result is characteristic-free and we get the

SUPERSTANDARD BASIS THEOREM. *The superstandard tableaux form a integral basis for the $\mathbb{Z}$-module of tableaux.*

Here is another example of a nontrivial invariant.

EXAMPLE 3.8. *(The Pfaffian of a skew-symmetric matrix)*
Consider a skew-symmetric tensor

$$t(x_1, x_2, \ldots, x_{2r}) \quad = \quad \sum_{1 \leq i < j \leq 2r} A_{ij}\, x_i\, x_j$$

of degree $n = 2$ in an even number of $d = 2r$ variables. We can think of t as a skew-symmetric $2r \times 2r$-matrix $(A_{ij})_{1 \leq i,j \leq 2r}$. The determinant of that matrix is the square of a polynomial $\mathrm{Pf}(t) := \det[(A_{ij})]^{\frac{1}{2}}$ in the A_{ij}, which is known as the *Pfaffian* of t. We have

$$\mathrm{Pf}(t) \quad = \quad \sum_{\sigma} \mathrm{sign}(\sigma)\, A_{\sigma_1\sigma_2} A_{\sigma_3\sigma_4} \cdots A_{\sigma_{2r-1}\sigma_{2r}}$$

where the sum is over all permutations σ of $\{1, 2, \ldots, 2r\}$ such that $\sigma_1 < \sigma_2$, $\sigma_3 < \sigma_4, \ldots, \sigma_{2r-1} < \sigma_{2r}$.

The Pfaffian is an invariant of the skew-symmetric tensor t which has the following umbral representation

$$\langle\, U \mid [\alpha_1^{(2)}\, \alpha_2^{(2)} \, \ldots \, \alpha_r^{(2)}]\,\rangle \quad = \quad r! \cdot \mathrm{Pf}(t).$$

Here $\alpha_1, \alpha_2, \ldots, \alpha_r$ are positive letters belonging to t. To verify this umbral representation, we compute the expansion

$$\begin{aligned} & [\alpha_1^{(2)}\, \alpha_2^{(2)} \, \ldots \, \alpha_r^{(2)}] \\ = \; & (\alpha_1^{(2)}\, \alpha_2^{(2)} \, \ldots \, \alpha_r^{(2)} \mid 1\,2\,3 \, \ldots \, 2r) \\ = \; & r! \cdot \sum_{\sigma} \mathrm{sign}(\sigma)\, (\alpha_1^{(2)} \mid \sigma_1\sigma_2)(\alpha_2^{(2)} \mid \sigma_3\sigma_4) \; \cdots \; (\alpha_r^{(2)} \mid \sigma_{2r-1}\sigma_{2r}) \\ = \; & r! \cdot \sum_{\sigma} \mathrm{sign}(\sigma)\, (\alpha_1 \mid \sigma_1)(\alpha_1 \mid \sigma_2)(\alpha_2 \mid \sigma_3)(\alpha_2 \mid \sigma_4) \; \cdots \; (\alpha_r \mid \sigma_{2r-1})(\alpha_r \mid \sigma_{2r}) \end{aligned}$$

where the sum is over all permutations σ with $\sigma_1 < \sigma_2,\ \sigma_3 < \sigma_4,\ \ldots, \sigma_{2r-1} < \sigma_{2r}$. Applying the umbral operator U to that expansion we obtain $\mathrm{Pf}(t)$. □

Let us close this section by stating without proof the umbral representation for the invariant in Example 3.4 which characterizes when six lines in projective 3-space form a line complex.

EXAMPLE 3.9. *(Umbral representation for the line complex invariant)*

As in Example 3.4, we consider six skew-symmetric tensors $t_1, t_2, \ldots, t_6$ of step 2 in 4 variables. Their joint invariant $\mathcal{B}(t_1, t_2, t_3, t_4, t_5, t_6)$ is the 6×6-determinant of their Plücker coordinates.

The following umbral representation in terms of superbrackets or tableaux has been proved by T. McMillan and N. White.

$$\mathcal{B}(t_1, t_2, t_3, t_4, t_5, t_6) \quad = \quad \langle\, U \mid \begin{bmatrix} \alpha & \alpha & \delta & \epsilon \\ \beta & \beta & \delta & \phi \\ \gamma & \gamma & \epsilon & \phi \end{bmatrix} \quad - \quad \begin{bmatrix} \alpha & \beta & \delta & \delta \\ \alpha & \gamma & \epsilon & \epsilon \\ \beta & \gamma & \phi & \phi \end{bmatrix} \rangle$$

Here $\alpha, \beta, \gamma, \delta, \epsilon, \phi$ are positive letters belonging to $t_1, t_2, t_3, t_4, t_5, t_6$ respectively. Notice that the two tableaux in the above expression are not superstandard. Using the super-straightening algorithm, this umbral representation can be rewritten as a sum of four super-standard tableaux. For details we refer to McMillan's thesis [12]. □

4. Some remarks.

We will briefly summarize what has been achieved so far. A syntactic characterization has been derived for arbitrary invariants of symmetric and skew-symmetric tensors: The desired invariants are precisely the expressions of the form

$$\langle\, U \mid \text{some bracket polynomial } \rangle .$$

In other words, the subalgebra generated by all brackets in the superalgebra Super$[L \mid P]$ is mapped onto the ring of invariants under the umbral operator U. This generalizes to covariants.

The brackets in these umbral representations consist of certain symbols associated with the tensors in question. With symmetric tensors we associate anticommutative symbols $\alpha_1, \alpha_2, \dots$, giving the well-known anticommutative bracket

$$[\alpha_1, \alpha_2, \dots, \alpha_d] \quad = \quad (\operatorname{sign} \sigma)\, [\alpha_{\sigma_1}, \alpha_{\sigma_2}, \dots, \alpha_{\sigma_\alpha}].$$

With skew-symmetric tensors we associate commutative symbols $\alpha_1, \alpha_2, \dots$, giving a new type of commutative bracket

$$[\alpha_1, \alpha_2, \dots, \alpha_d] \quad = \quad [\alpha_{\sigma_1}, \alpha_{\sigma_2}, \dots, \alpha_{\sigma_\alpha}].$$

On the first encounter the above switch of signs and the notion of positive and negative symbols may seem rather odd and unnatural. Yet, it turns out that one is forced to introduce these notions if one wants to develop a symbolic method also for skew-symmetric tensors. One question that remains open is whether and how this switch of sign can be understood in some "philosophical" sense. Is there an adequate geometric or physical interpretation for superbrackets and the supcralgebra ?

Let us remark that the umbral operator U acting on the superalgebra is a special case of a more general class of operators, called *Schur functors* [1]. If we want to further generalize our results to other classes of tensors, we are lead to the study of Schur functors.

It turns out that the superalgebra is also useful for deriving tensor algebraic identities. As an example we give a new and very simple proof for a classical determinant identity due to Bazin. Using bracket notation for $d \times d$-determinants, Bazin's identity reads as follows.

BAZIN'S THEOREM. *Let $k \leq d$, and let $a_1, \dots, a_d, b_1, \dots, b_k \in K^d$. Then*

$$\begin{aligned}
&\sum_\sigma \pm [b_{\sigma_1} \widehat{a_1} a_2 \dots a_d][b_{\sigma_2} a_1 \widehat{a_2} \dots a_d] \dots [b_{\sigma_k} a_1 a_2 \dots \widehat{a_k} \dots a_d] \\
&= \quad [b_1 \dots b_k a_{k+1} \dots a_d][a_1 a_2 \dots a_d]^{k-1}
\end{aligned}$$

where the sum is over all permutations of $\{1, 2, \dots, k\}$ and where each summand has a suitable sign.

In order to prove Bazin's theorem, we consider the superalgebra generated by the positively signed symbols $L = \{\alpha, \gamma_1, \ldots, \gamma_k\}$ in dimension d. We claim that the following two tableaux are equal in that algebra up to an integer constant:

$$\begin{bmatrix} \alpha & \gamma_2 & \cdots & \cdots & \gamma_2 \\ \alpha & \gamma_3 & \cdots & \cdots & \gamma_3 \\ \vdots & \vdots & & & \vdots \\ \alpha & \gamma_k & \cdots & \cdots & \gamma_k \\ \alpha & \underbrace{\gamma_1 \cdots \gamma_1}_{d-k} & \gamma_2 & \cdots & \gamma_k \end{bmatrix} \quad = \quad c \cdot \begin{bmatrix} \overbrace{\alpha \ldots \alpha}^{k} & \overbrace{\gamma_1 \cdots \gamma_1}^{d-k} \\ \gamma_2 \quad \cdots & \cdots \quad \gamma_2 \\ \vdots & \vdots \\ \gamma_{k-1} \quad \cdots & \cdots \quad \gamma_{k-1} \\ \gamma_k \quad \cdots & \cdots \quad \gamma_k \end{bmatrix}.$$

The tableau on the right hand side is the only possible superstandard tableau with respect to the linear order $\alpha < \gamma_1 < \gamma_2 < \cdots < \gamma_k$. Hence the tableau on the left hand side and the tableau on the right hand side are equal up to a constant $c \in \mathbb{Z}$.

Now suppose that the symbol γ_1 belongs to the decomposable skew-symmetric tensor

$$a_{k+1} \ldots a_d \qquad \text{(the exterior product of } d-k \text{ linear forms),}$$

that the symbols $\gamma_2, \gamma_3, \ldots, \gamma_k$ belong to the decomposable tensor $a_1 \ldots a_d$, and that and the symbol α belongs to the tensor $b_1 b_2 \ldots b_k$. Applying the umbral operator U on both sides of the above superalgebra identity, we obtain Bazin's identity up to the global constant $c \in \mathbb{Z}$. Plugging in unit vectors for the a_i, we can easily see that $c = 1$. □

Another important application of invariant theory in superalgebras is the classification of canonical forms of skew-symmetric tensors. As an example consider skew-symmetric tensors of step 3 in dimension 6. Using superalgebra methods, it is proved in [8, Chapter 5] that every such tensor

$$t \quad \in \quad \mathrm{Ext}^3_K(\{x_1, x_2, \ldots, x_6\}).$$

can be represented as

I. $t = abc$, or

II. $t = abc + ade$, or

III. $t = abc + aef + bde$, or

IV. $t = abc + def$,

for some suitable linear forms $a, b, c, d, e, f \in \mathrm{Ext}^1_K(\{x_1, x_2, \ldots, x_6\})$.

Let us close by pointing out an important area for future research. It has been remarked by Dieudonné already in 1971 that "it might be worthwhile to push the 19th century computations a little further along, with the help of modern computers" [6, Preface]. With the very recent exciting developments in computer algebra, most notably B. Buchberger's *Gröbner bases method* [4], it is now possible to efficiently implement many existing algorithms from invariant theory.

Some classical methods provide basic tools for new applicable symbolic algorithms. The straightening algorithm, for example, is a crucial subroutine in N. White's *Cayley factorization* which rewrites projectively invariant polynomials in terms of synthetic geometric constructions. It would therefore be interesting to analyze the computational complexity of invariant theory algorithms, both from a practical and a worst-case point of view. We suggest to further explore these and other aspects of what might be called *Computational Invariant Theory*.

REFERENCES

[1] K. AKIN, D.A. BUCHSBAUM AND J. WEYMAN, *Schur functors and Schur complexes*, Advances in Mathematics 44 (1982) 207–278..

[2] R.S. BALL, *A treatise on the theory of screws*, Cambridge University Press, 1900.

[3] M. BARNABEI, A. BRINI AND G.-C. ROTA, *On the exterior calculus of invariant theory*, Journal of Algebra 96 (1985) 120-160.

[4] B. BUCHBERGER, *Gröbner bases - an algorithmic method in polynomial ideal theory*, Chapter 6 in N.K. Bose (ed.): "Multidimensional Systems Theory", D. Reidel Publisher, 1985.

[5] J. DÉSARMIEN, J.P.S. KUNG AND G.-C. ROTA, *Invariant theory, Young bitableaux, and combinatorics*, Advances in Math. 27 (1978) 63–92.

[6] J.A. DIEUDONNÉ AND J.B. CARRELL, *Invariant Theory - Old and New*, Academic Press, New York, 1971.

[7] P. DOUBILET, G.-C. ROTA AND J.A. STEIN, *On the foundations of combinatorial theory IX: Combinatorial methods in invariant theory*, Stud. Appl. Math 53 (1976) 185–216.

[8] F.D. GROSSHANS, G.-C. ROTA AND J.A. STEIN, *Invariant theory and superalgebras*, AMS Regional Conference Series 69, Providence, R.I., 1987.

[9] D. HILBERT, *Über die Theorie der algebraischen Formen*, Math. Annalen 36 (1890) 473–534.

[10] G. KEMPF, *Computing invariants*, in S.S. Koh (ed.): Invariant Theory, Springer Lecture Notes 1278, Heidelberg, 1987.

[11] J.P.S. KUNG AND G.-C. ROTA, *The invariant theory of binary forms*, Bull. Amer. Math. Soc. 10 (1984) 27–85.

[12] T. MCMILLAN, *Ph.D. Dissertation*, University of Florida, in preparation.

[13] G.-C. ROTA AND J.A. STEIN, *Symbolic method in invariant theory*, Proc. Natl. Acad. Sci. USA 83 (1986) 844-847.

[14] B. STURMFELS AND N. WHITE, *Gröbner bases and invariant theory*, Advances in Math., to appear.

[15] M. SWEEDLER, *Hopf algebras*, Benjamin, Reading, Massachussetts, 1969.

[16] N. WHITE, *The bracket of 2-extensors*, Congressus Numerantium 40 (1983) 419–428.

[17] N. WHITE, *Multilinear Cayley factorization*, in: Symbolic Computations in Geometry, I.M.A. Preprint # 389, University of Minnesota, 1988.

[18] N. WHITE AND W. WHITELEY, *The algebraic geometry of stresses in frameworks*, SIAM J. Alg. Discrete Math. 4 (1983) 481–511.

IMPLEMENTATION OF THE STRAIGHTENING ALGORITHM OF CLASSICAL INVARIANT THEORY

NEIL WHITE*

Abstract. The straightening algorithm for bracket polynomials or Young tableaux has many possible variations. We examine the choices that are involved in implementing a straightening algorithm, describe some particular variations which we found to be relatively efficient, and provide a comparison of their performances on a number of instances of the problem. We also draw connections to the more general Gröbner basis normal form algorithms. Finally, we use one of the variations to straighten an invariant of Turnbull and Young which specifies when ten points lie on a common quadric surface in projective three-space.

1. The Straightening Algorithm. Let $a, b, \ldots, z$ be N points (or unspecified vectors) in a vector space V of dimension r over an arbitrary field K. The First Fundamental Theorem of Invariant Theory [4,5,13] states that the invariants of the general linear group acting on V are the homogeneous polynomials in the *brackets*, which are determinants of r of the vectors. Homogeneity here refers to the usual $\mathbf{N}$ grading of polynomials. The invariants of the projective linear group are the bracket polynomials which satisfy a stronger homogeneity condition, namely that they are homogeneous with respect to the $\mathbf{N}^N$-grading induced on a bracket monomial M by the number of occurrences of each point in M. Of course, V may have infinite cardinality, but a bracket polynomial involves only a finite number of points, and for notational convenience we are assuming that they are always contained in the given set of points. Each bracket monomial will be abbreviated as a *tableau*, or rectangular $d \times r$ array of points, where each bracket has its points listed as one row of the tableau, and d is the $\mathbf{N}$-degree of the monomial.

The straightening algorithm is a procedure for writing any bracket polynomial, or linear combination of tableaux, as a linear combination of certain tableaux called *standard tableaux*. It is well known that the standard tableaux form a linear basis of the vector space of bracket polynomials. The reader is assumed to have some familiarity with the straightening algorithm; for an introduction see [10].

The following six items are required for the straightening algorithm.

1. A linear ordering on the points.

We will assume that alphabetical ordering on $a, b, \ldots, z$ has been chosen. The definition of standard tableau is stated in terms of this ordering: a tableau is standard if the points in each row are written in strictly increasing order from left to right, and the points in each column are written in non-decreasing order from top to bottom. Actually, by anti-symmetry of the brackets, we may rewrite any non-zero tableau so that each row is in strictly increasing order, up to a sign change on coefficient of the entire tableau. We will henceforth assume that all tableaux are so rewritten whenever appropriate.

---

*Institute for Mathematics and its Applications, University of Minnesota, Minneapolis, MN 55455; U.S.A. and Department of Mathematics, University of Florida, Gainesville, FL 32611; U.S.A.

2. A linear ordering on the brackets, or rows.

The usual orderings are lexicographic or reverse lexicographic ordering induced by the order in Item 1. We will now define these on arbitrary n-sequences of elements from a linearly ordered set S. We say that $(x_1, x_2, \ldots, x_n) < (y_1, y_2, \ldots, y_n)$ in *lexicographic order induced by the order on* S if there exists i, $1 \leq i \leq n$, such that $x_i < y_i$ in S, and for all j, $1 \leq j < i$, $x_j = y_j$. We say that $(x_1, x_2, \ldots, x_n) < (y_1, y_2, \ldots, y_n)$ in *reverse lexicographic order induced by the order on* S if there exists i, $1 \leq i \leq n$, such that $x_i < y_i$ in S, and for all j, $i < j \leq n$, $x_j = y_j$.

Since multiplication of brackets is commutative, the rows of a tableau may be interchanged freely, hence we may assume that the rows are always written in order. For example, in lexicographic order, this implies that we may assume that every tableau has non-decreasing order in its first column, as well as strictly increasing rows.

3. A linear ordering on tableaux.

This is usually obtained by thinking of each tableau as a sequence of d rows, and using either lexicographic or dual reverse lexicographic order induced by the order on the rows in Item 2. *Dual* here means that $(x_1, x_2, \ldots, x_n) > (y_1, y_2, \ldots, y_n)$ replaces $(x_1, x_2, \ldots, x_n) < (y_1, y_2, \ldots, y_n)$.

4. A class of syzygies which gives a monotone improvement in the tableau ordering and which is sufficient to correct any violation of standardness.

A *syzygy* is a non-trivial bracket polynomial which is identically zero. For our purposes, a syzygy will always be solved for its greatest term (in the tableau ordering), and used to substitute for that term. The resulting monotone improvement will guarantee termination of the straightening algorithm, because there are always a finite number of tableaux less than a given tableau.

The most commonly used syzygy for straightening algorithm purposes is the *van der Waerden syzygy*, which is

$$\sum sgn(\sigma)[x_1, x_2, \ldots, \sigma(x_q), \sigma(x_{q+1}), \ldots, \sigma(x_r)][\sigma(y_1), \sigma(y_2), \ldots, \sigma(y_q), y_{q+1}, \ldots, y_r] = 0.$$

The sum is over combinations σ of q of the symbols $y_1, \ldots, y_q, x_q, \ldots, x_r$ chosen for insertion into the first q places in the second bracket and the remaining $r-q+1$ symbols inserted into the last $r-q+1$ places in the first bracket. The sign of σ is just the usual sign of σ when σ is thought of as a permutation. If $x_q > y_q$, then $[x_1, \ldots, x_d][y_1, \ldots, y_d]$ is the greatest tableau in the syzygy in either lexicographic or dual reverse lexicographic order on tableaux, and, furthermore, the violation of standardness by x_q and y_q in column q is corrected in all other terms of the syzygy. The above syzygy will be abbreviated by the following underline notation:

$$\begin{bmatrix} x_1 & x_2 & \cdots & \underline{x}_q & \underline{x}_{q+1} & \cdots & \underline{x}_r \\ \underline{y}_1 & \underline{y}_2 & \cdots & \underline{y}_q & y_{q+1} & \cdots & y_r \end{bmatrix}.$$

That all such syzygies are identically zero is easily seen, for the syzygy itself may be regarded as an alternating multilinear form in $r+1$ vector variables from a vector space of dimension r.

Another useful syzygy is the *multiple syzygy*,

$$[x_1, x_2, \ldots, x_{r-q}, x_{r-q+1}, \ldots, x_r][y_1, y_2, \ldots, y_r] \; - \\ \sum [x_1, x_2, \ldots, x_{r-q}, y_{i_1}, \ldots, y_{i_q}][y_1, y_2, \ldots, y_r \leftarrow x_{r-q+1}, \ldots, x_r] \quad = \quad 0,$$

where the arrow denotes substitution of $x_{r-q+1}, \ldots, x_r$ for $y_{i_1}, \ldots, y_{i_q}$ in order, and the sum is over all q-combinations $i_1 < i_2 < \ldots < i_q$ from $1, \ldots, r$. This syzygy will be abbreviated

$$\begin{bmatrix} x_1 & x_2 & \cdots & x_{r-q} & \underline{x}_{r-q+1} & \cdots & \underline{x}_r \\ \underline{y}_1 & \underline{y}_2 & \cdots & \underline{y}_{r-q} & \underline{y}_{r-q+1} & \cdots & \underline{y}_r \end{bmatrix}.$$

Because of the special case of van der Waerden syzygy we are using, as above, the notation for van der Waerden and multiple syzygies will not coincide except in the particular situation of the last element of one bracket and the entire next bracket being underlined. However, in this case, the two types of syzygies are easily seen to coincide, and this syzygy is usually referred to as an *ordinary syzygy*. The ordinary syzygies are the well-known Grassmann-Plücker relations, and in general they can interchange any fixed element of one bracket with variable elements of another bracket.

5. A strategy to choose which violation of standardness in a given tableau T will be corrected.

One typical choice is to find the first pair of consecutive rows of T, starting from the top, which contain a violation, and choose the first column from the left containing a violation in those two rows. This will be referred to as the *first-first* strategy. Another choice is the analogous *last-last* strategy. Other strategies may involve choosing a violation in non-consecutive rows.

6 A strategy to choose which syzygy in Item 4 will be used to correct the violation in Item 5.

The straightening algorithm can now be described as a whole. The input bracket polynomial is first broken into its homogeneous components with respect to the $\mathbf{N}^N$-grading. The algorithm is then applied to each component separately. The terms of the homogeneous polynomial are sorted according to the ordering in Item 3. The greatest non-standard tableau T is selected, and one of its violations corrected according to Items 5 and 6. This means that a syzygy is applied which rewrites T as a linear combination of smaller tableaux, each of the same $\mathbf{N}^N$-grade as T. Thus the greatest non-standard tableau remaining is smaller than previously, and since there are only a finite number of tableaux of a given $\mathbf{N}^N$-grade, the algorithm must terminate, outputing a linear combination of standard tableaux.

As shown in [10], the straightening algorithm is a special case of the normal form algorithm with respect to a Gröbner basis. On the other hand, the terminology of standard and non-standard terms can be very useful when working with Gröbner bases. Items 1 and 2 correspond to the choice of an ordering on the variables in Gröbner basis theory. Two steps are needed to specify the ordering, since the variables are the brackets, which have another level of combinatorial structure, namely, the points. Item 3, the tableau order, is the admissable order on monomials. The choice of syzygies in Item 4 is simply the choice of a particular Gröbner basis, which is far from unique. A reduced Gröbner basis, which is unique, corresponds to a syzygy for each non-standard tableau of two rows, which expresses that tableau directly as a linear combination of standard tableaux. In the straightening algorithm, it is generally not practical to compute the reduced Gröbner basis ahead of time. The strategies in Items 5 and 6 amount to a strategy for an order in which to reduce terms in the normal form algorithm. As we shall see, there are many ways to choose such a strategy in general, and the choice may have a significant effect on the efficiency of the algorithm, even if a reduced Gröbner basis is used. This problem seems not to have been sufficiently studied in Gröbner basis theory.

2. Implementation of the Straightening Algorithm. There is a wide variety of choices in the orderings, syzygies, and strategies of a straightening algorithm. All must give the same output for a given input, since the standard tableaux form a linear basis for all tableaux. What effect do these different choices make on the efficiency of the algorithm?

First we observe that any bracket straightening algorithm is inherently exponential. The number of standard tableaux of a given shape is given by the well-known hook-length formula [6], which is factorial in the number of entries of the tableau. An input of a single non-standard tableau may require most of these standard tableaux to appear in its output.

We implemented a number of variations of the straightening algorithm, and will now describe some of the relatively more efficient ones. All were written in FORTRAN and run on an Apollo Domain 3000. All of them keep two separate stacks of tableaux, the first having both standard and non-standard tableaux sorted in tableau order, and the second having only standard tableaux which are greater in tableau order than any tableau in the first stack. The program then looks at the greatest tableau in the first stack, and if it is standard, moves it to the second stack, but if non-standard, applies the appropriate syzygy and returns all tableaux resulting from the syzygy to the first stack by a mergesort. All of the variations have a limit of 5000 tableaux in each of the two stacks. All of the variations discussed below use lexicographic order on tableaux, since a variation using dual reverse lexicographic order may be converted to lexicographic order by rotating all tableaux by 180°, reversing the order on the points, switching from reverse lexicographic on rows to lexicographic or vice-versa, and suitably modifying the syzygies and strategies used. The variations discussed also use only lexicographic order on the rows, since a few variations using reverse lexicographic order on rows and lexicographic order on tableaux were found not to be competitive. The two simplest variations, using only van der Waerden syzygies or else only multiple syzygies, and lexicographic order on both rows and tableaux, were both found not to be competitive.

We now list five variations which were found to be relatively competitive. In the following descriptions, $a, b, \ldots$ are the first points in order, and $\ldots, y, z$ the last points. Let T be the tableau being currently examined, that is, the greatest tableau in the first stack, which we may assume is non-standard.

A. If the first pair of consecutive rows of T which have a violation have one in the last column but not in the second column, use an ordinary syzygy on the last element of the former row and the entire latter row. If there are violations in the second and last columns, use an ordinary syzygy on the first element of the former row and the entire latter row. If there is no violation in the last column of the two rows, then correct the first violation in those two rows using a van der Waerden syzygy (a first-first strategy). For example,

$$\begin{bmatrix} a & b & \dots & \underline{z} \\ \underline{b} & \underline{c} & \dots & \underline{y} \end{bmatrix},$$

$$\begin{bmatrix} \underline{a} & d & \dots & z \\ \underline{b} & \underline{c} & \dots & \underline{y} \end{bmatrix},$$

$$\begin{bmatrix} a & \dots & \underline{k} & \dots & \underline{y} \\ \underline{b} & \dots & \underline{j} & \dots & z \end{bmatrix}.$$

B. Examine the first two consecutive rows of T which have a violation. Let the s-th column be the first column having a violation in those two rows. If the last column or two of the last three columns have a violation in those two rows, then use a multiple syzygy on the s-th through last points in the former row, and the entire latter row. Otherwise use a van der Waerden syzygy on the violation in the s-th column (a first-first strategy). For example,

$$\begin{bmatrix} a & \dots & \underline{k} & \dots & \underline{z} \\ \underline{b} & \dots & \underline{j} & \dots & \underline{y} \end{bmatrix},$$

$$\begin{bmatrix} a & \dots & \underline{k} & \dots & \underline{w} & \underline{x} & \underline{y} \\ \underline{b} & \dots & \underline{j} & \dots & \underline{u} & \underline{w} & \underline{z} \end{bmatrix},$$

$$\begin{bmatrix} a & \dots & \underline{k} & \dots & \underline{y} \\ \underline{b} & \dots & \underline{j} & \dots & z \end{bmatrix}.$$

C. This is the same as B, except that if those two rows have a violation in the last column, say z over y, then the last column is searched for the least element situated below z, and a multiple syzygy is performed as in B on the row of z and that least element. Thus we may be correcting a violation in non-consecutive rows.

D. This is the same as C, except that only the last column non-consecutive row multiple syzygies and the van der Waerden syzygies are used, not the multiple syzygies when there are violations in two of the last three columns.

E. Find the last violation in the first pair of consecutive rows having a violation, and do van der Waerden syzygies.

In Table 1, we describe the performance of the five above variations on seventeen different inputs. For each input, we list r, the dimension or row length, d, the **N**-degree or column length, and N, the number of distinct points among the rd entries in each tableau. This is followed by the number of terms of the input, the number of terms of the output, and the performance of each of the five variations. This performance is measured by the number of syzygies performed times the average size of the first stack of tableaux. This number is approximately proportional to the execution time, excluding time for input, output, and an initial sort. An hour of execution time on the Apollo 3000 corresponded to approximately 2.2 million, a minute to about 30,000. In the table, 'e' denotes that the stack size was exceeded, 'm' denotes million and 'k' denotes thousand. The best performance for each input is indicated in boldface.

TABLE 1

Performance of Five Variations

	r	d	N	In	Out	A	B	C	D	E
1.	5	2	10	12	16	709	**185**	**185**	709	347
2.	4	3	12	1	6	24k	1557	**7**	**7**	22k
3.	5	3	12	1	115	13k	16k	9k	9k	**2.6k**
4.	3	6	9	9	0	**13k**	**13k**	21k	21k	52k
5.	5	4	15	1	2677	6m	**1.2m**	1.3m	1.5m	9m
6.	6	3	14	1	508	159k	115k	**109k**	147k	158k
7.	6	2	12	1	104	**603**	1206	1206	**603**	802
8.	5	4	13	1	78	4.6k	1.8k	**1.6k**	**1.6k**	6.6k
9.	5	4	14	1	314	95k	29k	**7.3k**	**7.3k**	65k
10.	3	6	10	2	27	**303**	**303**	**303**	**303**	372
11.	4	3	12	9	3	**171**	3.6k	7.5k	2.8k	240
12.	3	5	15	24	526	368k	368k	467k	467k	**365k**
13.	4	4	16	27	2111	e	28m	28m	**16m**	e
14.	3	6	10	2	245	75k	75k	**59k**	**59k**	79k
15.	8	3	14	1	60	3k	**181**	**181**	**181**	5k
16.	8	3	15	1	295	31k	9.7k	**9.2k**	14k	36k
17.	8	3	18	1	2531	e	**2.042m**	**2.042m**	2.044m	e

We note that the relative performances of the five variations, as shown in Table 1, are quite erratic. The best overall were perhaps C and D, but they were beaten badly by each of the others on certain inputs. If time but not space were the only consideration, then a parallel implementation of all five of these (and perhaps several more) would be the fastest straightening algorithm. We do not claim to be exhaustive in our investigation, and it is possible that better variations exist.

3. The Turnbull-Young Invariant. The Turnbull-Young invariant [12] is a 240-term bracket polynomial with $r = 5$, $d = 4$, and $N = 10$, which is zero precisely when the 10 points lie on a common quadric surface in projective 3-space. This polynomial was used as input for our variations C and D. The stack size was exceeded by C, but D produced the output after about two days. This output appears below; this is probably the first time that this invariant has been straightened. A numerical factor of 20 appears, as predicted by Turnbull and Young. A very interesting question is whether there exists a theorem analogous to Pascal's theorem for 6 points on a planar conic, which would give an incidence property of ten points equivalent to their lying on a common quadric. Such a property would amount to a Cayley factorization (see [15]) of the Turnbull-Young invariant, or of some bracket multiple of it. The straightened output of this invariant was input into another program which attempts to find such a Cayley factorization, by identifying certain points and re-straightening. This program exceeded the stack size in one of these later straightenings, but it had proceeded far enough that it appears very unlikely that the Turnbull-Young invariant can be Cayley factored directly. Sturmfels and Whiteley [11] show that some bracket multiple of any such invariant must be Cayley factorable, hence there exists a projective geometric incidence theorem. However, this theorem may be so complex as to be uninteresting.

TABLE 2

Straightened form of the Turnbull-Young Invariant

[0136] [0247] [1258] [3459] [6789] (-20)
[0135] [0247] [1268] [3469] [5789] (20)
[0135] [0246] [1278] [3479] [5689] (-20)
[0134] [0257] [1268] [3569] [4789] (-20)
[0134] [0256] [1278] [3579] [4689] (20)
[0134] [0256] [1258] [3479] [6789] (-20)
[0134] [0256] [1257] [3489] [6789] (20)
[0134] [0246] [1258] [3579] [6789] (20)
[0134] [0246] [1257] [3589] [6789] (-20)
[0134] [0245] [1268] [3579] [6789] (-20)
[0134] [0245] [1267] [3589] [6789] (20)
[0134] [0245] [1256] [3789] [6789] (20)
[0134] [0236] [1258] [4579] [6789] (-20)
[0134] [0236] [1257] [4589] [6789] (20)
[0134] [0235] [1268] [4579] [6789] (20)
[0134] [0235] [1267] [4589] [6789] (-20)
[0134] [0235] [1256] [4789] [6789] (-20)
[0134] [0234] [1256] [5789] [6789] (20)
[0126] [0347] [1358] [2459] [6789] (20)
[0125] [0347] [1368] [2469] [5789] (-20)
[0125] [0346] [1378] [2479] [5689] (20)
[0124] [0357] [1368] [2569] [4789] (20)
[0124] [0356] [1378] [2579] [4689] (-20)
[0124] [0356] [1358] [2479] [6789] (20)
[0124] [0356] [1357] [2489] [6789] (-20)
[0124] [0346] [1358] [2579] [6789] (-20)
[0124] [0346] [1357] [2589] [6789] (20)
[0124] [0345] [1368] [2579] [6789] (20)
[0124] [0345] [1367] [2589] [6789] (-20)
[0124] [0345] [1356] [2789] [6789] (-20)
[0124] [0236] [1358] [4579] [6789] (-20)
[0124] [0236] [1357] [4589] [6789] (20)
[0124] [0235] [1368] [4579] [6789] (20)
[0124] [0235] [1367] [4589] [6789] (-20)
[0124] [0235] [1356] [4789] [6789] (-20)
[0124] [0234] [1356] [5789] [6789] (20)
[0123] [0245] [1468] [3579] [6789] (-20)
[0123] [0245] [1467] [3589] [6789] (40)
[0123] [0245] [1457] [3689] [6789] (-20)
[0123] [0245] [1456] [3789] [6789] (20)
[0123] [0245] [1378] [4679] [5689] (-20)
[0123] [0245] [1368] [4679] [5789] (-20)
[0123] [0245] [1367] [4589] [6789] (-40)
[0123] [0245] [1347] [5689] [6789] (-40)
[0123] [0237] [1458] [4569] [6789] (20)
[0123] [0236] [1457] [4589] [6789] (-40)
[0123] [0235] [1467] [4589] [6789] (20)
[0123] [0235] [1457] [4689] [6789] (20)
[0123] [0234] [1567] [4689] [5789] (-20)
[0123] [0234] [1567] [4589] [6789] (-40)
[0123] [0234] [1467] [5689] [5789] (20)
[0123] [0157] [2468] [3469] [5789] (20)
[0123] [0157] [2458] [3469] [6789] (-20)
[0123] [0156] [2478] [3479] [5689] (-20)
[0123] [0156] [2458] [3479] [6789] (20)
[0123] [0147] [2468] [3569] [5789] (-20)
[0123] [0147] [2458] [3569] [6789] (20)
[0123] [0147] [2368] [4569] [5789] (20)
[0123] [0146] [2478] [3579] [5689] (20)
[0123] [0146] [2458] [3579] [6789] (20)
[0123] [0146] [2378] [4579] [5689] (-20)
[0123] [0146] [2358] [4579] [6789] (-40)
[0123] [0145] [2478] [3679] [5689] (-20)
[0123] [0145] [2468] [3679] [5789] (-20)
[0123] [0145] [2468] [3579] [6789] (-40)
[0123] [0145] [2467] [3689] [5789] (20)
[0123] [0145] [2458] [3679] [6789] (40)
[0123] [0145] [2457] [3689] [6789] (-20)
[0123] [0145] [2456] [3789] [6789] (40)
[0123] [0145] [2378] [4679] [5689] (20)
[0123] [0145] [2368] [4679] [5789] (20)
[0123] [0145] [2368] [4579] [6789] (40)

TABLE 2 Continued

[0124] [0136] [2357] [4589] [6789] (-20)	[0123] [0145] [2367] [4689] [5789] (-20)
[0124] [0135] [2367] [4589] [6789] (20)	[0123] [0145] [2358] [4679] [6789] (-20)
[0124] [0135] [2356] [4789] [6789] (20)	[0123] [0145] [2357] [4689] [6789] (20)
[0124] [0135] [2348] [5679] [6789] (20)	[0123] [0145] [2356] [4789] [6789] (-60)
[0124] [0135] [2347] [5689] [6789] (-20)	[0123] [0145] [2348] [5679] [6789] (40)
[0124] [0135] [2346] [5789] [6789] (20)	[0123] [0145] [2346] [5789] [6789] (60)
[0124] [0135] [2345] [6789] [6789] (-20)	[0123] [0145] [2345] [6789] [6789] (-60)
[0124] [0134] [2358] [5679] [6789] (-20)	[0123] [0137] [2458] [4569] [6789] (-20)
[0124] [0134] [2357] [5689] [6789] (20)	[0123] [0136] [2458] [4579] [6789] (20)
[0124] [0134] [2356] [5789] [6789] (-40)	[0123] [0136] [2457] [4589] [6789] (20)
[0123] [0457] [1468] [2569] [3789] (-20)	[0123] [0135] [2467] [4589] [6789] (-20)
[0123] [0456] [1478] [2579] [3689] (20)	[0123] [0135] [2458] [4679] [6789] (-20)
[0123] [0356] [1458] [2479] [6789] (-20)	[0123] [0134] [2568] [4579] [6789] (40)
[0123] [0356] [1457] [2489] [6789] (20)	[0123] [0134] [2567] [4689] [5789] (20)
[0123] [0346] [1457] [2589] [6789] (-20)	[0123] [0134] [2467] [5689] [5789] (-20)
[0123] [0345] [1468] [2579] [6789] (20)	[0123] [0134] [2456] [5789] [6789] (20)
[0123] [0345] [1458] [2679] [6789] (-20)	[0123] [0127] [3468] [4569] [5789] (20)
[0123] [0345] [1457] [2689] [6789] (20)	[0123] [0127] [3458] [4569] [6789] (20)
[0123] [0345] [1456] [2789] [6789] (-20)	[0123] [0126] [3478] [4579] [5689] (-20)
[0123] [0257] [1468] [3469] [5789] (-20)	[0123] [0126] [3458] [4579] [6789] (-20)
[0123] [0257] [1458] [3469] [6789] (20)	[0123] [0126] [3457] [4589] [6789] (-40)
[0123] [0256] [1478] [3479] [5689] (20)	[0123] [0125] [3478] [4679] [5689] (20)
[0123] [0256] [1457] [3489] [6789] (-20)	[0123] [0125] [3468] [4679] [5789] (20)
[0123] [0247] [1468] [3569] [5789] (20)	[0123] [0125] [3467] [4689] [5789] (-20)
[0123] [0247] [1368] [4569] [5789] (-20)	[0123] [0125] [3467] [4589] [6789] (60)
[0123] [0247] [1358] [4569] [6789] (-20)	[0123] [0124] [3567] [4689] [5789] (-20)
[0123] [0246] [1478] [3579] [5689] (-20)	[0123] [0124] [3567] [4589] [6789] (-60)
[0123] [0246] [1457] [3589] [6789] (-20)	[0123] [0124] [3478] [5679] [5689] (-20)
[0123] [0246] [1378] [4579] [5689] (20)	[0123] [0124] [3468] [5679] [5789] (-20)
[0123] [0246] [1358] [4579] [6789] (20)	[0123] [0124] [3467] [5689] [5789] (40)
[0123] [0246] [1357] [4589] [6789] (40)	[0123] [0124] [3457] [5689] [6789] (-20)
[0123] [0245] [1478] [3679] [5689] (20)	[0123] [0123] [4567] [4689] [5789] (20)
[0123] [0245] [1468] [3679] [5789] (20)	[0123] [0123] [4567] [4589] [6789] (60)

4. Acknowledgements. I wish to thank Bernd Sturmfels for programming an early version of the straightening algorithm, which provided several helpful subroutines. I also thank Joel Stein for help in preparing the input of the Turnbull-Young invariant.

REFERENCES

[1] M. Barnabei, A. Brini and G.-C. Rota, *On the exterior calculus of invariant theory*, J. Algebra, 96 (1985), 120–160.

[2] C. De Concini and C. Procesi, *A characteristic free approach to invariant theory*, Advances in Math., 21 (1976), 330–354.

[3] J. Désarménien, J. Kung and R.-C. Rota, *Invariant theory, Young tableaux, and Combinatorics*, Advances in Math., 27 (1978), 63–92.

[4] J. Dieudonné and J. Carrell, *Invariant Theory, Old and New*, Academic Press (1971).

[5] P. Doubilet, G.-C. Rota and J. Stein, *On the foundations of combinatorial theory: IX, Combinatorial methods in invariant theory*, Studies in Appl. Math., 53, (1974) 185–216.

[6] J. S. Frame, G. de B. Robinson, and R. Thrall, *The hook graphs of* S_n, Can. J. Math., 6 (1954), 316–324.

[7] W.V.D. Hodge and D. Pedoe, *Methods of Algebraic Geometry*, Cambridge University Press (1947).

[8] C. Procesi, *A Primer in Invariant Theory*, Brandeis Lecture Notes 1, September (1982).

[9] B. Sturmfels, *Computational Synthetic Geometry*, Ph.D. Dissertation, University of Washington, Seattle (1987).

[10] B. Sturmfels and N. White, *Gröbner bases and invariant theory*, to appear, Advances in Math.

[11] B. Sturmfels and W. Whiteley, *On the synthetic factorization of homogeneous invariants*, in Symbolic Computations in Geometry, Institute for Mathematics and Its Applications, University of Minnesota Preprint Series no. 389.

[12] H. Turnbull and A. Young, *Linear invariants of ten quaternary quadrics*, Trans. Camb. Phil. Soc., 23 (1926), 265–301.

[13] H. Weyl, *The Classical Groups – Their Invariants and Representations*, Princeton University Press (1939).

[14] N. White, *The bracket ring of a combinatorial geometry. I*, Transactions Amer. Math. Soc. 202 (1975), 79–103.

[15] N. White, *Multilinear Cayley factorization*, in Symbolic Computations in Geometry, Institute for Math and Its Appl., Preprint no. 389, Minneapolis.

[16] A. Young, *On quantitative substitutional analysis* (3rd paper), Proc. London Math. Soc., Ser. 2, **28** (1928) 255–292..

CANONICAL FORMS OF BINARY FORMS: VARIATIONS ON A THEME OF SYLVESTER

JOSEPH P.S. KUNG*

Abstract. In 1851, J.J. Sylvester gave two proofs of his theorem that a generic binary form of odd degree $n = 2p+1$ can be expressed as the sum of $p+1$ n^{th} powers of linear forms. The ideas in the chronologically first proof led to S. Gundelfinger's general theorem giving conditions in terms of the vanishing of certain covariants related to Wronskians for a form of degree n to be expressible as a sum of k n^{th} powers. The ideas in the second proof led to the theory of apolarity. In this survey, we trace the historical development of these ideas. In particular, we describe Sylvester's two proofs and Gundelfinger's own proof of his theorem. Most of this paper is elementary and assumes no prior knowledge of classical invariant theory.

Key words. Polynomials, binary forms, canonical forms, classical invariant theory.

AMS(MOS) subject classifications. (1985): 11C08, 15A72, 05A40, 01A55.

1. Waring's Problem for Binary Forms. The method of "completing the square" allows us to write a quadratic polynomial in the form $a(x-h)^2+k$. Over the complex numbers, this method yields an expression of a quadratic as a sum of two squares, $(\alpha x+\beta)^2+\gamma^2$. This suggests the following question, reminiscent of Waring's problem in number theory:

How many n^{th} powers are needed to express a polynomial of degree n as a sum of nth powers?

This question was the starting point of the theory of canonical forms for binary forms in classical invariant theory (see [2, Chap. 12] and [4, Chap. 11] for contemporaneous accounts). The purpose of this paper is to give a historically oriented account of the two main theorems in this area, Sylvester's theorem and Gundelfinger's theorem. Throughout this paper, we shall work over the field of complex numbers.

A *binary form f of degree n in the variables x and y* is a homogeneous polynomial of degree n in x and y. Thus

$$\begin{aligned} f(x,y) &= \sum_{i=0}^{n} \binom{n}{i} a_i x^i y^{n-i} \\ &= a_n x^n + \binom{n}{1} a_{n-1} x^{n-1} y + \binom{n}{2} a_{n-2} x^{n-2} y^2 + \cdots + a_0 y^n. \end{aligned}$$

Note that binomial coefficients are built into the coefficients of f; this is convenient for working with powers $(\alpha x+\beta y)^n$ of linear forms. The numbers a_i are called the *(normalized) coefficients* of f.

*Department of Mathematics, University of North Texas, Denton, Texas 76203. The author was supported by National Science Foundation Grant DMS-8722431.

By homogenizing a polynomial $p(x)$ of degree n, we can obtain a form $f(x,y) = y^n p(x/y)$. Conversely, from a form $f(x,y)$, we can obtain a polynomial $f(x,1)$ by setting $y = 1$. Thus, polynomials and binary forms are equivalent mathematical objects. A form $f(x,y)$ of degree n can be factored into a product of n linear forms by factoring the polynomial $f(x,1)$. Two nonzero linear forms $\alpha x + \beta y$ and $\gamma x + \delta y$ are said to be *distinct* if $\alpha x + \beta y$ is not a constant multiple of $\gamma x + \delta y$.

The set of all binary forms of degree n forms a vector space under addition and scalar multiplication. We shall denote this vector space by $\mathfrak{F}_n$. The dimension of $\mathfrak{F}_n$ equals $n+1$. The general linear group $GL(2)$ acts on $\mathfrak{F}_n$ as linear changes of variables: if (c_{ij}) is a nonsingular 2×2 matrix, then $f(x,y)$ is sent to $\overline{f}(\overline{x},\overline{y})$, where

$$x = c_{11}\overline{x} + c_{12}\overline{y} \quad \text{and} \quad y = c_{21}\overline{x} + c_{22}\overline{y}.$$

Expanding and grouping terms in

$$\begin{aligned}\overline{f}(\overline{x},\overline{y}) &= \sum_{i=0}^{n} \binom{n}{i} \overline{a}_i \overline{x}^i \overline{y}^{n-i} \\ &= \sum_{i=0}^{n} \binom{n}{i} a_i (c_{11}\overline{x} + c_{12}\overline{y})^i (c_{21}\overline{x} + c_{22}\overline{y})^{n-i}.\end{aligned}$$

we conclude that (c_{ij}) acts linearly on the coefficients by

$$\overline{a}_k = \sum_{m=0}^{n} \left[\sum_{i=m-n+k}^{\min(m,k)} \binom{k}{i} \binom{n-k}{m-i} c_{11}^i c_{12}^{m-i} c_{21}^{k-i} c_{22}^{n-k-m+i} \right] a_m. \tag{1.1}$$

Since a sum of m n^{th} powers remains a sum of m n^{th} powers under linear changes of variables, one expects $GL(2)$ to play some role in the theory of canonical forms. As we shall see, it does, but in a very hidden manner.

2. Sylvester's "Remarkable Discovery". In 1851, Sylvester made the (in his own words) "remarkable discovery" that a generic binary form of odd degree $n = 2p+1$ can be written as the sum of
$p + 1 n^{th}$ powers of linear forms. Sylvester gave two proofs of this result and both proofs contributed basic ideas to the theory of canonical forms. We shall first describe the chronologically second proof which appeared in [18].

Suppose $f(x,y) = \sum_{i=0}^{n} \binom{n}{i} a_i x^i y^{n-i}$ can in fact be written as the sum of $p + 1$ n^{th} powers: that is,

$$f(x,y) = (\alpha_1 x + \beta_1 y)^n + (\alpha_2 x + \beta_2 y)^n + \cdots + (\alpha_{p+1} x + \beta_{p+1} y)^n. \tag{2.1}$$

Writing $\lambda_i = \beta_i/\alpha_i$, expanding, and equating coefficients, we obtain, for $i = 0, 1, 2, \ldots, n$, the equations:

$$a_i = \alpha_1^n \lambda_1^{n-i} + \alpha_2^n \lambda_2^{n-i} + \cdots + \alpha_{p+1}^n \lambda_{p+1}^{n-i} \tag{E_i}$$

Eliminating $\alpha_1, \alpha_2, \ldots, \alpha_{p+1}$ successively from the first $p+2$ equations $E_0, \ldots, E_{p+1}$, we obtain the equation

$$E_0 - s_p E_1 + s_{p-1} E_2 - \cdots \pm s_0 E_{p+1},$$

which works out to be

$$a_0 s_{p+1} - a_1 s_p + a_2 s_{p-1} - \cdots \pm a_{p+1} s_0 = 0;$$

here $s_{p+1} = 1$, $s_p = \lambda_1 + \lambda_2 + \cdots + \lambda_{p+1}, s_{p-1} = \Sigma \lambda_i \lambda_j, \ldots, s_0 = \lambda_1 \lambda_2 \ldots \lambda_{p+1}$ are the elementary symmetric functions in $\lambda_1, \lambda_2, \ldots \lambda_{p+1}$. An example may make this computation clearer. Take $n = 3$. Then the three equations E_0, E_1, E_2 are

$$\begin{aligned} a_0 &= {\alpha_1} 3 {\lambda_1}^3 + {\alpha_2}^3 {\lambda_2} 3, \\ a_1 &= {\alpha_1} 3 \lambda_1^2 + {\alpha_2}^3 {\lambda_2}^2, \\ a_2 &= {\alpha_1}^3 \lambda_1 + {\alpha_2}^3 \lambda_2. \end{aligned}$$

The equation $E_0 - (\lambda_1 + \lambda_2) E_1 + \lambda_1 \lambda_2 E_2$ can be obtained by adding together the equations

$$\begin{aligned} a_0 &= {\alpha_1}^3 {\lambda_1}^3 + {\alpha_2}^3 {\lambda_2}^3, \\ -\lambda_1 a_1 &= -{\alpha_1}^3 {\lambda_1}^3 - {\alpha_2}^3 \lambda_1 {\lambda_2}^2, \\ -\lambda_2 a_1 &= -{\alpha_1}^3 \lambda_2 {\lambda_1}^2 - {\alpha_2}^3 {\lambda_2}^3, \\ \lambda_1 \lambda_2 a_2 &= {\alpha_1}^3 {\lambda_1}^2 \lambda_2 + {\alpha_2}^3 \lambda_1 {\lambda_2}^2. \end{aligned}$$

From this, we see that terms on the right hand side occur in pairs with opposite signs and hence, cancel to yield zero. It is not hard to show, by induction, that this pairing and cancellation occur in general.

Applying the same argument to the equations $E_i, \ldots, E_{i+p+1}$, we obtain the equations: for $i = 0, 1, \ldots, p$,

$$a_i s_{p+1} - a_{i+1} s_p + a_{i+2} s_{p-1} - \cdots \pm a_{i+p+1} s_0 = 0. \tag{2.2}$$

This gives a triangular system of linear equations in the unknowns s_j. By definition, the numbers s_j are the unnormalized coefficients of the form

$$r(x, y) = (x + \lambda_1 y)(x + \lambda_2 y) \ldots (x + \lambda_{p+1} y);$$

hence, the ratios λ_i can be found by solving

$$x^{p+1} + s_p x^p + s_{p-1} x^{p-1} + \cdots + s_0 = 0,$$

Once the ratios λ_i are known, α_i (and hence β_i) can be found by substituting $\beta_i = \alpha_i \lambda_i$ into Equation (2.1), equating coefficients, solving the resulting system of linear equations in ${\alpha_i}^n$, and taking n^{th} roots.

All this can be done provided that the polynomial $r(x,1)$ has distinct roots. From the fact that a determinant with two identical rows is zero, Sylvester obtained the following explicit formula for a scalar multiple of the form $r(x,y)$:

$$J(x,y) = \begin{vmatrix} y^{p+1} & -xy^p & \dots & (-x)^{p+1} \\ a_n & a_{n-1} & & a_p \\ a_{n-1} & a_{n-2} & & a_{p-1} \\ & \vdots & & \\ a_{p+1} & a_p & \dots & a_0 \end{vmatrix} \tag{2.3}$$

The form J is called the *canonizant* of f.

Summarizing, Sylvester proved the following theorem.

THEOREM 2.1 (SYLVESTER). *Let f be a binary form of odd degree $n = 2p+1$. If the form J can be factored into distinct linear factors, then there exist $\alpha_1, \alpha_2, \dots, \alpha_{p+1}, \beta_1, \beta_2, \dots, \beta_{p+1}$ such that*

$$f(x,y) = \sum_{i=1}^{p+1} (\alpha_i x - \beta_i y)^n.$$

Note that J has distinct linear factors if and only if its discriminant $D[J]$ is nonzero. Thus, the conclusion in Theorem 2.1 holds for all binary forms of degree n in the complement of the algebraic subset consisting of the zeroes of the polynomial $D[J]$ in the space $\mathfrak{F}_n$. This makes precise the word "generic" used earlier.

In the latter part of [18], Sylvester raised the problem of finding a suitable canonical form for binary forms of even degree. This area is surveyed in [8].

3. Apolarity. From Equation (2.2), one is led naturally to the apolar covariant. Let $f(x,y) = \sum_{i=0}^{n} \binom{n}{i} a_i x^i y^{n-i}$ and $g(x,y) = \sum_{j=0}^{m} \binom{m}{j} b_j x^j y^{m-j}$ be forms of degree n and m, where $n \geq m$. Their *apolar covariant* $\{f,g\}$ is the form $\sum_{k=0}^{n-m} \binom{n-m}{k} c_k x^k y^{n-m-k}$ of degree $n-m$ whose normalized coefficients c_k are defined by

$$c_k = \sum_{l=0}^{m} (-1)^{m-l} \binom{m}{l} a_{l+k} b_{m-l}.$$

Because the coefficients c_k are linear in the coefficients a_i and b_j, the apolar covariant is a bilinear map from $\mathfrak{F}_n \times \mathfrak{F}_m$ to $\mathfrak{F}_{n-m}$.

Here is where the $GL(2)$-action comes in. Up to a scalar multiple, the apolar covariant $\{f,g\}$ is the unique bilinear map from $\mathfrak{F}_n \times \mathfrak{F}_m$ to $\mathfrak{F}_{n-m}$ which is "invariant" under $GL(2)$ in the following sense. The group $GL(2)$ acts on polynomials in the coefficients $a_0, a_1, \dots, a_n, b_0, b_1, \dots, b_m$ and variables x and y of f and g by sending $I(f,g) = I(a_i, b_j, x, y)$ to $I(\overline{f}, \overline{g}) = I(\overline{a}_i, \overline{b}_j, \overline{x}, \overline{y})$, where $\overline{a}_i$ and $\overline{b}_j$ are given by Equation (1.1). A polynomial $I(f,g)$ is a (*relative*) *covariant* if for all matrices (c_{ij}) in $GL(2)$, $I(\overline{f}, \overline{g}) = [\det(c_{ij})]^k I(f,g)$ for some nonnegative integer k. The apolar covariant $\{f,g\}$ is covariant in this sense; indeed,

$$\{\overline{f}, \overline{g}\} = [\det\,(c_{ij})]^m \{f,g\}.$$

Moreover, up to a scalar multiple, the apolar covariant $\{f,g\}$ is the unique bilinear covariant map from $\mathfrak{F}_n \times \mathfrak{F}_m$ to $\mathfrak{F}_{n-m}$ (see [9, Lemma 5.1]).

Two binary forms f and g are said to be *apolar* if their apolar covariant $\{f,g\}$ is identically zero. The set of binary forms of degree p apolar to a given form h is a subspace of the vector space $\mathfrak{F}_p$. We shall denote this subspace by $h^{\perp}$.

PROPOSITION 3.1. *Let g be a nonzero binary form of degree m and let n be a positive integer such that $n \geq m$. Then the subspace $g^{\perp}$ in $\mathfrak{F}_n$ has dimension m. If g can be factored into the product of m distinct linear forms $\mu_i x - \nu_i y$, $i = 1, 2, \ldots, m$, then the nth powers:*

$$(\mu_i x - \nu_i y)^n, \quad i = 1, 2, \ldots, m,$$

form a basis for $g^{\perp}$.

Proof. Let $g(x,y) = \sum_{j=0}^{m} \binom{m}{j} b_j x^j y^{m-j}$ and let $f(x,y) = \sum_{i=0}^{n} \binom{n}{i} a_i x^i y^{n-i}$ be a form of degree n apolar to g. Equating coefficients in $\{f,g\} \equiv 0$, we obtain the following system of linear equations: for $l = 0, 1, \ldots, n-m$,

$$b_m a_l - \binom{m}{1} b_{m-1} a_{l+1} + \binom{m}{2} b_{m-2} a_{l+2} - \cdots \pm b_0 a_{m+l} = 0.$$

Since these equations are upper triangular, they determine a subspace in $\mathfrak{F}_n$ of dimension $(n+1) - (n-m+1) = m$.

Now suppose that $\mu x - \nu y$ is a factor of $g(x,y)$. The form $(\mu x - \nu y)^n$ has normalized coefficients $a_k = \mu^k (-\nu)^{n-k}$. Since

$$b_m \mu^l \nu^{n-l} + \binom{m}{1} b_{m-1} \mu^{l+1} \nu^{n-l-1} + \cdots + b_0 \mu^{l+m} \nu^{n-l-m} = (-1)^{n-l} \mu^l \nu^{n-l-m} g(\nu, \mu) = 0,$$

$(\mu x - \nu y)^n$ is apolar to $g(x,y)$.

Finally, observe that if $(\mu_i x - \nu_i y)$ are distinct linear forms, the nth powers $(\mu_i x - \nu_i y)^n$ are linearly independent because the $m \times m$ determinant $|\mu_i^{n-j+1} \nu_i^{j-1}|_{1 \leq i,j \leq m}$ is a nonzero multiple of a van der Monde determinant. □

Next, we describe the subspace $g^{\perp}$ when g has repeated linear factors.

PROPOSITION 3.2. *Suppose*

$$g(x,y) = a(\mu_1 x - \nu_1 y)^{m_1} (\mu_2 x - \nu_2 y)^{m_2} \ldots (\mu_s x - \nu_s y)^{m_s},$$

where $\mu_i x - \nu_i y$ are distinct linear factors. Then, the forms

$$x^j (\mu_i x - \nu_i y)^{n-j}, \quad i = 1, 2, \ldots, s, \quad j = 0, 1, \ldots, m_i - 1$$

form a basis for the subspace $g^{\perp}$.

Proof. Use the fact that because

$$[n(n-1)\ldots(n-j+1)] x^j (\mu x - \nu y)^{n-j} = (\partial^j / \partial \mu^j)(\mu x - \nu y)^n,$$

the normalized coefficients of $[n(n-1)\ldots(n-j+1)] x^j (\mu x - \nu y)^{n-j}$ are given by $a_k = (-1)^{n-k} (\partial^j / \partial \mu^j)(\mu^k \nu^{n-k})$. □

PROPOSITION 3.3. *Let $f(x,y) = \sum_{i=0}^{n} \binom{n}{i} a_i x^i y^{n-i}$ and let m be a positive integer such that $m \leq n$. Then the subspace $f^{\perp}$ in $\mathfrak{F}_m$ has dimension $m - r + 1$, where r is the rank of the $(n-m+1) \times (m+1)$ matrix*

$$[a_{i+j}]_{0 \leq i \leq n-m,\, 0 \leq j \leq m}$$

Proof. Using the same notation as in Proposition 3.1, $\{f, g\} \equiv 0$ is equivalent to

$$\sum_{j=0}^{m} (-1)^{m-j} \binom{m}{j} a_{j+l} b_{m-j} = 0, \quad l = 0, 1, \ldots, n-m. \tag{3.1}$$

As these linear equations have the same rank as the matrix $[a_{i+j}]$, $f^{\perp}$ has dimension $m - r + 1$. □

Next, we state explicitly the connection between apolarity and sums of powers.

PROPOSITION 3.4. *Let f be a binary form of degree n. Then f can be written as the sum of m or fewer n^{th} powers of linear forms if and only if there exists a binary form g of degree m with m distinct linear factors apolar to f.*

From the foregoing results, we can prove Theorem 2.1 by observing that when $n = 2p+1$, $m = p+1$, the matrix $[a_{i+j}]$ in Proposition 3.3 has rank at most $p+1$. Hence, if f is a binary form of degree $2p+1$, there exists a nonzero binary form of degree $p+1$ apolar to f. Theorem 2.1 now follows from Proposition 3.4.

4. Partial Derivatives. The first proof of Sylvester given in [17] is (at least in hindsight) the first step towards the general theorem of Gundelfinger (see §5), In this section, we shall describe the ideas behind this proof.

We follow Sylvester by discussing the proof for a septic form $f(x,y) = a_7 x^7 + 7a_6 x^6 y + \cdots + a_0 y^7$. Suppose that f can be written as a sum of four 7th powers, that is,

$$f(x,y) = S^7 + T^7 + U^7 + V^7,$$

where $S = \alpha_1 x + \beta_1 y$, $T = \alpha_2 x + \beta_2 y$, $U = \alpha_3 x + \beta_3 y$, and $V = \alpha_4 x + \beta_4 y$. Consider the 4×4 determinant

$$J = \{1/7!\}^4 |\partial^6 f / \partial x^{6-i-j} \partial y^{i+j}|_{0 \leq i,j \leq 3}.$$

(This determinant arises from a general theory of eliminating variables from systems of polynomial equations which Sylvester developed in [16].) Observing that if $Z = \alpha x + \beta y$,

$$\partial^6 Z^7 / \partial x^{6-i-j} \partial y^{i+j} = \{7!\} \alpha^{6-i-j} \beta^{i+j} Z,$$

we obtain

$$J = |\alpha_1^{6-i-j} \beta_1^{i+j} S + \alpha_2^{6-i-j} \beta_2^{i+j} T + \alpha_3^{6-i-j} \beta_3^{i+j} U + \alpha_4^{6-i-j} \beta_4^{i+j} V\ |_{0 \leq i,j \leq 3}.$$

Next, we shall show that J is a scalar multiple of $STUV$. This follows from the next lemma.

LEMMA 4.1. *Let* $T_{ij} = \sum_{k=1}^{m} A_k \alpha_k^{e-c_i-d_j} \beta_k^{c_i+d_j}$, *where* $e, c_1, c_2, \ldots, c_n, d_1, d_2, \ldots, d_n$ *are integers. Then the determinant* $|T_{ij}|_{1 \leq i,j \leq n}$ *is zero when* $n > m$. *When* $n = m$, $|T_{ij}|$ *equals* $C(\alpha_1, \ldots, \alpha_n, \beta_1, \ldots, \beta_n) A_1 \ldots A_n$, *where* $C(\alpha_1, \ldots, \alpha_n, \beta_1, \ldots, \beta_n) A_1 \ldots A_n$ *is a rational function in* $\alpha_1, \ldots, \alpha_n, \beta_1, \ldots, \beta_n$

Proof. The $n \times n$ matrix $[T_{ij}]$ is the product of the $n \times m$ matrix $[(\beta_k/\alpha_k)^{c_i}]$, the $m \times m$ diagonal matrix diag $[A_k \alpha_k^e]$, and the $m \times n$ matrix $[(\beta_k/\alpha_k)^{d_j}]$. □

Evaluating J directly, we obtain

$$J = \begin{vmatrix} a_7x + a_6y & a_6x + a_5y & a_5x + a_4y & a_4x + a_3y \\ a_6x + a_5y & a_5x + a_4y & a_4x + a_3y & a_3x + a_2y \\ a_5x + a_4y & a_4x + a_3y & a_3x + a_2y & a_2x + a_1y \\ a_4x + a_3y & a_3x + a_2y & a_2x + a_1y & a_1x + a_0y \end{vmatrix}$$

As in the first proof, we can now find the linear forms S, T, U, and V by factoring J.

Applying this argument to a binary form of odd degree, Sylvester obtained an alternate formula for the canonizant.

PROPOSITION 4.2. *A canonizant for the form* $f(x,y) = \sum_{i=0}^{n} \binom{n}{i} a_i x^i y^{n-i}$ *of odd degree* $n = 2p+1$ *is given by*

$$J(x,y) = \left| a_{n-i-j}x + a_{n-i-j-1}y \right|_{0 \leq i,j \leq p+1}.$$

5. Gundelfinger's Theorem. Using the first part of Lemma 4.1 and the argument in §4, one can prove that if a binary form of (odd or even) degree n is the sum of s n^{th} powers, then the determinant $|\partial^{2s} f / \partial x^{2s-i-j} \partial y^{i+j}|_{0 \leq i,j \leq s}$ is identically zero. In [5,6], Gundelfinger proved that this is also a sufficient condition.

We first define the kth *Gundelfinger covariant* $G_k[f]$ of a binary form f to be the $(k+1) \times (k+1)$ determinant

$$|\partial^{2k} f / \partial x^{2k-i-j} \partial y^{i+j}|_{0 \leq i,j \leq k}.$$

THEOREM 5.1 (GUNDELFINGER). *Let* f *be a binary form of degree* n. *There exists a nonzero binary form of degree* k *apolar to* f *if and only if the* k^{th} *Gundelfinger covariant* $G_k[f]$ *is identically zero.*

Note that $G_0[f]$ equals f and $G_1[f]$ is the Hessian

$$(\partial^2 f/\partial x^2)(\partial^2 f/\partial y^2) - (\partial^2 f/\partial x \partial y)^2.$$

Hence, Theorem 5.1 includes as a special case the theorem that a binary form f of degree n is a perfect n^{th} power if and only if its Hessian is zero. Moreover, since $G_k[f] \equiv 0$ for $k > n/2$, $G_{p+1}[f] \equiv 0$ when f has odd degree $2p+1$. Thus, Sylvester's theorem is also a special case of Theorem 5.1. If $f(x,y) = \sum_{i=0}^{n} \binom{n}{i} a_i x^i y^{n-i}$ is a form of even degree $n = 2p$, then

$$G_p[f] = \{n(n-1)\ldots(n-p+1)\}^{p+1} |a_{n-i-j}|_{0 \leq i,j \leq p}$$

and is a multiple of the *catalecticant* $|a_{n-i-j}|_{0\le i,j\le p}$. In this case, Theorem 5.1 yields the theorem that a generic binary form of even degree $n = 2p$ is a sum of p n^{th} powers if and only if the catalecticant is zero.[1] Finally note that for $k < n/2$, $G_k[f]$ has degree $(n-2k)(k+1)$.

6. Wronskians. In this section, we shall outline Gundelfinger's proof [5, 6] of his theorem. A similar but briefer account can be found in [4, pp. 233–235].

The proof proceeds in three steps.

LEMMA 6.1. *Let $m \le n$, g a binary form of degree m and f a binary form of degree n. Then g is apolar to f if and only if for some k, $0 \le k \le n-m$, g is apolar to all the partial derivatives $\partial^k f/\partial x^k, \partial^k f/\partial x^{k-1}\partial y, \partial^k f/\partial x^{k-2}\partial y^2, \ldots, \partial^k f/\partial y^k$ of order k.*

Proof. Let $f(x,y) = \sum_{i=0}^{n}\binom{n}{i}a_i x^i y^{n-i}$ and $g(x,y) = \sum_{j=0}^{m}\binom{m}{j}b_j x^j y^{m-j}$. Since

$$\partial^k f/\partial x^{k-s}\partial y^s = [n!/(n-k)!]\sum_{i=0}^{n-k}\binom{n-k}{i}a_{k-s+i}x^i y^{n-k-i},$$

the condition that

$$\{\partial^k f/\partial x^{k-s}\partial y^s, g\} \equiv 0$$

is equivalent to

$$\sum_{j=0}^{m}(-1)^{m-j}\binom{m}{j}a_{j+l}b_{m-j} = 0, \qquad l = k-s,\ k-s+1, \ldots, n-m-s.$$

We can now prove the lemma by observing that this is a subsystem of the system (3.1) of linear equations equivalent to $\{f,g\} \equiv 0$. □

LEMMA 6.2. *Let f be a binary form of degree n and k a nonnegative integer such that $n-2k \ge 0$. Then, $G_k[f] \equiv 0$ if and only if the $k+1$ partial derivatives $\partial^k f/\partial x^k, \partial^k f/\partial x^{k-1}\partial y, \partial^k f/\partial x^{k-2}\partial y^2, \ldots, \partial^k f/\partial y^k$ are linearly dependent.*

Proof. This lemma is an analogue of the theorem relating linear dependence of functions and their Wronskian and the same proof works here. □

LEMMA 6.3. *Let f, n, and k be as in Lemma 6.2. There exists a nonzero binary form g of degree k apolar to f if and only if the forms $\partial^k f/\partial x^k, \partial^k f/\partial x^{k-1}\partial y$, $\partial^k f/\partial x^{k-2}\partial y^2, \ldots, \partial^k f/\partial y^k$ are linearly dependent.*

Proof. Suppose g is a nonzero form of degree k apolar to f. By Lemma 6.1, g is apolar to all the partial derivatives $\partial^k f/\partial x^{k-i}\partial y^i$. However, by Proposition 3.1, $g^{\perp}$ has dimension k. Hence, the forms $\partial^k f/\partial x^{k-i}\partial y^i$ are linearly dependent.

---

[1] This theorem explains the term "catalecticant". A generic binary form of degree $n = 2p$ has $2p+1$ coefficients or parameters, but a sum of p n^{th} powers has only $2p$ parameters, one less than the generic form. As Dr. R.T. Tugger has kindly informed me, the term "catalectic" is used to describe the analogous situation in prosody when the last foot in a line of a poem has one less syllable than it should have.

Conversely, suppose that

$$\lambda_0 \partial^k f/\partial x^k + \lambda_1 \partial^k f/\partial x^{k-1}\partial y + \cdots + \lambda_k \partial^k f/\partial y^k = 0$$

is a nontrivial linear relation among the kth partial derivatives. Applying $\partial^{n-2k}/\partial x^{n-2k-i}\partial y^i$ to this relation, we obtain a triangular (hence linearly independent) system of linear relations $R_0, R_1, \ldots, R_{n-2k}$ between the $n-k^{th}$ partial derivatives of f. Thus, the subspace spanned by the $n-k^{th}$ partial derivatives of f has dimension d, where $d \leq (n-k+1)-(n-2k+1) = k$. Let $h_1, h_2, \ldots, h_d$ be d linearly independent $n-k^{th}$ partial derivatives of f. Since h_i has degree k, the condition that $\{h_i, g\} \equiv 0$ is equivalent to one linear equation. Hence, the subspace $\cap_{i=0}^{d} h_i{}^{\perp}$ in $\mathfrak{F}_k$ has positive dimension $k+1-d$. Because the apolar covariant is bilinear, $\cap_{i=0}^{d} h_i{}^{\perp}$ equals $\cap_{i=0}^{n-k}(\partial^{n-k} f/\partial x^{n-k-i}\partial y^i)^{\perp}$, which in turn equals $f^{\perp}$ by Lemma 6.1. □

This completes Gundelfinger's proof of Theorem 5.1. The canonizant can be obtained from a Gundelfinger covariant.

PROPOSITION 6.4. *Let f be a binary form of degree n and let s be the minimum integer for which $G_s[f] \equiv 0$. Then $G_{s-1}[f] = g(x,y)^{n-2s+2}$, where g is a form of degree s apolar to f.*

Proof. We shall only consider the generic case when g has distinct linear factors. The case of repeated factors can be handled by "confluence" or taking partial derivatives as in Proposition 3.2.

Let $g(x,y) = \prod_{k=1}^{s} (\alpha_k x + \beta_k y)$ be a form of degree s apolar to f chosen so that $f(x,y) = \sum_{k=1}^{s} (\alpha_k x + \beta_k y)^n$. Then,

$$\begin{aligned} &\partial^{2s-2} f/\partial x^{2s-i-j-2}\partial y^{i+j} \\ &= \sum_{k=1}^{s} [n!/(n-2s+2)!]\alpha_k^{2s-i-j-2}\beta_k{}^{i+j}(\alpha_k x + \beta_k y)^{n-2s+2}. \end{aligned}$$

We can now finish the proof by using Lemma 4.1. □

7. Umbral Methods. A natural and more direct way to proving Theorem 5.1 is to relate the vanishing of the Gundelfinger's covariants $G_k[f]$ of the form $f(x,y) = \sum_{i=0}^{n} \binom{n}{i} a_i x^i y^{n-i}$ to the rank of the matrix $[a_{i+j}]$. This was done in [7] using umbral methods and we shall briefly sketch the argument here. We shall assume that the reader is familiar with the version of the umbral calculus set forth in [9].

The basic result in [7] relates the coefficients of $G_k[f]$ to the subdeterminants of $[a_{i+j}]$.

LEMMA 7.1. *Let $d = (n-2s)(s+1)$ and let*

$$G_k[f] = \sum_{j=0}^{d} g_{j,k} x^j y^{d-j},$$

where the coefficients $g_{j,k}$ *are unnormalized. Then*

$$g_{j,k} = [n!/(n-2k)!]^{k+1} \sum \binom{n-2k}{l_0}\binom{n-2k}{l_1}\cdots\binom{n-2k}{l_k}\Gamma[l_0, l_1, \ldots, l_k],$$

where the summation is over all $(k+1)$*–tuples* $l_0, l_1, \ldots, l_k$ *of nonnegative integers such that* $l_0 + l_1 + \cdots + l_k = j$ *and*

$$\Gamma[l_0, l_1, \ldots, l_s] = |a_{2s-i-j+l_i}|_{0\le i,j\le k}.$$

Proof. By the commutation relations $\partial/\partial x \circ U = U \circ \partial/\partial u_2$ and $\partial/\partial y \circ U = -U \circ \partial/\partial u_1$,

$$G_k[f] = \left| < U|(-1)^{i+j}\partial^{2k}[l\alpha(i)\ u]^n/\partial u_2^{2k-i-j}\partial u_1^{i+j} > \right|_{0\le i,j\le k}.$$

where $\alpha(0),\ \alpha(1), \ldots, \alpha(k)$ are $k+1$ distinct Greek umbral letters. Hence,

(7.1) $$G_k[f] =$$

$$[n!/(n-2k)!]^{k+1} < U|\{\prod_{i=0}^{k}[\alpha(i)\ u]^{n-2k}\}\{|\alpha(i)_1^{2k-i-j}\alpha(i)_2^{i+j}|_{0\le i,j\le k}\} > . \ \Box$$

It follows from Lemma 7.1 that $G_k[f] \equiv 0$ if and only if the $(n-k+1)\times(k+1)$ rectangular matrix $[a_{i+j}]_{0\le i\le n-k,\ 0\le j\le k}$ has rank at most k. Hence, by Proposition 3.3, there exists a nonzero binary form of degree k apolar to f.

From Equation (7.1), we can umbrally represent the Gundelfinger covariants as bracket polynomials.

PROPOSITION 7.2. *Let* $\alpha, \beta, \ldots, \epsilon$ *be* $k+1$ *Greek umbral letters. Then,*

$$G_k[f] = [n!/(n-2k)!]^{k+1}[(k+1)!]^{-1} < U| \prod_{\gamma<\delta}[\gamma\ \delta]\prod_{\gamma}[\gamma\ u]^{n-2k} > .$$

Proof. Use the symmetrization argument on p. 65–67 of [9]. $\Box$

Lemma 7.1, together with Proposition 3.3 and the fact that a determinant with two identical rows is zero, yield an explicit formula for the canonizant. The canonizant is not a covariant in general.

PROPOSITION 6.5. *Let* $f(x,y) = \sum_{i=0}^{n}\binom{n}{i}a_i x^i y^{n-i}$, s *be the minimum number such that* $G_s[f] \equiv 0$, *and* t *be the minimum number such that the coefficient of* $x^t y^{s(n-2s+2)-t}$ *in* $G_{s-1}[f]$ *is nonzero. Then*

$$\begin{vmatrix} y^{s+1} & -xy^s & \cdots & (-x)^{s+1} \\ a_{2s-1+t} & a_{2s-2+t} & & a_{s-1+t} \\ a_{2s-2} & a_{2s-3} & & a_{s-2} \\ & \cdots & & \\ a_s & a_{s-1} & \cdots & a_0 \end{vmatrix}$$

is a nonzero form of degree s apolar to f

8. Recent Work and Open Problems. We first mention three recent papers. In [1], Dür developed a theory of apolarity in finite fields and use it to find decoding algorithms for Cauchy codes. Reichstein in [13, 14, 15] rediscovered some of the results in §3 with new methods. Using these very interesting methods, he obtained results about expressing n-ary cubics as sums of cubes. Finally, Lascoux [10] has given another proof in the style of Sylvester's first proof of Theorem 2.1 using symmetric functions.

There are many open problems in this area. I shall discuss three.

Arithmetic Invariants. An *arithmetic* invariant $a[f]$ of a form f is a function on forms taking values in nonnegative integers which is invariant (i.e. $a[f] = a[\overline{f}]$) under $GL(2)$. For example, the minimum degree $s[f]$ of a nonzero form apolar to f is an arithmetic invariant. It follows from Theorem 5.1 that the inverse image $\{f: \ s[f] = k\}$ equals the set

$$\{f : G_0[f] \neq 0, \ G_1[f] \neq 0, \ldots, G_{k-1}[f] \neq 0, \ G_k[f] = 0\};$$

hence, $\{f : s[f] = k\}$ is an finite intersection of algebraic sets or their complements. Call an arithmetic invariant $a[f]$ *polynomially definable* if the inverse images $\{f : a[f] = k\}$ can be obtained in a finite number of steps by taking unions, intersections, and complements of invariant algebraic sets.

Problem 8.1. Characterize polynomially definable arithmetic invariants.

An interesting example of an arithmetic invariant is the minimum number $m[f]$ of active multiplications needed to evaluate a form f. More precisely, a multiplication in an evaluation of a form f is said to be *active* (see [11]) if both multiplicands involve x and y. For example, $x^2 - y^2$ requires two active multiplications when evaluated as $xx - yy$ but only one when evaluated as $(x + y)(x - y)$. Since multiplications used in computing the $GL(2)$–action are not active, $m[f]$ is an arithmetic invariant. However, it seems improbable that $m[f]$ would be polynomially definable.

Locations of Zeroes. Apolarity is used in locating zeroes of polynomials in the complex plane. See [12, pp. 57–62] for an excellent account. An example is Grace's theorem [3]: Let $p(x)$ and $q(x)$ be polynomials of degree n such that their homogenization $y^n p(x/y)$ and $y^n q(x/y)$ are apolar. Then every disc $\{z : |z - a| \leq r\}$ in the complex plane that contains all the zeroes of one polynomial also contains at least one zero of the other. Is there a connection between canonical forms and location of zeroes? Szegö has touched on this question in [22].

Newton's rule for the discovery of imaginary zeroes. In the first chapter of *Arithmetica Universalis* [published anonymously, but with the authorship an open secret, in 1707 by W. Whiston; republished with minor changes by Newton in 1729], Newton gave a rule without any proof for finding a lower bound on the number of imaginary (i.e. non–real) zeroes of a real polynomial.

Conjecture 8.2 (Newton's Rule). Let $p(x) = \Sigma_{i=0}^{n} \binom{n}{i} a_i x^i$ be a polynomial with real coefficients a_i. Write down the sequence ${a_n}^2, a_{n-1}^2 - a_n a_{n-2}, a_{n-2}^2 - a_{n-1} a_{n-3},$

$\ldots, a_1^2 - a_2a_0,\ a_0^2$ formed by taking "discriminants" of three consecutive terms in $0,\ a_n,\ a_{n-1}, \ldots, a_0,\ 0$. Then, the number of imaginary zeroes is at least the number of sign changes in the sequence of discriminants.

Newton's rule has been proved (by MacLaurin, Campbell, and others; see [19]) in the weaker form: if there exists a sign change in the discriminant sequence, then $p(x)$ has at least one imaginary zero. Thus, the conjecture is verified for quadratics and cubics. Sylvester [19, 20] verified it for quartics and quintics. His proof for quintics is not easy and uses the fact that any generic quintic can be written as the sum of three (complex) fifth powers. He also proved an analogous rule for what he called "superlinear" polynomials.

THEOREM 8.3 (SYLVESTER'S RULE). *Suppose a real polynomial $p(x)$ of degree n can be put into the canonical form*

$$p(x) = \lambda_1(x + c_1)^n + \lambda_2(x + c_2)^n + \cdots + \lambda_s(x + c_s)^n$$

where λ_i and c_i are real and $c_1 < c_2 < \cdots < c_s$. Then the number of real zeroes is at most the number of sign changes in the sequence $\lambda_1, \lambda_2, \ldots, \lambda_s,\ (-1)^n\lambda_1$.

Further work in this area is needed.

Acknowledgement. I would like to thank David Richman for a careful reading of the first draft of this paper.

REFERENCES

[1] A. DÜR, *On the decoding of extended Reed–Solomon codes*, Discrete Math., to appear.

[2] E.B. ELLIOTT, *An introduction to the algebra of quantics*, 2nd edition, Oxford Univ. Press, Oxford, 1913.

[3] J.H. GRACE, *On the zeroes of a polynomial*, Proc. Cambridge Philos. Soc. 11 (1900–02), 352–357.

[4] J.H. GRACE AND A. YOUNG, *The algebra of invariants*, Cambridge Univ. Press, Cambridge, 1903.

[5] S. GUNDELFINGER, *Zur Theorie der binären Formen*, Göttinger Nachr. 12 (1883), 115–121.

[6] S. GUNDELFINGER, *Zur Theorie der binären Formen*, J. Reine Angew. Math. 100 (1886), 413–424.

[7] J.P.S. KUNG, *Gundelfinger's theorem on binary forms*, Stud. Appl. Math. 75 (1986), 163–170.

[8] J.P.S. KUNG, *Canonical forms for binary forms of even degree*, in S.S. Koh (ed.), *Invariant theory*, Lecture Notes in Math. 1278, Springer–Verlag, New York, 1987, pp. 52–61.

[9] J.P.S. KUNG AND G.-C. ROTA, *The invariant theory of binary forms*, Bull. Amer. Math. Soc. (New Series) 10 (1985), 27–85.

[10] A LASCOUX, *Forme canonique d'une forme binaire*, in S.S. Koh (ed.), *Invariant theory*, Lecture Notes in Math. 1278, Springer–Verlag, New York, 1987, pp. 44-51.

[11] T.S. MOTZKIN, *Evaluations of polynomials and evaluations of rational functions*, Bull. Amer. Math. Soc. 61 (1955), 163.

[12] G. PÓLYA AND G. SZEGÖ, *Problems and theorems in analysis II*, (= translation by C.E. Billigheimer of *Aufgaben und Lehrsätze aus der Analysis II*, 4th edn), Springer–Verlag, New York, 1976.

[13] B. REICHSTEIN, *On symmetric operators of higher degree and their applications*, Linear Alg. Appl. 75 (1986), 155–172.

[14] B. REICHSTEIN, *On expressing a cubic form as a sum of cubes of linear forms*, Linear Alg. Appl. 75 (1986), 91–122.

[15] B. REICHSTEIN, *An algorithm to express a cubic form as a sum of cubes of linear forms*, in F. Uhlig and R. Grone (eds.), *Current trends in matrix theory*, North–Holland, New York, 1987, pp. 273–283.

[16] J.J. SYLVESTER, *Sketch of a memoir on elimination, transformation and canonical forms*, Cambridge and Dublin Math. J. 6 (1851), 186–200 (= *Collected mathematical papers*, Vol. I, Paper 32).

[17] J.J. SYLVESTER, *An essay on canonical forms, supplement to a sketch of a memoir on elimination*, George Bell, Fleet Street, London, 1851 (= *Collected mathematical papers*, Vol. I, Paper 34).

[18] J.J. SYLVESTER, *On a remarkable discovery in the theory of canonical forms and of hyperdeterminants*, Math. Mag. 2 (1951), 391–410 (= *Collected mathematical papers*, Vol. I, Paper 41).

[19] J.J. SYLVESTER, *Algebraical researches, containing a disquisition on Newton's rule for the discovery of imaginary roots, and an allied rule applicable to a particular class of equations, together with a complete invariantive determination of the character of the roots of the general equation of the fifth degree, &c.*, Philosophical Trans. Roy. Soc. London 154(1964), 579–666 (= *Collected mathematical papers*, Vol. II, Paper 74, Part 1 and 3).

[20] J.J. SYLVESTER, *An inquiry into Newton's rule for the discovery of imaginary roots*, Proc. Roy. Soc. London 13(1863–4), 179–183 (= *Collected mathematical papers*, Vol. II, Paper 74, Part 2).

[21] J.J. SYLVESTER, *Collected mathematical papers*, Vol. I–IV, Cambridge Univ. Press, Cambridge, 1904–1912.

[22] G. SZEGÖ, *Bemerkungen zu einem Satz von J.H. Grace über die Wurzeln algebraischer Gleichungen*, Math. Z. 13 (1922), 28–55.

INVARIANT THEORY, EQUIVALENCE PROBLEMS, AND THE CALCULUS OF VARIATIONS*

PETER J. OLVER†

Abstract. This paper surveys some recent connections between classical invariant theory and the calculus of variations, stemming from the mathematical theory of elasticity. Particular problems to be treated include the equivalence problem for binary forms, covariants of biforms, canonical forms for quadratic variational problems, and the equivalence problem for particle Lagrangians. It is shown how these problems are interrelated, and results in one have direct applications to the other.

1. Introduction. My mathematical researches into elasticity and the calculus of variations over the past eight years have led to several surprising connections with classical invariant theory. My original motivation for pursuing classical invariant theory arose through a study of existence theorems for non-convex problems in the calculus of variations of interest in nonlinear elasticity. This theory, due to John Ball, relies on a complete classification of all null Lagrangians, which are differential polynomials whose Euler-Lagrange equations vanish identically, or, equivalently, can be written as a divergence. The basic classification tool is a transform, analogous to the Fourier transform from analysis, originally introduced by Gel'fand and Dikii, [8], in their study of the Korteweg-deVries equation, and developed by Shakiban, [31]. This reduced the original problem to a question about determinantal ideals, which had been answered in fairly recent work in commutative algebra; see Ball, Currie and Olver, [1]. I further noticed that, when the relevant functions involved were homogeneous polynomials, the transform coincided with the classical symbolic method of classical invariant theory, but had the advantage of being applicable even when the functions were not polynomials, leading to a "symbolic method" for the "invariant theory of analytic functions". Moreover, even in the classical case of polynomials, the transform provides a ready mechanism for determination of the expression for classical covariants and invariants in terms of partial derivatives of the form, a significant problem mentioned by Kung and Rota, [20]. However, I have already written a survey of these results and applications, [25], and, as I want to discuss several more recent connections between classical invariant theory and the calculus of variations, space limitations will preclude any further presentation of this range of ideas.

Following upon these results, in the summer of 1981, John Ball asked me whether there were any conservation laws for nonlinear elasticity beyond the classical conservation laws of energy, momentum, etc. I set out to answer his question, but soon realized that the linear theory was not in all that great shape. A detailed

*Research supported in part by NSF Grant DMS 86-02004.

†School of Mathematics, University of Minnesota, Minneapolis, MN USA 55455

study, [22], of the equations of linear isotropic elasticity in both two and three dimensions revealed new classes of conservation laws (despite claims in the literature to the contrary). The next logical step was to extend these results to anisotropic elasticity, but here a direct attack on the equations in physical coordinates proved to be too complicated. I hit upon the idea of employing some a priori change of variables, which would have the effect of placing the general quadratic variational problem into a much simpler canonical form, similar to the well-known canonical forms for ordinary polynomials. It turned out that the full power of classical invariant theory, modified to incorporate the theory of "biforms", was required to effect the classification of canonical forms for quadratic Lagrangians, and hence of linear elastic media, [28]. Importance consequences of this approach included the determination of "canonical elastic moduli", reducing the number of constants required to characterize and simplify the behavior of anisotropic elastic materials, [27], and a complete classification of conservation laws for arbitrary anisotropic planar elastic media, in which it was shown that there are always two infinite families of new conservation laws which depend on two arbitrary analytic functions of two complex variables, [29].

I still had not tried to tackle Ball's original question for nonlinear elasticity, and therefore started searching for an appropriate tool that would handle the nonlinear case as effectively as classical invariant theory had taken care of the linear case. Contemporaneously, I learned of a powerful method introduced by Elie Cartan for answering general (nonlinear) equivalence problems. By definition, an *equivalence* problem is to determine when two given objects, e.g. two polynomials, two differential equations, or two variational integrals, can be mapped into each other by an appropriate change of variables. In 1908, through his pioneering study of Lie pseudogroups, [3], Cartan proposed a fundamentally algorithmic procedure, based on the rapidly developing subject of differential forms, which would completely solve general equivalence problems, leading to necessary and sufficient conditions for equivalence of two objects. In spirit, Cartan's method is very much like classical invariant theory, in that it leads to certain functions of the original objects which are invariants of the problem, and so must have the same values for any two equivalent objects. But, even more, Cartan's theory tells you which invariants are really important as far as the equivalence problem is concerned, and gives the necessary and sufficient conditions for equivalence in terms of a finite number of these invariants. Although Cartan's method is extremely powerful, and received further developments in the 1930's for solving several equivalence problems of interest in differential equations and differential geometry, it never did catch on in the mathematical community at large. Most of the recent work can trace its inspiration back to a paper of R. Gardner in mathematical control theory, [7]. In the last few years, the method has had a number of successful applications to differential equations, [14], [16], calculus of variations, [17], differential operators and molecular dynamics, [18], etc., all of which have pointed to its increasing importance. However, the method still awaits a real popularization in the applied mathematical community as a straightforward, algorithmic, computational tool that will provide explicit and effective necessary and sufficient conditions for the solution of many equivalence problems of current mathematical and applied interest. I am convinced that there are even more profound, practical applications of Cartan's method to both pure and applied mathematics, not to mention physics and engineering, in the offing.

I still have not applied Cartan's method to nonlinear elasticity, although the relevant equivalence problem is now under investigation with Niky Kamran. However, while learning the method, I came across the remarkable observation that the fundamental equivalence problem of classical invariant theory, namely that of determining when two binary forms can be mapped to each other by a linear transformation, could be recast as a special case of a Cartan equivalence problem for a particular type of one-dimensional variational integral, a problem Cartan himself had solved in the 1930's, [5]. Even more surprising is the fact that the Cartan solution of the Lagrangian equivalence problem has, by this connection, profound consequences for classical invariant theory. In particular, a new solution to the equivalence problem for binary forms, based on just three of the associated covariants, as well as new results on symmetry groups and equivalence to monomials and sums of nth powers are direct results of this remarkable connection, cf. [26], [30].

This paper will give a brief overview of these connections between the calculus of variations and classical invariant theory; proofs and more detailed developments of results can be found in the cited literature. This survey begins with a general discussion of equivalence problems, illustrated by several examples from classical invariant theory and the calculus of variations and their inter-relationships. In section 3, we introduce and compare the basic concepts of invariants, covariants and other kinds of invariant objects which make their appearance in the solutions to the equivalence problems discussed, which are presented in section 4. Section 5 includes a discussion of symmetry groups and how they arise from Cartan's approach.

2. Equivalence Problems. The general equivalence problem is to determine when two geometric or algebraic objects are really the same object, re-expressed in different coordinate systems. Of course, there are two underlying questions that must be precisely answered before we can mathematically formulate an equivalence problem: 1. Exactly what do we mean by two objects being the "same"? 2. Which changes of coordinates are to be allowed? Once we have been more precise in the specification of our equivalence problem, we can begin the mathematical analysis of our problem. In this section we briefly present several different types of equivalence problems arising in classical invariant theory and the calculus of variations, and discuss their inter-relationships.

1. *The equivalence problem for forms*

By a *form*[1] of *degree* n, we mean a homogeneous polynomial function

$$f(x) = \sum a_I x^I, \tag{2.1}$$

defined for $\mathbf{x} = (x^1, \ldots, x^m)$ in $\mathbb{R}^m$ or $\mathbb{C}^m$, where the sum is over all multi-indices $I = (i_1, \ldots, i_m)$, with $|I| \equiv i_1 + \cdots + i_m = n$, and where $x^I \equiv (x^1)^{i_1} \cdot \ldots \cdot (x^m)^{i_m}$. In classical invariant theory, the appropriate changes of coordinates are provided by the general linear group $GL(m)$ (meaning either $GL(m, \mathbb{R})$ or $GL(m, \mathbb{C})$), which acts on the variables via the standard linear representation $\mathbf{x} \to A \cdot \tilde{\mathbf{x}}, A \in GL(m)$. The induced action on the forms, which takes $f(\mathbf{x})$ to

$$\tilde{f}(\tilde{\mathbf{x}}) = f(A \cdot \tilde{\mathbf{x}}) = \sum \tilde{a}_I \tilde{x}^I, \tag{2.2}$$

induces an action of $GL(m)$ on the coefficients $\mathbf{a} = (a_I)$ of the form, whose explicit expression is easy to derive, but not overly helpful. Two forms f and $\tilde{f}$ are called (real or complex) *equivalent* if they can be transformed into each other by a suitable element of $GL(m)$, so the basic equivalence problem here is to determine whether or not two given forms can be mapped into each other by a suitable linear transformation.

In the particular case of binary forms, meaning $m = 2$, so $f = f(x, y)$ depends on $\mathbf{x} = (x, y) = (x^1, x^2)$, we can replace $\mathbf{x}$ by the projective coordinate $p = x/y$, and write

$$g(p) = f(p, 1) \tag{2.3}$$

for the corresponding inhomogeneous polynomial. Now, on the projective line, the corresponding group of coordinate changes consists of the linear fractional transformations

$$\tilde{p} = \frac{ap + b}{cp + d}, \tag{2.4}$$

where $A = \begin{pmatrix} a & b \\ c & d \end{pmatrix} \in GL(2)$. Two n^{th} degree polynomials g and $\tilde{g}$ are *equivalent* if

$$g(p) = (cp + d)^n \tilde{g}(\tilde{p}) = (cp + d)^n \tilde{g}\left(\frac{ap + b}{cp + d}\right), \tag{2.5}$$

for some $A \in GL(2)$. The equivalence problem for inhomogeneous polynomials is to determine when two given polynomials can be mapped into each other by such a linear fractional transformation.

[1] Here we already encounter the first in a series of conflicting terminologies which will plague our attempts to unite these two fields. In classical invariant theory, a form means a homogeneous polynomial; in Cartan's equivalence method, differential forms are the key objects of interest. Needless to say, these are very different kind of objects. To distinguish them, we will always use "form" for a homogeneous polynomial, whereas "differential form" will always have "differential" in front of it for emphasis.

2. *The equivalence problem for biforms.*

A second important type of algebraic equivalence problem is provided by generalizing the considerations of part 1 to what will be called "biforms". By definition, a *biform* of *bidegree* (m,n) is a polynomial function

$$Q(\mathbf{x},\mathbf{u}) = \sum a_{IJ} x^I u^J, \tag{2.6}$$

depending on two vector variables $(\mathbf{x},\mathbf{u}) \in \mathbb{R}^p \times \mathbb{R}^q$ (or in $\mathbb{C}^p \times \mathbb{C}^q$), which for, fixed $\mathbf{u}$, is a homogeneous polynomial of degree m in $\mathbf{x}$, and, for fixed $\mathbf{x}$, is a homogeneous polynomial of degree n in $\mathbf{u}$. The appropriate changes of variable are the linear transformations in the Cartesian product group $GL(p) \times GL(q)$; the transformation $\mathbf{x} \to A \cdot \tilde{\mathbf{x}}, \mathbf{u} \to B \cdot \tilde{\mathbf{u}}$ maps the biform $Q(\mathbf{x},\mathbf{u})$ to the biform

$$\tilde{Q}(\tilde{\mathbf{x}},\tilde{\mathbf{u}}) = Q(A \cdot \tilde{\mathbf{x}}, B \cdot \tilde{\mathbf{u}}) = \sum \tilde{a}_{IJ} \tilde{x}^I \tilde{u}^J.$$

As with forms, two biforms are *equivalent* if they can be mapped to each other by a suitable group element, and we have a similar, but less well investigated, type of equivalence problem. (However, see Turnbull, [34], for a discussion of bilinear forms, and Weitzenböck, [35], for even more general types of polynomials.)

3. *Equivalence problems for variational integrals.*

We now consider some of the possible equivalence problems associated with a general problem from the calculus of variations. Consider the integral

$$\mathcal{L}[\mathbf{u}] = \int_\Omega L(\mathbf{x}, \mathbf{u}^{(n)}) dx. \tag{2.7}$$

Here the domain of integration is an open subset $\Omega \subset \mathbb{R}^p$, and the Lagrangian L is a smooth function of the independent variables $\mathbf{x} \in \Omega$, the dependent variables $\mathbf{u} \in \mathbb{R}^q$, and their derivatives up to some order n, denoted by $\mathbf{u}^{(n)}$. The typical calculus of variations problem is to find extremals (i.e. minimizers or maximizers) $\mathbf{u} = \mathbf{f}(\mathbf{x})$ for the integral $\mathcal{L}[\mathbf{u}]$ subject to suitable boundary conditions. Here we are more concerned with the integral itself, rather than the specific minimizers. There are at least four different versions of the notion of "equivalence of variational problems", depending on the type of changes of variables allowed, and the precise form the equivalence is to take. First, there are two immediately obvious possible choices of allowable coordinate changes:

1) The *fiber-preserving transformations*, in which the new independent variables depend only on the old independent variables, so the transformations have the form

$$\tilde{\mathbf{x}} = \phi(\mathbf{x}), \qquad \tilde{\mathbf{u}} = \psi(\mathbf{x},\mathbf{u}). \tag{2.8}$$

2) The general *point transformations*, in which an arbitrary change of independent and dependent variables is allowed, and the transformations have the form

$$\tilde{\mathbf{x}} = \phi(\mathbf{x},\mathbf{u}), \qquad \tilde{\mathbf{u}} = \psi(\mathbf{x},\mathbf{u}). \tag{2.9}$$

Other possible classes of coordinate transformations include contact transformations, linear transformations, volume-preserving transformations, etc., etc., but these two are sufficient for our purposes. Furthermore, there are two choices for deciding when two Lagrangians are equivalent:

a) *Standard Equivalence*: Here we require that the two variational problems agree on all possible functions $\mathbf{u} = \mathbf{f}(\mathbf{x})$. This implies that the two Lagrangians are related by the change of variables formula for multiple integrals:

$$L(\mathbf{x}, \mathbf{u}^{(n)}) = \tilde{L}(\tilde{\mathbf{x}}, \tilde{\mathbf{u}}^{(n)}) \cdot \det \mathbf{J}, \tag{2.10}$$

where $\mathbf{J} = (D_i \phi^j)$ is the Jacobian matrix of the transformation. (Here, we are treating $\mathcal{L}[\mathbf{u}]$ as an oriented integral; otherwise we should put an absolute value on the factor $\det \mathbf{J}$.)

b) *Divergence Equivalence:* Here we only require that the variational problems agree on extremals, or, equivalently, that the associated Euler-Lagrange equations are mapped directly to each other by the change of variables. A standard result, [23; Theorem 4.7], says that two Lagrangians have the same Euler-Lagrange equations if and only if they differ by a divergence, so the two Lagrangians must be related by the formula

$$L(\mathbf{x}, \mathbf{u}^{(n)}) = \tilde{L}(\tilde{\mathbf{x}}, \tilde{\mathbf{u}}^{(n)}) \cdot \det \mathbf{J} + \operatorname{Div} \mathbf{F}, \tag{2.11}$$

where $\mathbf{F}(\mathbf{x}, \mathbf{u}^{(m)}) = (F_1, \ldots F_p)$ is an arbitrary p-tuple of functions of $\mathbf{x}, \mathbf{u}$ and derivatives of $\mathbf{u}$.

Combining the two notions of equivalence with each of the two classes of coordinate transformations, we are led to four different equivalence problems for Lagrangians, such as the standard point transformation equivalence problem, the divergence fiber-preserving equivalence problem, etc. Depending on the context, each of these problems is important, and warrants a solution. To date, however, only some of the simpler cases, e.g. $p = 1$ or 2, $q = 1$, $n = 1$ or 2, have been looked at, and only the simplest case $p = q = n = 1$ has been solved completely, cf. [17].

4. *The equivalence problem for particle Lagrangians.*

Now we specialize the general discussion on equivalence problems in the calculus of variations to present one particular equivalence problem in detail. A *particle Lagrangian* is one involving only one independent variable, x, and we specialize to first order particle Lagrangians in one dependent variable also, u. Let $p \equiv \frac{du}{dx}$ denote the derivative variable. The equivalence problem is to determine when two first order scalar variational problems

$$\mathcal{L}[u] = \int L(x, u, p) dx, \quad \text{and} \quad \tilde{\mathcal{L}}[u] = \int \tilde{L}(\tilde{x}, \tilde{u}, \tilde{p}) d\tilde{x},$$

can be transformed into each other by a point transformation

$$\tilde{x} = \phi(x, u), \qquad \tilde{u} = \psi(x, u), \tag{2.12}$$

without any additional divergence terms. Let us see what this entails.

According to the chain rule, if $\tilde{x}, \tilde{u}$ are related to x, u according to (2.12), the change in the derivative p is given by a linear fractional transformation:

$$\tilde{p} = \frac{ap+b}{cp+d}, \tag{2.13}$$

where

$$a = \frac{\partial \psi}{\partial u}, \quad b = \frac{\partial \psi}{\partial x}, \quad c = \frac{\partial \phi}{\partial u}, \quad d = \frac{\partial \phi}{\partial x}. \tag{2.14}$$

Specializing the general transformation rule (2.10), we deduce that equivalent Lagrangians must be related by the basic change of variables formula

$$L(x,u,p) = (cp+d)\tilde{L}(\tilde{x},\tilde{u},\tilde{p}) \tag{2.15}$$

under (2.12), (2.13). Thus the original problem from the calculus of variations can be recast as a problem of determining when two functions of three variables are related by the formula (2.15) for some transformation of the form (2.12), (2.13). This is the basic problem solved by Cartan in [5].

A remarkable observation is that the equivalence condition (2.15) for particle Lagrangians and the equivalence condition (2.5) for binary forms are essentially the same! Indeed, if, given a nonhomogeneous polynomial $g(p)$ of degree n, we define the "Lagrangian"

$$L(p) = \sqrt[n]{g(p)} \tag{2.16}$$

then the relevant transformation rules are identical, and so the equivalence problem (2.5) for the polynomial $g(p)$ under the linear fractional transformation (2.4) is the *same* as the equivalence problem for the (x,u)-independent Lagrangian (2.16) under the transformation (2.12), (2.13), (2.15). Therefore, any solution to the Lagrangian equivalence problem immediately induces a solution to the equivalence problem for binary forms. This observation can be extended to forms in more variables, connecting the equivalence problem for forms with an equivalence problem for multi-particle Lagrangians, cf. [26].

5. *Equivalence problems for quadratic Lagrangians.*

Another important special class of equivalence problems from the calculus of variations comes from specializing the general considerations of part 3 to the special case of the divergence equivalence of quadratic variational problems

$$\mathcal{L}[\mathbf{u}] = \int \sum a_{IJ}^{\alpha\beta} \frac{\partial^n u^\alpha}{\partial x^I} \frac{\partial^n u^\beta}{\partial x^J} dx. \tag{2.17}$$

For simplicity, we assume that the coefficients $a_{IJ}^{\alpha\beta}$ are constants. These problems are motivated by the applications to linear elasticity, but they also arise in many other contexts. Mathematically, the relevant subclass of changes of variables that preseves the quadratic form of the Lagrangian are the linear change of variables $\mathbf{x} \to A \cdot \mathbf{x}, \mathbf{u} \to B \cdot \mathbf{u}$, already presented in our discussion of biforms. We define the symbol of a quadratic Lagrangian to be the biform

$$Q(\mathbf{x}, \mathbf{u}) = \sum a_{IJ}^{\alpha\beta} \, \mathbf{x}^I \mathbf{x}^J u^\alpha u^\beta, \tag{2.18}$$

bidegree $(2n, 2)$. It is not hard to show that, since we can add arbitrary divergences to our Lagrangian, the symbol is well-defined, independently of any quadratic divergence which might be added in. Moreover, except for an extra determinantal factor $\det A$ in (2.11), which can always be effectively eliminated by rescaling, and the replacement of A by A^{-1}, the transformation rules for quadratic Lagrangian and those for their symbols are exactly the same. Therefore, we deduce the important fact that quadratic Lagrangians are divergence equivalent under a linear change of variables if and only if their symbols are equivalent as biforms. Thus, this equivalence problem reduces to the previous algebraic equivalence problem.

3. Invariants. Of fundamental importance to the solution of any equivalence problem are certain functions, the invariants, whose values do not change under the change of variables apposite to the problem. These can be appropriately defined for all of the equivalence problems considered above. Additional invariant quantities are provided by relative invariants or covariants, whose values change only by some multiplicative factor, and invariant differential forms, which are the key to Cartan's approach. The terminology here is slightly complicated by the different use of the term "invariant" in the different subject areas. In classical invariant theory, the invariants are distinguished from the more general covariants by the fact that they do not involve the variables $\mathbf{x}$. Furthermore, both invariants and covariants are really relatively invariant functions, as they do change by some multiplicative factor under the action of the general linear group. What we will be calling invariants would be known as absolute invariants in the classical terminology. In Cartan's approach, the distinction between invariants and covariants blurs, and they are all called invariants for simplicity. Thus, the invariants of Cartan's approach would be labelled absolute covariants in the classical invariant theory approach, while the invariants and covariants of classical invariant theory are really relative invariants according to the Cartan terminology. (I hope that this doesn't cause undue confusion for the reader!) Absolute invariants for two equivalent objects must agree identically, while relative invariants only need agree up to a factor. However, the vanishing of a relative invariant is an invariant condition, that often carries important geometric information about the object.

There are two basic methods for constructing invariants. In classical invariant theory, the powerful symbolic method provides a ready means of constructing all the (relative) polynomial invariants and covariants of a form. Hilbert's Basis Theorem says that there are a finite number of polynomially independent covariants for a form of a given degree, but the precise number of independent covariants increases rapidly with the degree n of the form (although this is partially mitigated by the presence of many polynomial syzygies). Indeed, a complete system of covariants has been constructed only for binary forms of degrees $2, 3, 4, 5, 6$, and 8. Despite the constructive methods used to generate the covariants themselves, it is by no means clear which covariants play the crucial role in the equivalence problem. For example, in the case of a binary quartic, there are two important invariants, denoted by i and j, but it is the strange combination $i^3 - 27j^2$ (the discriminant) which provides the key to the classification of quartic polynomials, [12; page 292]. Symbolic techniques can also be applied to construct covariants of biforms, [35], [24], and there is an analogous version of Hilbert's Basis Theorem. However, explicit Hilbert bases for even the simplest biforms are not known.

Cartan's approach provides an alternative method for constructing invariants, this time as functions of the partial derivatives of the basic object. Moreover, Cartan's algorithm automatically determines which invariants are important for the equivalence problem and readily gives necessary and sufficient conditions for equivalence based on the fundamental invariants. (We also note that Maschke, [21], developed a symbolic method for use in equivalence problems in Riemannian geometry. See Tresse, [33], for a treatment of relative invariants of differential equations.) We will see how the two approaches bear on each other in the equivalence problem considered. One conclusion will be that, as far as the equivalence problem is concerned, Cartan's approach is certainly the more powerful of the two.

Consider first the equivalence problem for forms. A classical *covariant* of weight w is a function $J(\mathbf{a}, \mathbf{x})$ depending on the coefficients $\mathbf{a} = (a_I)$ of the form and the variables $\mathbf{x}$, which, except for a determinantal factor, does not change under the action of $GL(n)$ given in (2.1), (2.2):

$$J(\tilde{\mathbf{a}}, \tilde{\mathbf{x}}) = (\det A)^w \cdot J(\mathbf{a}, \mathbf{x}). \tag{3.1}$$

If $w = 0$, we call J an *absolute covariant*; these are the invariants in Cartan's terminology. If a covariant J does not depend on $\mathbf{x}$ it is called an *invariant*, although here we will use the term *relative invariant* (unless $w = 0$).

In the case of a binary form $f(x, y)$, we list some of the most important classical covariants. First, the *Hessian*

$$H = (f, f)^{(2)} = \frac{2}{n^2(n-1)^2}(f_{xx}f_{yy} - f_{xy}^2), \tag{3.2}$$

is a covariant of weight 2 and degree $2n - 4$. The Jacobians

$$T = (f, H) = \frac{1}{2n(n-2)}(f_x H_y - f_y H_x), \tag{3.3}$$

$$U = (H, T) = \frac{1}{6(n-2)^2}(H_x T_y - H_y T_x), \tag{3.4}$$

are covariants of weight 3 and 6 and degrees $3n-6$ and $5n-12$ respectively. In these formulae, the notation $(f, g)^{(k)}$ denotes the kth transvectant of f and g. A classical result states that the kth transvectant of any two covariants is again a covariant, and, moreover, all the polynomial covariants can be constructed using successive transvection, [9], [12].

Turning to the equivalence problem for biforms, we define a *covariant of biweight* (v, w) of a biform (2.6) to be a polynomial function $J(\mathbf{a}, \mathbf{x}, \mathbf{u})$, depending on the coefficients $\mathbf{a} = (a_{IJ})$ and the variables $\mathbf{x}, \mathbf{u}$, which satisfies:

$$J(\tilde{\mathbf{a}}, \tilde{\mathbf{x}}, \tilde{\mathbf{u}}) = (\det A)^v (\det B)^w J(\mathbf{a}, \mathbf{x}, \mathbf{u}), \quad \mathbf{A} = (A, B) \in GL(p) \times GL(q).$$

As with forms, *a relative invariant* is just a covariant which does not depend on the variables $\mathbf{x}$ or $\mathbf{u}$. If $v = w = 0$, then we have an absolute covariant.

For example, consider the first nontrivial case of a biform, a binary biquadratic

$$\begin{aligned} Q(\mathbf{x}, \mathbf{u}) = & a_{11}^{11} x^2 u^2 + 2a_{12}^{11} xyu^2 + a_{22}^{11} y^2 u^2 + 2a_{11}^{12} x^2 uv + 4a_{12}^{12} xyuv + \\ & + 2a_{22}^{12} y^2 uv + a_{11}^{22} x^2 v^2 + 2a_{12}^{22} xyv^2 + a_{22}^{22} y^2 v^2. \end{aligned} \tag{3.5}$$

Here $\mathbf{x} = (x^1, x^2) = (x, y)$ and $\mathbf{u} = (u^1, u^2) = (u, v)$, and we have incorporated some multi- nomial coefficients to conform with the references. Since Q is a quadratic function of $\mathbf{x}$ for each fixed $\mathbf{u}$, it is not hard to see that the discriminant

$$\Delta_{\mathbf{x}}(\mathbf{u}) = \frac{1}{4}(Q_{xx} Q_{yy} - Q_{xy}^2) \tag{3.6}$$

is a covariant of biweight (2,0). Similarly, the $\mathbf{u}$-discriminant

$$\Delta_{\mathbf{u}}(\mathbf{x}) = \frac{1}{4}(Q_{uu} Q_{vv} - Q_{uv}^2) \tag{3.7}$$

is a covariant of biweight $(0, 2)$. These discriminants have the usual properties enjoyed by the discriminant of an ordinary quadratic polynomial; for instance, $\Delta_{\mathbf{x}}(\mathbf{u}_0) = 0$ implies that $Q(\mathbf{x}, \mathbf{u}_0)$ is a perfect square, etc. There is a mixed bi-quadratic covariant of biweight (2,2), which has the explicit formula

$$C_2 = \frac{1}{4}(Q_{xu} Q_{yv} - Q_{xv} Q_{yu}). \tag{3.8}$$

The simplest relative invariant of Q is the quadratic expression

$$I_2(\mathbf{a}) = 2a_{11}^{11} a_{22}^{22} - 4a_{12}^{11} a_{12}^{22} + 2a_{22}^{11} a_{11}^{22} - 4a_{11}^{12} a_{22}^{12} + 4(a_{12}^{12})^2, \tag{3.9}$$

and has biweight $(2, 2)$. There is a single cubic invariant, namely

$$I_3(\mathbf{a}) = a_{11}^{11} a_{12}^{12} a_{22}^{22} - a_{11}^{11} a_{22}^{12} a_{12}^{22} - a_{12}^{11} a_{11}^{12} a_{22}^{22} + a_{12}^{11} a_{22}^{12} a_{11}^{22} + a_{22}^{11} a_{11}^{12} a_{12}^{22} - a_{22}^{11} a_{12}^{12} a_{11}^{22}. \tag{3.10}$$

Further invariants and covariants can be found by applying the technique of *composition* of covariants. If Q is any (bi)form, and J is a polynomial covariant for Q, then we can regard J itself as a (bi)form, whose coefficients are certain polynomial combinations of the coefficients of Q. Any covariant K, which depends directly on the coefficients of the new form J, is then, by composition, a covariant of the original biform Q, and is denoted by $K \circ J$. Thus, since the discriminant $\Delta_{\mathbf{u}}(\mathbf{x})$

of a binary biquadratic form is a binary quartic form in the variables $\mathbf{x} = (x, y)$, all the standard covariants of the binary quartic yield, under composition, covariants of the original biquadratic Q. Thus we have the Hessian of the $\mathbf{u}$-discriminant

$$H_{\mathbf{u}}(\mathbf{x}) = H \circ \Delta_{\mathbf{u}} = (\Delta_{\mathbf{u}}, \Delta_{\mathbf{u}})^{(2)}, \tag{3.11}$$

which is again a binary quartic in $\mathbf{x}$, and a covariant of biweight $(2, 4)$, as well as the two relative invariants

$$i_{\mathbf{u}} = i \circ \Delta_{\mathbf{u}} = (\Delta_{\mathbf{u}}, \Delta_{\mathbf{u}})^{(4)} \quad \text{and} \quad j_{\mathbf{u}} = j \circ \Delta_{\mathbf{u}} = (\Delta_{\mathbf{u}}, H_{\mathbf{u}})^{(4)}, \tag{3.12}$$

which have biweights (4,4) and (6,6) respectively. Similarly, the Hessian of the $\mathbf{x}$-discriminant

$$H_{\mathbf{x}}(\mathbf{u}) = H \circ \Delta_{\mathbf{x}} = (\Delta_{\mathbf{x}}, \Delta_{\mathbf{x}})^{(2)}, \tag{3.13}$$

has biweight (4,2), and the two relative invariants

$$i_{\mathbf{x}} = i \circ \Delta_{\mathbf{x}} = (\Delta_{\mathbf{x}}, \Delta_{\mathbf{x}})^{(4)} \text{ and } j_{\mathbf{x}} = j \circ \Delta_{\mathbf{x}} = (\Delta_{\mathbf{x}}, H_{\mathbf{x}})^{(4)}, \tag{3.14}$$

have biweights (4,4) and (6,6), respectively. The two Hessians $H_{\mathbf{u}}$ and $H_{\mathbf{x}}$ are easily seen to be different quartic polynomials in general (even if one identifies the variables $\mathbf{x}$ and $\mathbf{u}$). Remarkably, the i and j invariants of the two discriminants are the same invariants of the original biquadratic polynomial Q.

THEOREM 1, [24]. *Let Q be a binary biquadratic form. Let $\Delta_{\mathbf{x}}(\mathbf{u})$ and $\Delta_{\mathbf{u}}(\mathbf{x})$ be the two discriminants, which are quartic forms in $\mathbf{u}$ and $\mathbf{x}$ respectively. Then the invariants of these two quartic forms are the same:*

$$i_{\mathbf{x}} = i \circ \Delta_{\mathbf{x}} = i_{\mathbf{u}} = i \circ \Delta_{\mathbf{u}}, \qquad j_{\mathbf{x}} = j \circ \Delta_{\mathbf{x}} = j_{\mathbf{u}} = j \circ \Delta_{\mathbf{u}}.$$

The structure of the roots of the two discriminants is an important invariant of the biquadratic (3.5), and provides the key to the determination of canonical forms and the solution to the equivalence problem. (See below.) For instance, if $\Delta_{\mathbf{x}}(\mathbf{u})$ has two double roots in one coordinate system, then it has two double roots in every coordinate system. Since the discriminant of a quartic, whose vanishing indicates the presence of repeated roots, is given by $i^3 - 27j^2$, cf. [12; page 293], Theorem 1 immediately implies the following interesting interconnection between the root structures of the two discriminants.

COROLLARY 2. *The two discriminants $\Delta_{\mathbf{x}}(\mathbf{u})$ and $\Delta_{\mathbf{u}}(\mathbf{x})$ of a binary biquadratic either both have all simple roots, or both have repeated roots.*

Note that it is *not* asserted that $\Delta_{\mathbf{x}}(\mathbf{u})$ and $\Delta_{\mathbf{u}}(\mathbf{x})$ have identical root multiplicities! For example, the biquadratic form $Q = x^2u^2 + xyv^2$ has $\mathbf{u}$-discriminant $\Delta_{\mathbf{u}} = -4x^3y$, which has a triple root at 0 and a simple root at ∞, whereas the $\mathbf{x}$-discriminant $\Delta_{\mathbf{x}} = v^4$ has a quadruple root at ∞. We also note that since the ratio i^3/j^2 essentially determines the cross ratio of the four roots of the quartic, [9], [12], the cross ratios of the roots of the two discriminants $\Delta_{\mathbf{x}}(\mathbf{u})$ and $\Delta_{\mathbf{u}}(\mathbf{x})$ must be the same.

There are at least three possible ways to prove Theorem 1. One is to explicitly write out the invariants $i_{\mathbf{u}}, j_{\mathbf{u}}, i_{\mathbf{x}}$ and $j_{\mathbf{x}}$, and compare terms. This was the original version of the proof, and was effected on an Apollo computer using the symbolic manipulation language SMP. The explicit formula for $j_{\mathbf{u}}$ runs to two entire printed pages! A second approach is to write out the formulas for the invariants in terms of the partial derivatives of the biquadratic form. The final approach is to work entirely symbolically; this last proof is the easiest for hand computation, and can be found in [24].

We now turn to a discussion of the invariants for Lagrangian equivalence problems. For specificity we consider example 4 of section 2 - the standard equivalence problem for a particle Lagrangian under point transformations. We assume that we are at a point where neither L nor the second derivative L_{pp} vanishes. (In particular, we are excluding the elementary affine Lagrangians $a(x,u)p + b(x,u)$.) The simplest invariant of this problem is the rational differential function

$$I = \frac{(LL_{ppp} + 3L_pL_{pp})^2}{LL_{pp}^3}. \tag{3.15}$$

The next most complicated one is

$$J(p) = \frac{2L^2L_{pp}L_{pppp} - 2LL_pL_{pp}L_{ppp} + 6LL_{pp}^3 - 3L_p^2L_{pp}^2 - 3L^2L_{ppp}^2}{2LL_{pp}^3}. \tag{3.16}$$

Below we shall see how both of these formulas are found using the Cartan method.

The numerator and denominator of these two absolute invariants are relative invariants. Indeed, according to (2.15),

$$\tilde{L} = (cp+d)^{-1}L,$$

hence, differentiating using the chain rule and (2.14), we find

$$\tilde{L}_{\tilde{p}\tilde{p}} = (ad-bc)^{-2}(cp+d)^3L,$$
$$\tilde{L}\tilde{L}_{\tilde{p}\tilde{p}\tilde{p}} + 3\tilde{L}_{\tilde{p}}\tilde{L}_{\tilde{p}\tilde{p}} = (ad-bc)^{-3}(cp+d)^5(LL_{ppp} + 3L_pL_{pp}).$$

This proves the invariance of I directly, the corresponding result for J follows after one further differentiation.

So far we have been discussing just scalar-valued invariants. We can consider vector-valued invariants (or even more general quantities) when we have an induced action of the relevant coordinate changes on some other vector space. The most important class of such invariants are the invariant differential forms, where the exterior powers of the cotangent space have the standard induced action under coordinate changes. In the Cartan equivalence method, one begins by encoding the original equivalence problem into a problem involving the mapping of non-invariant differential one-forms on the manifold. For the standard particle Lagrangian equivalence problem under point transformations we are led to introduce the one-forms

$$\omega_1 = L(x,u,p)dx, \qquad \omega_2 = du - p\,dx, \tag{3.17}$$

the first of which is essentially the integrand, and the second of which is known as the *contact form*, which must be preserved (up to multiple) in order that the derivative p transform correctly, as in (2.13). We introduce the corresponding one-forms for the transformed Lagrangian $\tilde{L}$:

$$\tilde{\omega}_1 = \tilde{L}(\tilde{x},\tilde{u},\tilde{p})d\tilde{x}, \qquad \tilde{\omega}_2 = d\tilde{u} - \tilde{p}\,d\tilde{x}.$$

Then we have the following reformulation of our basic equivalence problem.

LEMMA 3. *Two nonvanishing Lagrangians L and $\tilde{L}$ are equivalent if and only if there exist functions $A(x,u,p)$, $B(x,u,p)$, with $B \neq 0$, and a diffeomorphism $(\tilde{x},\tilde{u},\tilde{p}) = \Phi(x,u,p)$ such that the one-forms are related according to*

$$\Phi^*(\tilde{\omega}_1) = \omega_1 + A\omega_2, \qquad \Phi^*(\tilde{\omega}_2) = B\omega_2. \tag{3.18}$$

under the pull-back map Φ^*.

Now that we have recast the original equivalence problem into an equivalence problem involving differential forms, we are ready to apply the Cartan algorithm. Unfortunately, space considerations will preclude a discussion of the details of the algorithm, and we refer the interested reader to the references [4], [7], [15], [17] for the details. Suffice it to say that the method is completely algorithmic, to the extent that it can be programmed onto a symbolic manipulation computer package, of which several exist. Barring complications, the final result of the Cartan method is to produce a list of invariant one-forms, which can then be used to produce scalar invariants and the complete solution to the equivalence problem.

For instance, in the Lagrangian equivalence problem, if the Lagrangian does not depend on x or u, the invariant forms resulting from Cartan's method are

$$\begin{aligned} \theta_1 &= \pm\sqrt{|LL_{pp}|}(du - p\,dx), \\ \theta_2 &= (L - pL_p)dx + L_p du, \\ \theta_3 &= \pm\sqrt{\left|\frac{L_{pp}}{L}\right|}dp. \end{aligned} \tag{3.19}$$

(If L does depend on x and u, there are much more complicated expressions for the invariant forms, [17].) As the reader can check directly, as long as $L \neq 0$, $L_{pp} \neq 0$, (again we exclude the trivial affine Lagrangians), these differential forms constitute an invariant "coframe", or basis for the cotangent space at each point of of the (x,u,p)-space, meaning that under the point transformations (2.12), (2.13), with the Lagrangians matching up as in (2.15), the corresponding differential forms satisfy the invariance conditions

$$\tilde{\theta}_i = \theta_i, \qquad i = 1,2,3. \tag{3.20}$$

(The ambiguous $\pm$ signs in the coframe are unavoidable and stem from the ambiguity of the square root.) The differential form θ_2 is the well-known Cartan form from the calculus of variations, also known as Hilbert's invariant integral.

Once we have determined an invariant coframe, there is a straightforward method for producing invariant scalar-valued functions, such as those in (3.15), (3.16). It relies on the fact that the exterior derivative operation is invariant under coordinate changes, so we can differentiate the invariant coframe elements to determine new invariant forms. We can evaluate each $d\theta_i$ directly, and rewrite the resulting two-forms in terms of wedge products of the invariant coframe:

$$d\theta_i = \sum C^i_{jk} \theta_j \wedge \theta_k. \tag{3.21}$$

These are known as the *structure equations* for our problem. It is readily seen using (3.20) that the so-called *torsion coefficients* C^i_{jk} must all be scalar invariants of the problem, i.e. $\tilde{C}^i_{jk} = C^i_{jk}$.

For the Lagrangian equivalence problem, with $L = L(p)$, the structure equations take the explicit form

$$\begin{aligned} d\theta_1 &= -\tfrac{1}{2} I_0 \theta_1 \wedge \theta_3 + \theta_2 \wedge \theta_3, \\ d\theta_2 &= \mp \theta_1 \wedge \theta_3, \\ d\theta_3 &= 0. \end{aligned} \tag{3.22}$$

Hence, the only nonconstant torsion coefficient is the invariant

$$I_0 = \pm \frac{L L_{ppp} + 3 L_p L_{pp}}{\sqrt{|L| |L_{pp}|^3}}.$$

If we square I_0 to eliminate the ambiguous sign, we recover the earlier invariant (3.15).

Further scalar invariants are found by re-expressing the exterior derivatives of the invariants appearing in the structure equations in terms of the invariant coframe, leading to the "derived invariants". In our case, since L only depends on p, we find

$$dI_0 = J\theta_3,$$

where the derived invariant

$$J = \pm \sqrt{\left| \frac{L}{L_{pp}} \right|} \frac{\partial I_0}{\partial p}$$

is the same invariant given in (3.16). This process can lead to higher and higher order derived invariants; for instance the equation $dJ = K\theta_3$ leads to the second order derived invariant $K = \pm\sqrt{|L/L_{pp}|} \cdot J_p$, etc., etc. For the full Lagrangian equivalence problem, there are not one but three fundamental invariants appearing in the corresponding structure equations, and a host of interesting derived invariants, cf. [17].

4. Solution of Equivalence Problems. The invariant quantities play a fundamental role in the resolution of any equivalence problem. Basically, one tries to characterize the equivalence of two objects in terms of the associated invariants. The Cartan approach is especially efficacious in this regards, in that it readily identifies which invariants are of fundamental importance for the equivalence problem, and, moreover, through the powerful Cartan-Kähler existence theorem for exterior systems of differential equations, provides *necessary and sufficient conditions* for equivalence based on these invariants. Roughly, the key to the complete solution of the equivalence problem is the functional relationship between the invariants which appear in the structure equations and their corresponding derived invariants, as discussed above. (Here I am glossing over several complications, including structures of higher order, and structures that require prolongation.) These functional relations, which lead to the concept of a determining function for the equivalence problem, are the principal objects of interest. It is best if we illustrate this with a particular example - the standard particle Lagrangian equivalence problem in the special case when the Lagrangian only depends on the derivative variable p. In this case, there is one fundamental invariant appearing in the structure equations (3.22), namely $I_0(p)$, or, better, its square $I(p)$. If I happens to be constant, then its value must remain unchanged under the point transformations (2.12), and so must have the same value for both Lagrangians. Otherwise, if $I(p)$ is not constant, we express the derived invariant $J(p)$, cf. (3.16), in terms of I, leading to an equation of the form $J = F(I)$. The scalar function F is called the *determining function* for our equivalence problem, since it effectively determines the equivalence class of a given Lagrangian. Since F may well turn out to be a multiply-valued function, it is better to view the invariants I and J as parametrizing a curve in $\mathbb{C}^2$, which we may identify with the "graph" of the determining function F.

DEFINITION 4. Let $L(p)$ be a complex-analytic Lagrangian depending only on the derivative variable p. The *universal curve* corresponding to L is the complex curve

$$\mathcal{C} \equiv \{(I(p), J(p)) : p \in \mathbb{C}\} \subset \mathbb{C}^2. \tag{4.1}$$

(If I is constant, so $J = 0$, then $\mathcal{C}$ reduces to a single point.)

The universal curve is an invariant for the Lagrangian, so that two equivalent Lagrangians have identical universal curves. The Cartan method shows that, moreover, barring the trivial affine Lagrangians, the universal curve provides the complete necessary and sufficient conditions for the solution to the Lagrangian equivalence problem.

THEOREM 5. *Let $L(p)$ and $\tilde{L}(\tilde{p})$ be two complex analytic Lagrangians which are not affine functions of p. Then L and $\tilde{L}$ are equivalent under a complex analytic change of variables if and only if their universal curves are identical: $\mathcal{C} = \tilde{\mathcal{C}}$.*

In particular, if curve degenerates to a point, the invariants I and $\tilde{I}$ are both constant, and they must the same: $I = \tilde{I}$. For a real-valued Lagrangian, the theorem is essentially the same, except that one must add in the additional condition that the sign of the second derivatives L_{pp} must match that of $\tilde{L}_{\tilde{p}\tilde{p}}$ in order that the resulting change of variables be real, cf. [30].

According to the remarks in section 2.4, the equivalence problem for binary forms is a special case of the general Lagrangian equivalence problem, when the Lagrangian is the nth root of a polynomial of degree n. We can therefore translate Theorem 5 into the language of classical invariant theory by evaluating the invariants I and J directly in terms of known covariants of the binary form f. In each of the covariants (3.2), (3.3), (3.4), we can replace x and y by the homogeneous coordinate p to find corresponding covariants of the polynomial $g(p)$; we use the same symbols for these covariants. A simple exercise in differentiation using the formula $f(x,y) = y^n g(x/y)$ will prove the following formula:

$$L_{pp} = \frac{n-1}{2} L^{1-2n} H. \tag{4.2}$$

Note that (4.2) provides a simple proof of the classical fact that a binary form is the nth power of a linear form if and only if its Hessian is identically 0. Indeed, if $H \equiv 0$, then the Lagrangian L must be an affine function of p, i.e. $L = ap + b$, which implies that $g(p) = (ap+b)^n$.

Assume that this is not the case, i.e. H does not vanish identically. Then further differentiations prove that

$$I = \frac{8(n-2)^2}{n^3(n-1)} \frac{T^2}{H^3}, \qquad J = -\frac{12(n-2)^2}{n-1} \frac{gU}{H^3}. \tag{4.3}$$

Discarding the inessential constants, we deduce the following solution to the equivalence problem for binary forms.

THEOREM 6. *Let $f(x,y)$ be a binary form of degree n. Let H, T, U be the covariants defined by (3.2), (3.3), (3.4). Suppose that the Hessian H is not identically 0, so f is not the nth power of a linear form. Define the fundamental absolute rational covariants*

$$I^* = \frac{T^2}{H^3}, \qquad J^* = \frac{fU}{H^3}, \tag{4.4}$$

which are both covariants of weight 0 and degree 0. The functions I^* and J^* parametrize a rational curve $\mathcal{C}^*$ in the projective plane $\mathbb{CP}^2$, called the universal curve associated with the binary form f. (If I^* is constant, the curve reduces to a single point.) Then two binary forms f and $\tilde{f}$ are equivalent under the general linear group $GL(2,\mathbb{C})$ if and only if their universal curves are identical: $\mathcal{C}^* = \tilde{\mathcal{C}}^*$.

Therefore a complete solution to the complex equivalence problem for binary forms depends on merely two absolute rational covariants – I^* and J^*! Buchberger, [2], has developed a computationally effective method based on the idea of a Gröbner basis which can be used to eliminate the parameter p from the definition of the universal curve, and thereby give an implicit formula for the curve. This could provide a computationally effective method to find the universal curve associated with a binary form and explicitly solve the equivalence problem in a form amenable to symbolic computation. Clebsch, [6], gives another solution to the equivalence problem for binary forms based on the absolute invariants of the forms, assuming the existence of suitable linear or quadratic covariants. However, his approach is not applicable to all forms, whereas the Cartan approach is valid in all cases. In [30], the relationship between the Clebsch and Cartan approaches is explained, and the special role of the null forms made clear.

When the universal curve does not degenerate to a single point, the linear fractional transformations (2.13) themselves mapping equivalent Lagrangians, or equivalent binary forms, to each other can also be explicitly determined using the universal curve. Let $\mathbf{I}^* = (I^*, J^*) : \mathbb{CP}^1 \to \mathcal{C}^*$ denote the map parametrizing the universal curve.

THEOREM 7. *Let f and $\tilde{f}$ be equivalent binary forms, so that the universal curves are the same $\mathcal{C}^* = \tilde{\mathcal{C}}^*$. If the curve does not degenerate to a single point, then the implicit equation*

$$\tilde{\mathbf{I}}^*(\tilde{p}) = \mathbf{I}^*(p), \tag{4.5}$$

which has a discrete set of solutions, determines all the linear fractional transformations mapping f to $\tilde{f}$.

In other words, we can explicitly determine all the linear fractional transformations mapping f to $\tilde{f}$ by solving the equations

$$I^*(p) = \tilde{I}^*(\tilde{p}), \qquad J^*(p) = \tilde{J}^*(\tilde{p}).$$

Of course, the second of these two equations merely serves to delineate the appropriate branch of the universal curve, and so rule out spurious solutions to the first equation which map between different branches.

Another (related) approach to the solution of an equivalence problem is to solve the more difficult canonical form problem, which is to find a complete list of simple, canonical forms for the given type of objects. The invariants and covariants will then determine which of the canonical forms a given object is equivalent to. This appears to be difficult to do for general Cartan equivalence problems, since there are uncountably many different equivalence classes. However, for many binary forms, canonical forms are well known, cf. [12]. In the case of biforms, the solution to the equivalence problem for binary biquadratics presented in [28] depends on the enumeration of all possible canonical forms. These are listed in table 1. It can be shown that every binary biquadratic is complex-equivalent to exactly one of the 20 classes of canonical forms. (It is not quite uniquely equivalent to one of the canonical forms, since there are certain discrete automorphisms taking one of the canonical forms to another in the same class for the first four classes.). The real equivalence problem has the same sort of classification, except there are various subclasses given by placing $\pm$ signs in front of the squared terms x^2u^2, etc. The different canonical forms are basically distinguished by the root structure of the associated discriminants $\Delta_{\mathbf{x}}$, $\Delta_{\mathbf{u}}$ discussed above. Thus the first class corresponds to those biforms whose discriminants have four simple complex roots. (Note that according to Corollary 2, if one discriminant has all simple roots, so does the other.) Some of the classes need more sophisticated invariants to distinguish them. For instance, both case 13 and case 17 have discriminants with two pairs of double roots, but they are distinguished by the fact that the invariant I_3, cf. (3.10), vanishes for case 17, but is nonzero for case 13. See [28] for the full details of this classification.

Table 1: Canonical Forms for Binary Biquadratics.

1.	$x^2u^2 + y^2v^2 + \alpha(x^2v^2 + y^2u^2) + 2\beta xyuv,$	$(1+\alpha^2-\beta^2)^2 \neq 4\alpha^2, \alpha \neq 0,$
2.	$x^2u^2 + y^2v^2 + \alpha(x^2v^2 + y^2u^2) + 2\beta xyuv,$	$(1+\alpha^2-\beta^2)^2 = 4\alpha^2, \alpha \neq 0,$
3.	$x^2u^2 + y^2v^2 + 2\beta xyuv,$	$\beta^2 \neq 1,$
4.	$x^2u^2 + y^2v^2 + 2xyuv,$	
5.	$x^2u^2 + y^2v^2 + y^2u^2 + 2\beta xyuv,$	$\beta^2 \neq 1,$
6.	$x^2u^2 + y^2v^2 + y^2u^2 + 2xyuv,$	
7.	$x^2u^2 - x^2v^2 + xyu^2 + xyuv^2,$	
8.	$x^2u^2 - y^2u^2 + x^2uv + y^2uv,$	
9.	$x^2u^2 + y^2v^2 + y^2uv + 2xyuv,$	
10.	$x^2u^2 + y^2uv,$	
11.	$x^2u^2 + xyuv^2,$	
12.	$x^2uv + xyu^2,$	
13.	$x^2u^2 + y^2u^2 + xyuv,$	
14.	$x^2u^2 + x^2v^2 + xyuv,$	
15.	$x^2u^2 + y^2u^2,$	
16.	$x^2u^2 + x^2v^2,$	
17.	$x^2u^2 + xyuv,$	
18.	$x^2u^2,$	
19.	$xyuv,$	
20.	$0.$	

Each of these canonical biforms corresponds to a canonical form for a first order planar quadratic Lagrangian, i.e. $p = q = 2, n = 1$, in (2.17). In linear elasticity, the physically important Legendre-Hadamard strong ellipticity condition, [10], [27], requires that the symbol of the Lagrangian be a positive definite real biform, i.e. $Q(\mathbf{x}, \mathbf{u}) > 0$ for $\mathbf{x}, \mathbf{u} \neq 0$. Therefore, the only equivalence classes of interest there are the first two, case 1 corresponding to an anisotropic elastic material, case 2 to an isotropic material. The corresponding Lagrangian takes the form

$$u_x^2 + v_y^2 + \alpha(u_y^2 + v_x^2) + 2\beta u_x v_y.$$

This particular Lagrangian can be shown to be just a rescaled version of the stored energy function for an orthotropic elastic material, which, in three dimensions, is an anisotropic material with three reflectional planes of material symmetry, cf. [10].

(A block of wood is a good example of such a material.) The constants α and β are called the canonical elastic moduli of the material. Thus, the invariant theory produces the surprising new result that every linear, planar elastic material is equivalent to an orthotropic material, determined by just two canonical elastic moduli (as opposed to the standard six moduli for a planar anisotropic material described in any text book on linear elasticity). Complex variables methods, well-known for orthotropic materials, but considerably more cumbersome for more general anisotropic materials, can now be readily employed for the analysis of physical problems.

5. Symmetry Groups. For any equivalence problem, the symmetry group of an individual object is the group of "self-equivalences" - i.e. the group consisting of all allowable changes of variables which leave the given object unchanged. For example, in the case of a binary form, the symmetry group is the subgroup consisting of all matrices $\begin{pmatrix} a & b \\ c & d \end{pmatrix} \in GL(2)$ such that

$$f(ax + by, cx + dy) = f(x, y).$$

Similarly, for one of our Lagrangian equivalence problems, the symmetry group of a given Lagrangian is the group of all transformations (either fiber-preserving or point transformations, depending the allowed changes of variables) which map the Lagrangian to itself (either with or without a divergence, depending on the notion of equivalence used). For instance, if the Lagrangian is $L = p^2$, then the scaling group $(x, u) \to (\lambda x, \lambda^2 u)$, $\lambda > 0$, is a symmetry group for the fiber-preserving equivalence problem (and hence for any of the more complicated equivalence problems). The transformation group $(x, u) \to (x, u + \varepsilon x)$, $\varepsilon \in \mathbb{R}$, maps L to $(p+\varepsilon)^2 = p^2 + 2\varepsilon p + \varepsilon^2$, and so is not a symmetry group of the standard Lagrangian equivalence problem; however, the two extra terms can be written as a divergence, $2\varepsilon p + \varepsilon^2 = D(2\varepsilon u + \varepsilon^2 x)$, and hence this group *is* a symmetry group of the divergence equivalence problem. Similarly, in the case of the quadratic Lagrangian or biforms, we can talk about linear symmetries. For instance, the "isotropic" case 2 of table 1 is distinguished from the more generic "anisotropic" case 1 by the fact that it possesses an additional one-parameter rotational symmetry group beyond the obvious scaling symmetry common to all quadratic Lagrangians. Thus, two-dimensional isotropic elastic materials are distinguished from more general anisotropic materials by this additional one-parameter symmetry group.

One of the great strengths of the Cartan method is that it immediately provides the dimension of the symmetry group of any given object.

THEOREM 8. *Suppose we have solved an equivalence problem on an n dimensional space by constructing an invariant coframe. Let r be the rank of the invariant coframe, meaning the number of functionally independent invariants appearing among the torsion coefficients C^i_{jk} and their derived invariants. Then the symmetry group of the problem is an $n - r$ dimensional Lie group.*

(Here we have phrased the theorem so as to ignore additional complications which can arise when there are infinite pseudo-groups of symmetries, or when the problem must be "prolonged".) See Hsu and Kamran, [14], for an application of this result to the study of symmetry groups of ordinary differential equations. Here we use Theorem 8 to provide a complete determination of the possible symmetry groups for our Lagrangian equivalence problem.

THEOREM 9. *Let $L(p)$ be a Lagrangian which depends only on p. Then the two-parameter translation group $(x, u) \to (x + \delta, u + \varepsilon)$, $\delta, \varepsilon \in \mathbb{C}$, is always a symmetry group. If L is an affine function of p, then L possesses an infinite- dimensional Lie pseudogroup of symmetries depending on two arbitrary functions. If L is not an affine function of p, and the invariant I is constant, then L admits an additional one-parameter group of symmetries. If the invariant I is not constant, then the symmetry group of L is generated by only the translation group and, possibly, discrete symmetries.*

Indeed, the rank of the invariant coframe (3.19) is one, unless the invariant I is constant, in which case there are no non-constant invariants for the problem, and the rank is zero. (The affine case does not follow from the equivalence method results as presented above, since we explicitly excluded this case from consideration, but is easily verified by direct computation.) It is easy to see that, except in the case when the Lagrangian is an affine function of p, the symmetry groups of a binary form f and the corresponding Lagrangian (2.16) differ only by the translation group in (x, u). Therefore, Theorem 8 immediately implies the following theorem on symmetries of binary forms.

THEOREM 10. *Let $f(x, y)$ be a binary form of degree n.*

i) *If $H \equiv 0$, then f admits a two-parameter group of symmetries.*
ii) *If $H \not\equiv 0$, and I^* is constant, then f admits a one-parameter group of symmetries.*
iii) *If $H \not\equiv 0$, and I^* is not constant, then f admits at most a discrete symmetry group.*

(Case i) is proved by direct computation, using the fact that f is the nth power of a linear form, and hence equivalent to $\pm x^n$.)

In the case when the invariant I^* is not constant, so the universal curve $\mathcal{C}^*$ is really a curve, we can combine Theorems 7 and 10 to determine the cardinality of the discrete symmetry group of a binary form.

THEOREM 11. *Let f be a binary form with non-constant invariant I^*. Let d denote the covering degree of the universal curve $\mathbf{I}^* : \mathbb{CP}^1 \to \mathcal{C}^*$, i.e. the number of points in the inverse image $\mathbf{I}^{*-1}\{z\}$ of a generic point $z \in \mathcal{C}^*$. Then the symmetry group of f is a finite group consisting of d elements.*

Indeed, the symmetries will be determined by all solutions to the implicit equation

$$\mathbf{I}^*(\tilde{p}) = \mathbf{I}^*(p). \tag{5.1}$$

Moreover, since I^* and J^* are rational functions, the degree cannot be infinite, so we deduce that a binary form cannot have an infinite, discrete symmetry group.

A second consequence of Theorem 11 is the interesting result that any binary form with constant invariant I^* is equivalent to a monomial. The proof is elementary once the symmetry generator is placed in Jordan canonical form, cf. [30].

THEOREM 12. *A binary form f is complex-equivalent to a monomial, i.e. to $x^k y^{n-k}$, if and only if the covariant T^2 is a constant multiple of H^3, or, equivalently, its universal curve degenerates to a single point.*

Another new result which follows directly from the Cartan approach is a complete characterization of those binary forms which can be written as the sum of two n^{th} powers.

THEOREM 13. *A binary form of degree $n \geq 3$ is complex-equivalent to a sum of two n^{th}powers, i.e. to $x^n + y^n$, if and only if the invariant I^* is not constant, and its universal curve is an affine subspace of $\mathbb{CP}^2$ of the explicit form*

$$J^* = -\frac{n}{3n-6}(I^* + \frac{1}{2}). \tag{5.2}$$

This is equivalent to the condition that the covariants f, H, T, U are related by the equation

$$(3n-6)fU + nT^2 + \frac{1}{2}nH^3 = 0. \tag{5.3}$$

(Incidentally, there are other binary forms whose universal curves are also affine subspaces of $\mathbb{C}^2$, one example being a binary quartic whose roots are in equianharmonic ratio. An interesting open problem is to characterize all such binary forms.)

There is a classical theorem, due to Gundelfinger, cf. [11], [19], which gives an alternative generic test for determining how many binary forms of degree n can be written as the sum of k n^{th} powers. It would be interesting to find a relationship between Gundelfinger's result and the criterion in Theorem 13 in the case of two n^{th}powers. Another interesting line of investigation would be to see how the universal curve distinguishes between sums of three or more n^{th} powers, although this appears to be much more difficult as it is no longer determined by a single-valued function of I^*.

Acknowledgments.

I would like to thank Dennis Stanton for his effort in organizing such a stimulating conference. I would also like to express my thanks to Gian-Carlo Rota for encouraging my forays into classical invariant theory. Also, I would like to thank Bernd Sturmfels for alerting me to the possibility of using Gröbner bases to explicitly determine the universal curve of a binary form.

REFERENCES

[1] Ball, J.M., Currie, J.C. and Olver, P.J., *Null Lagrangians, weak continuity, and variational problems of arbitrary order*, J. Func. Anal., 41 (1981), pp. 135–174.

[2] Buchberger, B., *Applications of Gröbner bases in non-linear computational geometry*, in *Scientific Software, J.R. Rice, ed., IMA Volumes in Mathematics and its Applications, vol. 14*, Springer-Verlag, New York, 1988.

[3] Cartan, E., *Les sous-groupes des groupes continus de transformations*, Oeuvres Complétes, part. II, vol. 2, Gauthiers-Villars, Paris (1953), pp. 719–856..

[4] Cartan, E., *Les problémes d'èquivalence*, in *Oeuvres Complétes, part. II, vol. 2*, Gauthiers-Villars, Paris, 1952, pp. 1311–1334.

[5] Cartan, E., *Sur un problème d'équivalence et la théorie des espaces métriques généraléss*, in *Oeuvres Complètes, part. III vol. 2*, Gauthiers-Villars, Paris, 1955, pp. 1131–1153..

[6] Clebsch, A.,, *Theorie der Binären Algebraischen Formen*, B.G. Teubner, Leipzig, 1872.

[7] Gardner, R.B., *Differential geometric methods interfacing control theory*, in *Differential Geometric Control Theory, R.W. Brockett et. al., eds.*, Birkhauser, Boston, 1983, pp. 117–180.

[8] Gel'fand, I.M. and Dikii, L.A., *Asymptotic behavior of the resolvent of Sturm- Liouville equations and the algebra of the Korteweg-deVries equation*, Russ. Math. Surveys 30 (1975), 77-113.

[9] Grace, J.H. and Young, A., *The Algebra of Invariants*, Cambridge Univ. Press, Cambridge, 1903.

[10] Green, A.E. and Zerna, W., *Theoretical Elasticity*, The Clarendon Press, Oxford, 1954.

[11] Gundelfinger, S., *Zur Theorie der binären Formen*, J. Reine Angew. Math. 100 , (1886), 413-424.

[12] Gurevich, G.B., *Foundations of the Theory of Algebraic Invariants*, P. Noordhoff Ltd., Groningen, Holland, 1964.

[13] Hilbert, D., *Über die vollen Invariantensysteme*, in *Ges. Abh. II*, Springer-Verlag, Berlin, 1933, pp. 287-344.

[14] Hsu, L. and Kamran, N., *Classification of second-order ordinary differential equations admitting Lie groups of fiber-preserving symmetries*, Proc. London Math. Soc. (to appear).

[15] Kamran, N., *Contributions to the study of the equivalence problem of Elie Cartan and its applications to partial and ordinary differential equations*, preprint, 1988..

[16] Kamran, N., Lamb, K.G. and Shadwick, W.F., *The local equivalence problem for $d^2y/dx^2 = F(x,y,dx/dy)$ and the Painlevé transcendants*, J. Diff. Geom. 22 (1985), 139-150.

[17] Kamran, N., and Olver, P.J., *The equivalence problem for particle Lagrangians*, preprint, 1988.

[18] Kamran, N., and Olver, P.J., *Equivalence of differential operators*, preprint, 1988.

[19] Kung, J.P.S., *Canonical forms for binary forms of even degree*, in *Invariant Theory, S.S. Koh, ed., Lecture Notes in Math, vol. 1278*, Springer-Verlag, New York, 1987, pp. 52-61.

[20] Kung, J.P.S. and Rota, G.-C., *The invariant theory of binary forms*, Bull. Amer. Math. Soc., 10 (1984), pp. 27–85.

[21] Maschke, H., *A symbolic treatment of the theory of invariants of quadratic differential quantics of n variables*, Trans. Amer. Math. Soc. 4 (1903), 445–469.

[22] Olver, P.J., *Conservation laws in elasticity. II. Linear homogeneous isotropic elastostatics*, Arch Rat. Mech. Anal. 85 (1984), 131–160.

[23] Olver, P.J., *Applications of Lie Groups to Differential Equations*, Graduate Texts in Mathematics, vol. 107, Springer-Verlag, New York, 1986.

[24] Olver, P.J., *Invariant theory of biforms*, preprint, 1986.

[25] Olver, P.J., *Invariant theory and differential equations*, in *Invariant Theory*, S.S. Koh, ed., Lecture Notes in Mathematics, vol. 1278, Springer-Verlag, New York, 1987, pp. 62–80.

[26] Olver, P.J., *Classical invariant theory and the equivalence problem for particle Lagrangians*, Bull. Amer. Math. Soc. 18 (1988), 21-26.

[27] Olver, P.J., *Canonical elastic moduli*, J. Elasticity 19 (1988), 189-212.

[28] Olver, P.J., *The equivalence problem and canonical forms for quadratic Lagrangians*, Adv. Appl. Math. 9 (1988), 226-257.

[29] Olver, P.J., *Conservation laws in elasticity. III. Planar anisotropic elastostatics*, Arch. Rat. Mech. Anal. (to appear).

[30] Olver, P.J., *Classical invariant theory and the equivalence problem for particle Lagrangians. I. Binary forms*, Adv. in Math (to appear).

[31] SHAKIBAN, C., *A resolution of the Euler operator II*, Math. Proc. Camb. Phil. Soc. 89 (1981), 501-510.

[32] STROH, E., *Ueber eine fundamentale Eigenschaft des Ueberschiebungsprocesses und deren Verwerthung in der Theorie der binären Formen*, Math. Ann., 33 (1889), pp. 61–107.

[33] TRESSE, M.A., *Détermination des Invariants Ponctuels de l'Èquation Différentielle Ordinaire du Second Ordre* $y'' = \omega(x, y, y')$, S. Hirzel, Leipzig, 1896.

[34] TURNBULL, H.W., *The invariant theory of a general bilinear form*, Proc. London Math. Soc. 33 (1932), 1–21.

[35] WEITZENBÖCK, R., *Invariantentheorie*, P. Noordhoff, Groningen, 1923.

A SURVEY OF INVARIANT THEORY APPLIED TO NORMAL FORMS OF VECTORFIELDS WITH NILPOTENT LINEAR PART

R. CUSHMAN† AND J.A. SANDERS‡

This paper is a survey of an application of invariant theory to finding a normal form of a nonlinear differential equation

$$\frac{dx}{dt} = \mathbf{X}(x) \qquad x \in \mathbb{R}^{n+1} \tag{1}$$

near an equilibrium point which we assume is at 0. In other words, we assume that the vectorfield $\mathbf{X}$ vanishes at 0.

Most of the results in this survey can be found in various papers by one or both of the authors [1–9]. However the existence of a nice decomposition of the normal form module (see §3) is based on a recent theorem of Sturmfels and White [10]. This result was vaguely formulated in [4], stated as a precise conjecture when this paper was presented (by R. Cushman) at the IMA workshop in invariant theory, and essentially proved by Sturmfels and White. It is a pleasure to thank Professors Dennis Stanton and Willard Miller for the invitation (to R. Cushman) to attend the workshop. The authors also wish to thank Professors Bernd Sturmfels and Neil White for making their preprint [10] available to them. In §4 we give a few remarks on the history of the techniques used in normal form theory of vectorfields with nonsemisimple linear part.

§1. Introduction to normal form theory.

The basic idea behind normal form theory of the vectorfield $\mathbf{X}$ is to find a succession of co–ordinate changes which leave 0 fixed and simplify $\mathbf{X}$. Because the process of finding a normal form is entirely *algebraic*, we will assume that $\mathbf{X}$ is a formal power series

$$\mathbf{X}(x) = Ax + \mathbf{X}_2(x) + \cdots + \mathbf{X}_\ell(x) + \dots . \tag{2}$$

where A is a nonzero $(n+1) \times (n+1)$ matrix and $\mathbf{X}_\ell \in \mathbf{P}_\ell(\mathbb{R}^{n+1}, \mathbb{R}^{n+1})$, that is, $\mathbf{X}_\ell$ is an $n+1$–vector of homogeneous polynomials on $\mathbb{R}^{n+1}$ of degree ℓ.

If $\mathbf{X}$ is the linear vectorfield A, then there is a real linear co–ordinate change $x = Py$ such that A is in real canonical form B. In the new co–ordinates the vectorfield $\mathbf{X}$ is the linear vectorfield

$$\dot{y} = By = P^{-1}\mathbf{X}(Py) = (P^{-1}AP)y$$

†Mathematics Institute, Rijksuniversiteit Utrecht, 3508 TA Utrecht, the Netherlands
‡Department of Mathematics and Computer Science, Free University, 1081 HV Amsterdam, the Netherlands

which is in normal form. When $\mathbf{X}$ is nonlinear the process of finding a co–ordinate change φ which brings $\mathbf{X}$ into normal form is more difficult. Let's try to simplify the quadratic terms $\mathbf{X}_2$ of $\mathbf{X}$ by a near identity co–ordinate change

$$x = \varphi(y) = y + \varphi_2(y).$$

where $\varphi_2 \in \mathbf{P}_2(\mathbf{R}^{n+1}, \mathbf{R}^{n+1})$. In the new co–ordinates $\mathbf{X}$ becomes

$$\frac{dx}{dt} = D\varphi(y)\frac{dy}{dt} = \mathbf{X}(\varphi(y));$$

in other words,

$$\frac{dy}{dt} = \mathbf{Y}(y) = (I + D\varphi_2(y))^{-1}\mathbf{X}(y + \varphi_2(y)).$$

Up to quadratic terms, the formal power series for $\mathbf{Y}$ is

$$\begin{aligned} \mathbf{Y}(y) &= (I - D\varphi_2(y) + \dots)(A(y + \varphi_2(y)) + \mathbf{X}_2(y + \varphi_2(y)) + \dots) \\ &= Ay + \big(A\varphi_2(y) - D\varphi_2(y)Ay + \mathbf{X}_2(y)\big) + \dots \\ &= Ay + \big(\mathbf{X}_2(y) - ad_A(\varphi_2(y))\big) + \dots . \end{aligned} \tag{3}$$

Note that (3) defines ad_A and that for $\ell \geq 2$, ad_A is a linear mapping of $\mathbf{P}_\ell(\mathbf{R}^{n+1}, \mathbf{R}^{n+1}) = \mathbf{P}_\ell$ into itself. In particular for $\ell = 2$, we can find a subspace $\mathbf{C}_2$ of $\mathbf{P}_2$ which is a complement to the image of ad_A on $\mathbf{P}_2$, that is,

$$\mathbf{P}_2 = \mathbf{C}_2 \oplus im\ ad_A \mid \mathbf{P}_2. \tag{4}$$

Decomposing $\mathbf{X}_2$ along (4) gives

$$\mathbf{X}_2 = \mathbf{X}_2' + \mathbf{X}_2'' \quad \text{where} \quad \mathbf{X}_2' \in \mathbf{C}_2 \quad \text{and} \quad \mathbf{X}_2'' \in im\ ad_A|\mathbf{P}_2.$$

Next choose φ_2 so that

$$ad_A\ \varphi_2 = \mathbf{X}_2''.$$

This determines the co–ordinate change φ. Up to quadratic terms, the transformed vectorfield

$$\mathbf{Y}(y) = Ay + \mathbf{X}_2'(y) + \dots \tag{5}$$

is in normal form. We can now bring the cubic terms in $\mathbf{Y}$ into normal form by a co–ordinate change $\Phi = I + \Phi_3$ where Φ_3 is suitably chosen. Repeating this process we obtain a normal form

$$\widetilde{\mathbf{X}}(x) = Ax + \widetilde{\mathbf{X}}_2(x) + \dots + \widetilde{\mathbf{X}}_\ell(x) + \dots$$

where $\widetilde{\mathbf{X}}_\ell \in \mathbf{C}_\ell$ and $\mathbf{C}_\ell$ is a complement to the image of ad_A on $\mathbf{P}_\ell$.

Because the process of bringing $\mathbf{X}$ into normal form $\widetilde{\mathbf{X}}$ is complicated, describing the terms in $\widetilde{\mathbf{X}}$ is not easy. This we will try to do for vectorfields with nilpotent linear term. In other words from now on we will assume that

$$\mathbf{X}(x) = Nx + \mathbf{X}_2(x) + \cdots + \mathbf{X}_\ell(x) + \ldots \qquad x \in \mathbf{R}^{n+1}$$

is a formal power series vectorfield where N is a nonzero $(n+1)\times(n+1)$ nilpotent matrix. By a preliminary linear change of co–ordinates we may assume that N is a sum of m blocks N_{n_i} of size $n_i + 1$ where $n_1 \le n_2 \le \cdots \le n_m$. Here

$$\mathbf{R}^{n+1} = \sum_{i=1}^{m} \oplus \mathbf{R}^{n_i+1} \tag{6}$$

and the matrix of N_{n_i} is

$$N|\mathbf{R}^{n_i+1} = \begin{pmatrix} 0 & & & & & \\ 1 & 0 & & & & \\ & 2 & 0 & & & \\ & & \cdot & \cdot & & \\ & & & \cdot & \cdot & \\ & & & & n_i & 0 \end{pmatrix}.$$

To describe the normal form of $\mathbf{X}$, we must find a complement $\mathbf{C}_\ell$ to the image of ad_N on $\mathbf{P}_\ell$. A good way to do this is to embed N into a subalgebra of $g\ell(n+1,\mathbf{R})$ which is isomorphic to $s\ell_2(\mathbf{R})$. For an arbitrary nilpotent matrix in $g\ell(n+1,\mathbf{R})$, not necessarily in normal form, this embedding is possible by the Jacobson–Morosov theorem [12]. In our case, we define $M, H \in g\ell(n+1,\mathbf{R})$ by setting

$$M_{n_i} = M|\mathbf{R}^{n_i+1} = \begin{pmatrix} 0 & n_i & & & \\ & 0 & n_{i-1} & & \\ & & 0 & \ddots & \\ & & & & 1 \\ & & & & 0 \end{pmatrix}$$

and

$$H_{n_i} = H|\mathbf{R}^{n_i+1} = \begin{pmatrix} n_i & & & & & \\ & n_i - 2 & & & & \\ & & n_i - 4 & & & \\ & & & \cdot & & \\ & & & & \cdot & \\ & & & & & -n_i \end{pmatrix}.$$

It is easy to check that $\{M, N, H\}$ work, because

$$[H, M] = HM - MH = 2M, \ [H, N] = -2N, \ [M, N] = H.$$

Note that $\{M_{n_i}, N_{n_i}, H_{n_i}\}$ form an $n_i + 1$ dimensional irreducible representation of $s\ell_2(\mathbf{R})$ [13]. A short calculation shows that $\{ad_M, ad_N, ad_H\}$ is a basis of a subalgebra of the Lie algebra $\mathcal{X}_{n+1}$ of polynomial vectorfields on $\mathbf{R}^{n+1}$ under Lie bracket which is isomorphic to $s\ell_2(\mathbf{R})$. Because ad_M, ad_N, and ad_H map $\mathbf{P}_\ell$ into itself,

$$\{ad_M|\mathbf{P}_\ell, ad_N|\mathbf{P}_\ell, ad_H|\mathbf{P}_\ell\}$$

defines a finite dimensional representation of $s\ell_2(\mathbf{R})$. Using the explicit form of the matrices $\{M_{n_i}, N_{n_i}, H_{n_i}\}$ of the irreducible n_i+1 dimensional $s\ell_2(\mathbf{R})$ representation, it is easy to check that

$$\mathbf{R}^{n_i+1} = im\ N_{n_i} \oplus ker\ M_{n_i}.$$

Since every finite dimensional $s\ell_2(\mathbf{R})$ representation is a sum of irreducible representations, it follows that

$$\mathbf{P}_\ell = im\ ad_N|\mathbf{P}_\ell \oplus ker\ ad_M|\mathbf{P}_\ell. \tag{8}$$

Thus a complement $\mathbf{C}_\ell$ to the image of ad_N on $\mathbf{P}_\ell$ is $ker\ ad_M|\mathbf{P}_\ell$. In other words, the vectorfield

$$\widetilde{\mathbf{X}}(x) = Nx + \widetilde{\mathbf{X}}_2(x) + \cdots + \widetilde{\mathbf{X}}_\ell(x) + \dots \tag{9}$$

is in *normal form* if and only if for every $\ell \geq 2, \widetilde{\mathbf{X}}_\ell \in ker\ ad_M|\mathbf{P}_\ell$, where $\{M, N, H\}$ form a subalgebra of $g\ell(n+1, \mathbf{R})$ which is isomorphic to $s\ell_2(\mathbf{R})$. Of course this definition does not depend on the choice of basis of $\mathbf{R}^{n+1}$ made above.

We now give an invariant theoretic way of describing this normal form. Consider the one parameter subgroups $\mathcal{U}, \mathcal{U}^-, \mathcal{T}$ of $G\ell(n+1, \mathbf{R})$ generated by M, N and H, that is,

$$\mathcal{U} = \{\exp sM | s \in \mathbf{R}\},\ \mathcal{U}^- = \{\exp sN | s \in \mathbf{R}\},\ \mathcal{T} = \{\exp sH | s \in \mathbf{R}\}.$$

Then $\{\mathcal{U}, \mathcal{U}^-, \mathcal{T}\}$ generate a closed connected Lie subgroup $\mathcal{G}$ of $G\ell(n+1, \mathbf{R})$ which is isomorphic to $S\ell_2(\mathbf{R})$, since the tangent space of the identity element of $\mathcal{G}$ is spanned by $\{M, N, H\}$. Define an action of the unipotent group $\mathcal{U}$ on vectorfields $\mathcal{X}_{n+1}$ by

$$(u \cdot \mathbf{X})(x) = u^{-1}\mathbf{X}(ux) \qquad u \in \mathcal{U}. \tag{10}$$

Since

$$\frac{d}{ds}\Big|_{s=0}(\exp sM \cdot \mathbf{X}) = ad_M \mathbf{X}$$

and $\mathcal{U}$ is connected, it follows that $\mathbf{X} \in ker\ ad_M$ if and only if $u \cdot \mathbf{X} = \mathbf{X}$ for every $u \in \mathcal{U}$, that is, $\mathbf{X}$ is a $\mathcal{U}$–invariant vectorfield. Thus $\mathbf{X}$ is in normal form if and only if $\mathbf{X} - N \in \mathcal{X}^{\mathcal{U}}_{n+1}$, the set of $\mathcal{U}$–invariant polynomial vectorfields on $\mathbf{R}^{n+1}$. $\mathcal{U}$ acts on $P(\mathbf{R}^{n+1}, \mathbf{R}) = P$, the space of polynomials on $\mathbf{R}^{n+1}$, by

$$(u \cdot p)(x) = p(u^{-1}x) \qquad u \in \mathcal{U}. \tag{11}$$

Let $\mathcal{R} = P^{\mathcal{U}}$ be the ring of $\mathcal{U}$–invariant polynomials on $\mathbf{R}^{n+1}$. Thus $p \in \mathcal{R}$ if and only if

$$ad_M(p) = \frac{d}{ds}\Big|_{s=0}(\exp\ sM \cdot p) = 0.$$

Because

$$ad_M(p\mathbf{X}) = (ad_M p)\mathbf{X} + p(ad_M \mathbf{X})$$

for every $p \in \mathcal{R}$ and $\mathbf{X} \in \mathcal{X}^{\mathcal{U}}_{n+1}, \mathcal{X}^{\mathcal{U}}_{n+1}$ is a module over $\mathcal{R}$ which we call *normal form module* of $\mathbf{X}$.

<u>Example</u> 1. To make the above discussion more concrete we will compute the normal form of

$$\mathbf{X}(x,y) = \begin{pmatrix} 0 & 0 \\ 1 & 0 \end{pmatrix}\begin{pmatrix} x \\ y \end{pmatrix} + \mathbf{X}_2(x,y) + \dots \ .$$

First we embed $N = \left(\begin{smallmatrix} 0 & 0 \\ 1 & 0 \end{smallmatrix}\right)$ into an $s\ell_2(\mathbf{R})$ by setting

$$M = \begin{pmatrix} 0 & 1 \\ 0 & 0 \end{pmatrix} \quad \text{and} \quad H = \begin{pmatrix} 1 & 0 \\ 0 & -1 \end{pmatrix}.$$

Then the vectorfields

$$ad_N = -x\frac{\partial}{\partial y} \quad ad_M = -y\frac{\partial}{\partial x}\ , \ ad_H = -x\frac{\partial}{\partial x} + y\frac{\partial}{\partial y}$$

span an $s\ell_2(\mathbf{R})$, since

$$\begin{aligned}
[ad_H, ad_M] &= ad_H\ ad_M - ad_M ad_H = -y\frac{\partial}{\partial x} - y\frac{\partial}{\partial x} = 2ad_M \\
[ad_H, ad_N] &= -2ad_N \\
[ad_M, ad_N] &= ad_H.
\end{aligned}$$

Now $\mathbf{P}_\ell(\mathbf{R}^2, \mathbf{R}^2)$ has a basis

$$X_i = x^i y^{\ell-i}\frac{\partial}{\partial x} \qquad Y_i = x^i y^{\ell-i}\frac{\partial}{\partial y} \qquad i = 0, 1, \dots, \ell, \tag{12}$$

which are eigenvectors of ad_H, since

$$\begin{aligned}
ad_H X_i &= -ix^i y^{\ell-i}\frac{\partial}{\partial x} + (\ell - i)x^i y^{\ell-i}\frac{\partial}{\partial x} - x^i y^{\ell-i}\frac{\partial}{\partial x} \\
&= (\ell - 2i - 1)X_i
\end{aligned}$$

and

$$ad_H Y_i = (\ell - 2i + 1)Y_i.$$

Thus, on $\mathbf{P}_\ell$, ad_H has eigenvalues

$$\begin{aligned}
&\ell + 1, \ell - 1, \ \ell - 3, \dots, -(\ell - 1) \\
&\qquad\quad \ell - 1, \ \ell - 3, \dots, -(\ell - 1), -(\ell + 1).
\end{aligned}$$

Hence $\mathbf{P}_\ell$ is the sum of two $s\ell_2(\mathbf{R})$–irreducible subspaces of dimension $\ell+2$ and ℓ. Because $ker\ ad_M|\mathbf{P}_\ell$ is spanned by the eigenvector of ad_H in each irreducible subspace corresponding to the largest ad_H eigenvalue in that subspace, $\dim_{\mathbf{R}} ker ad_M|\mathbf{P}_\ell = 2$. From (11) and the eigenvalues of ad_H, it is easy to see that the ad_H eigenspace corresponding to the eigenvalue $\ell+1$ is spanned by X_0, which must lie in $ker\ ad_M$, and the eigenspace corresponding to the eigenvalue $\ell - 1$ is spanned by X_1 and Y_0. Since $ad_M(X_1 + Y_0) = 0, X_1 + Y_0$ is another basis vector for ker $ad_M|\mathbf{P}_\ell$. Therefore the normal form of $\mathbf{X}$ is

$$\widetilde{\mathbf{X}}(x,y) = \begin{pmatrix} 0 & 0 \\ 1 & 0 \end{pmatrix}\begin{pmatrix} x \\ y \end{pmatrix} + f_1(y)\begin{pmatrix} 1 \\ 0 \end{pmatrix} + f_2(y)\begin{pmatrix} x \\ y \end{pmatrix} \tag{13}$$

where f_1 and f_2 are formal power series beginning with quadratic and linear terms, respectively. Since $ker\ ad_M$ on $P_\ell = P_\ell(\mathbf{R}^2, \mathbf{R})$ is spanned by y^ℓ, the ring $\mathcal{R} = P^{\mathcal{U}} = \mathbf{R}[y]$. A basis for $\mathcal{X}_2^{\mathcal{U}}$ as an R–module is given by the vectorfields

$$\mathbf{X}_1(x,y) = (1,0) \quad \text{and} \quad \mathbf{X}_2(x,y) = (x,y).$$

In the appendix we list normal form modules for the vectorfields with nilpotent linear part given by

$$(2),\ (3),\ (4),\ (2,2),\ (5),\ (3,2),\ (2,2,2),\ \text{and}\ \ (3,3).$$

How do we verify that these candidate normal forms are indeed normal forms without decomposing the representation? We look at the case with nilpotent linear part (2) treated above. The problem is: how do we show that $\mathbf{X}_1 = (1,0)$ and $\mathbf{X}_2 = (x,y)$ is a basis of the $\mathcal{R}$–module $\mathcal{X}_2^{\mathcal{U}}$? First, we show that $\mathbf{X}_1$ and $\mathbf{X}_2$ are linearly independent over $\mathcal{R}$. This property of (13) we call *minimality*. Suppose that

$$\begin{pmatrix} 0 \\ 0 \end{pmatrix} = f_1(y)\begin{pmatrix} 1 \\ 0 \end{pmatrix} + f_2(y)\begin{pmatrix} x \\ y \end{pmatrix}$$

that is,

$$\begin{aligned} 0 &= f_1(y) + x f_2(y) \\ 0 &= f_2(y) y. \end{aligned}$$

From the second equation it follows that $f_2 = 0$. Substituting $f_2 = 0$ in the first equation gives $f_1 = 0$. Thus $\mathbf{X}_1$ and $\mathbf{X}_2$ are $\mathcal{R}$–independent. Second we must show that every element of $\mathcal{X}_2^{\mathcal{U}}$ can be written as $f_1\mathbf{X}_1 + f_2\mathbf{X}_2$ where $f_1, f_2 \in \mathcal{R}$. This property of (13) we call *completeness*. To prove completeness, we use a two–variable generating function

$$\mathcal{P}_2^2(u,t) = \sum_{\ell \geq 0} u^{\ell+1} t^\ell + \sum_{\ell \geq 1} u^{\ell-1} t^\ell = \frac{u+t}{1-ut}$$

where the exponent of u in the first sum is the ad_H eigenvalue of the vectorfield $y^\ell \frac{\partial}{\partial x}$ and the exponent of t in the first sum is its degree. A similar interpretation of the exponents of u and t hold in the second sum for the vectorfield $y^{\ell-1}x\frac{\partial}{\partial x} + y^\ell \frac{\partial}{\partial y}$. Since

$$\frac{\partial}{\partial u}(u\mathcal{P}_2^2(u,t))\big|_{u=1} = \frac{2}{(1-t)^2}$$

is the Poincaré series for $\mathbf{P}(\mathsf{R}^2, \mathsf{R}^2)$, we are finished with the proof of completeness, because for each ℓ, the coefficient of t^ℓ in each summand of $\frac{\partial}{\partial u}(u\mathcal{P}_2^2(u,t))\big|_{u=1}$ is the dimension of an irreducible $s\ell_2(\mathsf{R})$ summand, namely $\ell+2$ and ℓ; but $\ell+2+\ell = 2(\ell+1) = \dim_{\mathsf{R}} \mathbf{P}_\ell(\mathsf{R}^2, \mathsf{R}^2)$.

§2. The algebra of covariants of special equivariants. In this section we reduce the study of the structure of the normal form $\mathcal{R}$–module to the study of the structure of an R–algebra $\mathcal{C}$ of covariants of special equivariants. A nice discussion of the embedding of the ring $\mathcal{R}$ of $\mathcal{U}$–invariants into the algebra of covariants can be found in [11].

We begin by decomposing $\mathcal{X}_{n+1}^{\mathcal{U}}$ into a direct sum of $\mathcal{R}$–modules $\mathcal{X}_{n+1,n_i+1}^{\mathcal{U}}$ of special equivariants, which we now define. Since

$$\mathcal{X}_{n+1} = \mathbf{P}(\mathsf{R}^{n+1}, \mathsf{R}^{n+1}) = P(\mathsf{R}^{n+1}, \mathsf{R}) \otimes \mathsf{R}^{n+1},$$

every vectorfield $\mathbf{X}$ can be written as the sum of decomposable tensors $p \otimes v$ where $p \in P(\mathsf{R}^{n+1}, \mathsf{R})$ and $v \in \mathsf{R}^{n+1}$. Define an action of the unipotent group $\mathcal{U}$ on $P(\mathsf{R}^{n+1}, \mathsf{R}) \otimes \mathsf{R}^{n+1}$ by

$$u \cdot (p \otimes v) = u \cdot p \otimes u(v) \qquad u \in \mathcal{U} \tag{14}$$

for decomposable tensors and then extend by linearity for arbitrary tensors. Note that this action is the same as the action of $\mathcal{U}$ on $\mathcal{X}_{n+1}$ defined in (10). From the construction of M in §1 we see that M has the same Jordan block structure as N. Therefore the unipotent group $\mathcal{U}$ preserves the decomposition $\mathsf{R}^{n+1} = \sum_{i=1}^m \oplus \mathsf{R}^{n_i+1}$ into N–invariant and hence $\mathcal{U}$–invariant subspaces R^{n_i+1} Thus

$$\mathcal{X}_{n+1} = \sum_{i=1}^m \oplus P(\mathsf{R}^{n+1}, \mathsf{R}) \otimes \mathsf{R}^{n_i+1} = \sum_{i=1}^m \oplus \mathcal{X}_{n+1,n_i+1}. \tag{15}$$

Writing $v \in \mathsf{R}^{n+1}$ as

$$v = \sum_{i=1}^m v_i \qquad \text{where} \quad v_i \in \mathsf{R}^{n_i+1},$$

we see that every element of $\mathcal{X}_{n+1,n_i+1}$ can be expressed as a sum of decomposable tensors, $p \otimes v_i$ where $p \in P(\mathsf{R}^{n+1}, \mathsf{R})$ and $v_i \in \mathsf{R}^{n_i+1}$. The action of $\mathcal{U}$ on $\mathcal{X}_{n+1}$ induces a $\mathcal{U}$ action on $\mathcal{X}_{n+1,n_i+1}$ defined by

$$u \cdot (p \otimes v_i) = u \cdot p \otimes u(v_i) \qquad u \in \mathcal{U}. \tag{16}$$

for decomposable tensors and extending by linearity for arbitrary tensors. The space $\mathcal{X}^{\mathcal{U}}_{n+1,n_i+1}$ of $\mathcal{U}$–invariant elements of $\mathcal{X}_{n+1,n_i+1}$ is the space of *special* $\mathcal{U}$*–equivariants* from $\mathbf{R}^{n+1}$ to $\mathbf{R}^{n_i+1}$. The reason for the adjective "special" is that the action of $\mathcal{U}$ on $\mathbf{R}^{n_i+1}$ is irreducible. Clearly $\mathcal{X}_{n+1,n_i+1}$ is an $\mathcal{R}$ module. Since the $\mathcal{U}$–action on $\mathcal{X}_{n+1}$ preserves each $\mathcal{X}_{n+1,n_i+1}$, we have the $\mathcal{R}$–module direct sum decomposition

$$\mathcal{X}^{\mathcal{U}}_{n+1} = \sum_{i=1}^{n} \oplus \mathcal{X}^{\mathcal{U}}_{n+1,n_i+1} \, .$$

Thus we can concentrate our attention on the $\mathcal{R}$–module $\mathcal{X}^{\mathcal{U}}_{n+1,n_i+1}$ of special equivariants.

Next we embed $\mathcal{X}^{\mathcal{U}}_{n+1,n_i+1}$ into an algebra. To do this we need to recall some facts about the representation theory of the unipotent group

$$\mathcal{U}_1 = \{\begin{pmatrix} 1 & s \\ 0 & 1 \end{pmatrix} \in S\ell_2(\mathbf{R}) | s \in \mathbf{R}\}.$$

First, the unipotent group $\mathcal{U}_r = \{\exp sM_r | s \in \mathbf{R}\}$ where

$$M_r = \begin{pmatrix} 0 & r & & & \\ & 0 & r-1 & & \\ & & 0 & & \\ & & & \ddots & 1 \\ & & & & 0 \end{pmatrix}$$

is an $r+1$ irreducible representation ρ_r of $\mathcal{U}_1$ on $\mathbf{R}^{r+1}$. Second, let (ξ, η) be coordinates on $(\mathbf{R}^2)^*$ with respect to the standard dual basis of $\mathbf{R}^2$. Then the contragradient representation $\tilde{\rho}_1^*$ induced by ρ_1 on $P_r((\mathbf{R}^2)^*, \mathbf{R})$ is defined by

$$\begin{aligned} (\tilde{\rho}_1^*(u_1)p)(\xi,\eta) &= ((u_1^t)^{-1} \cdot p)(\xi,\eta) \qquad \text{(here $\cdot$ is defined in (11))} \\ &= p(u_1^t(\xi,\eta)) = p(\xi, s\xi + \eta) \end{aligned}$$

where $p \in P_r((\mathbf{R}^2)^*, \mathbf{R})$ and $u_1 = \left(\begin{smallmatrix} 1 & s \\ 0 & 1 \end{smallmatrix}\right)$ for some $s \in \mathbf{R}$. Third and most important, ρ_r acting on $\mathbf{R}^{r+1}$ is equivalent to $\tilde{\rho}_1^*$ acting on $P_r((\mathbf{R}^2)^*, \mathbf{R})$ via the mapping

$$j : \mathbf{R}^{r+1} \longrightarrow P_r((\mathbf{R}^2)^*, \mathbf{R}) : y = (y_0, \ldots, y_r) \longrightarrow \sum_{\ell=0}^{r} \binom{r}{\ell} y_\ell \xi^{r-\ell} \eta^\ell.$$

This amounts to showing that

$$j((\exp\ sM_r)y) = \tilde{\rho}_1^*(\begin{pmatrix} 1 & s \\ 0 & 1 \end{pmatrix})(j(y)), \tag{17}$$

which is a straightforward calculation. Thus the action of $\mathcal{U}$ on $\mathcal{X}_{n+1,n_i+1}$ given by (16) is the same as the action of $\mathcal{U}$ on $P(\mathbf{R}^{n+1}, \mathbf{R}) \otimes P_{n_i}((\mathbf{R}^2)^*, \mathbf{R})$ defined by the representation $\rho^* \otimes \tilde{\rho}_1^*$ of $\mathcal{U}_1$ where $\rho = \rho_{n_1} + \rho_{n_2} + \cdots + \rho_{n_m}$. On the graded algebra

$$\mathcal{A} = P(\mathbf{R}^{n+1} \oplus (\mathbf{R}^2)^*, \mathbf{R}) = \sum_{r,s} P_r(\mathbf{R}^{n+1}, \mathbf{R}) \otimes P_s((\mathbf{R}^2)^*, \mathbf{R}) = \sum_{r,s} \mathcal{A}_{r,s}$$

the representation $(\rho+\tilde{\rho}_1)^*$ gives rise to an action of $\mathcal{U}_1$ which preserves the grading. We can identify $\sum_r \mathcal{A}_{r,n_i}$ with $\mathcal{X}^{\mathcal{U}}_{n+1,n_i+1}$ using the map j.

$\mathcal{A}$ is not quite the algebra we want. Following the proof of Weitzenböck's theorem [14], we embed $\mathcal{A}$ into an algebra of $S\ell_2(\mathbf{R})$ covariants. Let σ_r be an $r+1$ dimensional irreducible representation of $S\ell_2(\mathbf{R})$ on $\mathbf{R}^{r+1}$ such that $\sigma_r|\mathcal{U}_1 = \rho_r$. Since σ_r is uniquely determined by ρ_r, we will also denote it by ρ_r. Thus on the algebra

$$\mathcal{C} = P(\mathbf{R}^{n+1} \oplus (\mathbf{R}^2)^*, \mathbf{R}) \otimes P((\mathbf{R}^2)^*, \mathbf{R}) = P(\mathbf{R}^{n+1} \oplus (\mathbf{R}^2)^* \oplus (\mathbf{R}^2)^*, \mathbf{R}) = \sum_{r,s,t} \mathcal{C}_{r,s,t}$$

we have an action of $S\ell_2(\mathbf{R})$ via the representation $(\rho+\tilde{\rho}_1+\tilde{\rho}_1)^*$, which preserves the grading by degree. Every term $\mathcal{C}_{r,s,t}$ in $\mathcal{C}$ is the set of homogeneous polynomials

$$p_{r,s,t} = p_{r,s,t}(x,(\xi,\eta),(X,Y))$$

in $P_r(\mathbf{R}^{n+1},\mathbf{R}) \otimes P_s((\mathbf{R}^2)^*,\mathbf{R}))$ of degree t in X,Y with $\binom{t}{i}X^{t-i}Y^i$ coefficient $p^{(i)}_{r,s} = p^{(i)}_{r,s}(x,(\xi,\eta))$ such that

$$p_{r,s,t}\big(\rho(g^{-1})(x), \rho_1(g)^t(\xi,\eta), \rho_1(g)^t(X,Y)\big) = p_{r,s,t}\big(x,(\xi,\eta),(X,Y)\big) \tag{18}$$

for every $g \in S\ell_2(\mathbf{R})$. In other words, $p_{r,s,t}$ is invariant under the $S\ell_2(\mathbf{R})$ action defined by the representation $(\rho+\tilde{\rho}_1+\tilde{\rho}_1)^*$. In particular, (18) holds for all $g \in \mathcal{U}_1$. Since $\begin{pmatrix}1&s\\0&1\end{pmatrix}^t\binom{0}{1} = \binom{0}{1}$ for every $s \in \mathbf{R}$, the vector $(X,Y)=(0,1)$ is invariant under $\rho_1(g)^t$ for every $g \in \mathcal{U}_1$. Substituting $(X,Y)=(0,1)$ into (18) with $g \in \mathcal{U}_1$ gives

$$p^{(t)}_{r,s}\big(\rho(g^{-1})(x), (\rho_1(g))^t(\xi,\eta)\big) = p^{(t)}_{r,s}(x,(\xi,\eta))$$

that is, $p^{(t)}_{r,s} \in \mathcal{A}_{r,s}$. In other words, we have a mapping

$$Cay: \mathcal{C}_{r,s,t} \longrightarrow \mathcal{A}_{r,s} : p_{r,s,t} \longrightarrow p^{(t)}_{r,s}$$

which is called the *Cayley correspondence*. To see that Cay is bijective, we argue as follows. Hold the variables $(x,(\xi,\eta))$ fixed. Then

$$j(p^{(0)}_{r,s},\ldots,p^{(t)}_{r,s}) = p_{r,s,t}$$

Hence the fact that

$$j \circ \rho_t(g) = \tilde{\rho}_1^*(g) \circ j,$$

for every $g \in \mathcal{U}_1$, continues to be valid when $g \in S\ell_2(\mathbf{R})$. Because ρ_t is an irreducible $t+1$ dimensional $S\ell_2(\mathbf{R})$ representation, it is uniquely determined by its top weight vector $p^{(t)}_{r,s}$. Thus Cay is bijective and $(Cay)^{-1}$ provides an embedding of $\mathcal{A}$ into $\mathcal{C}$. We call $\mathcal{C}$ the *algebra of covariants of special equivariants*.

We now compute the Poincaré series

$$\mathcal{P} = \mathcal{P}(\mathcal{C}) = \sum_{i,j,k\ge 0} (\dim_{\mathbf{R}} \mathcal{C}_{i,j,k}) t^i u^j v^k$$

of $\mathcal{C}$. Since $\mathcal{C}$ is the algebra of $S\ell_2(\mathbf{R})$–invariant polynomials on $\mathbf{R}^{n+1} \oplus (\mathbf{R}^2)^* \oplus (\mathbf{R}^2)^*$ under the $S\ell_2(\mathbf{R})$ action defined by the representation $\rho+\tilde{\rho}_1+\tilde{\rho}_1 = \tau$, applying Weyl's unitary trick to $\tau^{\mathbf{C}}$, the complexification of τ, we see that

$$\mathcal{P} = \sum_{i,j,k} \dim \mathcal{P}(\mathbf{C}^{n+1} \oplus (\mathbf{C}^2)^* \oplus (\mathbf{C}^2)^*, \mathbf{C})^{\tau^{\mathbf{C}}(SU(2))} t^i u^j v^k.$$

Applying Molien's formula [15] gives

$$\mathcal{P} = \int\limits_{SU(2)} \left[\det(1 - t\rho^{\mathbf{C}}(g)) \det(1 - u\tilde{\rho}_1^{\mathbf{C}}(g)) \det(1 - v\tilde{\rho}_1^{\mathbf{C}}(g))\right]^{-1} \, dg,$$

where dg is Haar measure on $SU(2)$ normalized so that $\int\limits_{SU(2)} dg = 1$.

By Weyl's integral formula [16], we get

$$\mathcal{P} = \frac{1}{2\pi i} \int\limits_{|z|=1} \frac{\frac{1}{2}(1 - z^2)(1 - z^{-2})}{\det(1 - t\rho^{\mathbf{C}}(z))(1 - uz)(1 - \frac{u}{z})(1 - vz)(1 - \frac{v}{z})} \, \frac{dz}{z}$$

$$\text{where} \quad \rho_n^{\mathbf{C}}(z) = \begin{pmatrix} z^n & & & \\ & z^{n-2} & & \\ & & \ddots & \\ & & & z^{-n} \end{pmatrix} \quad \text{and}$$

$$\det(1 - t\rho^{\mathbf{C}}(z)) = \prod_{i=1}^{m} \det(1 - t\rho_{n_i}^{\mathbf{C}}(z)) = D(z). \tag{19}$$

A bit of rearranging gives

$$\mathcal{P} = \frac{1}{2(1 - uv)} \cdot \frac{1}{2\pi i} \int\limits_{|z|=1} \frac{(1 - z^2)}{D(z)(1 - uz)(1 - vz)} \, \frac{dz}{z} + $$
$$+ \frac{1}{2(1 - uv)} \cdot \frac{1}{2\pi i} \int\limits_{|z|=1} \frac{(1 - z^{-2})}{D(z)(1 - \frac{u}{z})(1 - \frac{v}{z})} \, \frac{dz}{z} .$$

In the second integral replace z by $\frac{1}{z}$. Noting that $D(\frac{1}{z}) = D(z)$ and the orientation of $|z| = 1$ is reversed, we find that the second integral is equal to the first one. Hence we obtain

$$\mathcal{P} = \frac{1}{1 - uv} \cdot \frac{1}{2\pi i} \int\limits_{|z|=1} \frac{1 - z^2}{D(z)(1 - uz)(1 - vz)} \, \frac{dz}{z} . \tag{20}$$

Springer [17] gives a useful recursive procedure for evaluating this integral when $D(z)$ has simple zeroes.

Example 2. Here we show that when N is made up of two Jordan blocks $\begin{pmatrix} 0 & 0 \\ 1 & 0 \end{pmatrix}$, then the normal form module $\mathcal{X}_4^{\mathcal{U}}$ is given by

$$\begin{aligned} & f_1(p_1, p_2)\begin{pmatrix} \nu_1 \\ 0 \end{pmatrix} + f_2(p_1, p_2, p_3)\begin{pmatrix} \nu_2 \\ 0 \end{pmatrix} + f_3(p_1, p_2, p_3)\begin{pmatrix} \nu_3 \\ 0 \end{pmatrix} \\ & + g_1(p_1, p_2)\begin{pmatrix} 0 \\ \nu_1 \end{pmatrix} + g_2(p_1, p_2, p_3)\begin{pmatrix} 0 \\ \nu_2 \end{pmatrix} + g_3(p_1, p_2, p_3)\begin{pmatrix} 0 \\ \nu_3 \end{pmatrix} \end{aligned} \tag{21}$$

where $0 = \binom{0}{0}, \nu_1 = \binom{1}{0}, \nu_2 = \binom{x_1}{x_2}$ and $\nu_3 = \binom{x_3}{x_4}$. In addition, $p_1 = x_2, p_2 = x_4$ and $p_3 = x_2x_3 - x_1x_4$ and f_i, g_i for $i = 1, 2, 3$ are formal power series in the indicated variables. It is easy to check that all the terms in (21) lie in $ker\ ad_M$ where M is made up of two Jordan blocks $\begin{pmatrix} 0 & 1 \\ 0 & 0 \end{pmatrix}$. Note that (21) does *not* say that the vectors $\binom{\nu_i}{0}, \binom{0}{\nu_i}$ for $i = 1, 2, 3$ form a basis of the normal form module over the ring of formal power series in p_1, p_2, and p_3, because f_1 and g_1 are functions of p_1, p_2 alone. To prove minimality and completeness of (21) it is enough to show this for the first two components of (21), which is the normal form module for the $\mathcal{U}$-equivariants $\mathcal{X}^{\mathcal{U}}_{4,2}$. First we show minimality. Set

$$\binom{0}{0} = f_1(p_1, p_2)\binom{1}{0} + f_2(p_1, p_2, p_3)\binom{x_1}{x_2} + f_3(p_1, p_2, p_3)\binom{x_3}{x_4}. \tag{22}$$

Then the second component of (22) reads

$$\begin{aligned} 0 &= f_2(p_1, p_2, p_3)x_2 + f_3(p_1, p_2, p_3)x_4 \\ &= f_2(p_1, p_2, p_3)p_1 + f_3(p_1, p_2, p_3)p_2. \end{aligned}$$

Since p_1, p_2, p_3 are algebraically independent, there is a formal power series $g = g(p_1, p_2, p_3)$ such that

$$f_2 = -gp_2 = -gx_4 \quad \text{and} \quad f_3 = gp_1 = gx_2.$$

Substituting these expressions into the first component of (22) gives

$$\begin{aligned} 0 &= f_1(p_1, p_2) + g(p_1, p_2, p_3)(x_2x_3 - x_1x_4) \\ &= f_1(p_1, p_2) + g(p_1, p_2, p_3)p_3 \end{aligned}$$

But p_1, p_2 and p_3 are algebraically independent. Therefore $f_1 = g = 0$, which by definition of g gives $f_2 = f_3 = 0$. Thus we have shown minimality of (22). To prove completeness we compute the Poincaré series $\mathcal{P}^{2,2}$ of the algebra of covariant special equivariants containing $\mathcal{X}^{\mathcal{U}}_{4,2}$. Using (20) we have

$$\mathcal{P}^{2,2} = \frac{1}{1-uv}\,\frac{1}{2\pi i}\int\limits_{|z|=1} \frac{(1-z^2)}{(1-tz)^2(1-\frac{t}{z})^2(1-uz)(1-vz)}\,\frac{dz}{z}$$

The integrand has only one pole at $z = t$ inside $|z| = 1$ (since $|t| < 1$, $|u| < 1$ and $|v| < 1$) and it is of second order. Computing the residue at $z = t$ by brute force gives

$$\mathcal{P}^{2,2} = \frac{1-uvt^2}{(1-uv)(1-t^2)(1-ut)^2(1-vt)^2}. \tag{23}$$

The coefficient of the constant term in the Taylor series of $\mathcal{P}^{2,2}$ as a function of v about 0 is

$$\mathcal{P}^{2,2}_1 = \frac{1}{(1-t^2)(1-ut)^2}$$

which is the two variable generating function of the ring $\mathcal{R}$ of $\mathcal{U}$–invariant polynomials on $\mathbf{R}^4$, because

generator	m=eigenvalue of ad_H	r=degree	$u^m t^r$
$p_1 = x_2$	1	1	ut
$p_2 = x_4$	1	1	ut
$p_3 = x_2x_3 - x_1x_4$	0	2	t^2

Table 1

The coefficient of v^1 in $\mathcal{P}^{2,2}$ is

$$\begin{aligned} \mathcal{P}_2^{2,2} &= \frac{u + 2t - ut^2}{(1-t^2)(1-ut)^2} \\ &= \frac{u}{(1-ut)^2} + \frac{2t}{(1-t^2)(1-ut)^2}, \end{aligned} \tag{24}$$

which is the two variable generating function for the $\mathcal{R}$–module of special equivariants $\mathcal{X}_{4,2}^{\mathcal{U}}$, because the numerator of $\mathcal{P}_2^{2,2}$ comes from table 2.

generator	m=eigenvalue of ad_H	r=degree	$u^m t^r$
$\binom{1}{0}$	1	0	u
$\binom{x_1}{x_2}$	0	1	t
$\binom{x_3}{x_4}$	0	1	t

Table 2

This proves the completeness of (22). Thus every element of the $\mathcal{R}$–module $\mathcal{X}_{4,2}^{\mathcal{U}}$ can be written *uniquely* in the form

$$f_1(p_1, p_2)\begin{pmatrix}1\\0\end{pmatrix} + f_2(p_1, p_2, p_3)\begin{pmatrix}x_1\\x_2\end{pmatrix} + f_3(p_1, p_2, p_3)\begin{pmatrix}x_3\\x_4\end{pmatrix}$$

for some formal power series f_1, f_2, f_3. This completes the proof that the normal form module of $\mathcal{X}^{\mathcal{U}}$ is given by (21).

§3. **The Stanley decomposition.** In this section we will show that the normal form $\mathcal{R}$–module $\mathcal{X}_{n+1}^{\mathcal{U}}$ has a Stanley decomposition. To explain what this means, let $\{p_1, \ldots, p_\ell\}$ be a finite set of algebraically independent generators of $\mathcal{R}$. (Such a set exists by a theorem of Weitzenböck). $\mathcal{X}_{n+1}^{\mathcal{U}}$ has a *Stanley decomposition* if there is a finite collection $\mathcal{F}$ of subsets of $\{1, \ldots, \ell\}$ and homogeneous polynomial vectorfields $v_\alpha \in \mathcal{X}^{\mathcal{U}}_{n+1}$ with $\alpha \in \mathcal{F}$ such that for every $\mathbf{X} \in \mathcal{X}^{\mathcal{U}}_{n+1}$ there are *uniquely* determined polynomials f_α in the variables $\mathbf{p}_\alpha = \{p_i | i \in \alpha\}$ so that

$$\mathbf{X} = \sum_{\alpha \in \mathcal{F}} f_\alpha(\mathbf{p}_\alpha) v_\alpha. \tag{25}$$

Look at the appendix for some examples of Stanley decompositions of normal form modules.

Our argument makes heavy use of a result of Sturmfels and White [10] which states that every finitely generated k–algebra $\mathbf{A}$ has a Stanley decomposition. We recall the relevant definition here. Without loss of generality we may assume that $\mathbf{A} = k[z_1 \ldots, z_n] \big/_I$ where I is an ideal given by a finite number of generators. A Stanley decomposition for $\mathbf{A}$ is a splitting of $\mathbf{A}$ into a direct sum of k vector spaces of the form

$$\sum_{\alpha \in \mathcal{F}} \oplus z^\alpha k[\mathbf{Z}_\alpha] \tag{26}$$

where $\mathcal{F}$ is a finite subset of $\{0, 1, \ldots, n\}^n, z^\alpha = z_1^{\alpha_1} \ldots z_n{}^{a_n}$, and $\mathbf{Z}_\alpha$ is a finite subset of $\{z_1, \ldots, z_n\}$. In order to understand example 4, given at the end of this section, we give a very brief description of the ideas in the proof of the Sturmfels–White theorem. Their argument begins with the the observation that if I is an ideal generated by monomials, then there is a combinatorial construction for the Stanley decomposition of $\mathbf{A}$. In the case of a general ideal I, it is shown that the algorithm to construct a Gröbner basis for I preserves the Stanley decomposition. Thus the k–algebras $k[z_1 \ldots, z_n] \big/_I$ and $k[z_1, ., z_n] \big/_{init\ I}$ have the same Stanley decomposition. Because *init* I is the ideal generated by the largest monomials (with respect to an admissible ordering) in a Gröbner basis for I, we are finished.

We now apply this result to the $\mathcal{R}$–algebra $\mathcal{C}$ of covariants of special equivariants in which the $\mathcal{R}$–module $\mathcal{X}^{\mathcal{U}}_{n+1,n_i+1}$ is embeded . First we note that $\mathcal{C}$ is finitely generated, $\mathbf{R}$-algebra, being the ring of polynomials invariant under an $S\ell_2(\mathbf{R})$ action given by the representation $\rho + \tilde{\rho}_1 + \tilde{\rho}_1$; and $S\ell_2(\mathbf{R})$ is a reductive group.

Let $\mathbf{R}[z_1, \ldots, z_\ell] \big/_{\mathrm{I}}$ be a presentation of the $\mathbf{R}$-algebra $\mathcal{C}$ with $\tilde{q}_i \in \mathcal{C}$ the image of z_i under the presentation. Each $\tilde{q}_i$ is a homogeneous polynomial in $x, (\xi, \eta), (X, Y)$. By Sturmfels and White, $\mathcal{C}$ has a Stanley decomposition, that is, there is a finite set $\mathcal{F}$ of subsets of $\{0, \ldots, \ell\}^\ell$ such that every covariant $\tilde{C} \in \mathcal{C}$ may be written uniquely as

$$\tilde{C} = \tilde{C}\big(x, (\xi, \eta), (X, Y)\big) = \sum_{\alpha \in \mathcal{F}} \tilde{q}^\alpha C_\alpha\big(\tilde{q}_{I(\alpha)}\big) \tag{27}$$

where C_α is a polynomial in $\{\tilde{q}_i | i \in I(\alpha)\}$ with $I(\alpha) \subseteq \{1, \ldots, \ell\}$ and $\tilde{q}^\alpha = \tilde{q}_1^\alpha \cdots \tilde{q}_\ell^{\alpha_\ell}$. Next write the ordered set of generators $\{\tilde{q}_1, \cdots, \tilde{q}_\ell\}$ as the disjoint union of two ordered sets $\{\tilde{r_1}, \ldots, \tilde{r_k}\}$ and $\{\tilde{s_1}, \ldots, \tilde{s}_{\ell-k}\}$ where each $\tilde{r_j}$ is a homogeneous polynomial in $\big(x, (X, Y)\big)$; while each $\tilde{s_j}$ is a homogeneous polynomial in $(x,(X,Y),(\xi,\eta))$ with nonzero degree in (ξ,η). We may suppose that the (ξ,η) degree of $\tilde{C}$ is n_i, since $\mathfrak{X}^{\mathfrak{U}}_{n+1,n_i+1}$ is embedded in $\sum_{r,s\geq 0} \oplus C_{r,s,n_i}$. Apply the Cayley correspondence to (27). Namely, set $(X,Y)=(0,1)$. Then (27) becomes

$$C = C(x,(X,Y)) = \sum_{\alpha\in\mathcal{F}} q^{\alpha} C_{\alpha}(q_{I(\alpha)}) \tag{$*$}$$

where $q_i(x,(\xi,\eta)) = \tilde{q}_i(x,(\xi,\eta),(0,1))$ for $i=1,\ldots,\ell$. For each $\alpha\in\mathcal{F}$, $I(\alpha)$ can be written as the disjoint union of two subsets $J(\alpha)\subseteq\{1,\ldots,k\}$ and $K(\alpha)\subseteq\{k+1,\ldots,\ell\}$ where $i\in J(\alpha)$ if $q_i=r_i$ is a homogeneous polynomial in x alone and $i\in K(\alpha)$ if $q_i = s_{i-k}$ is a homogeneous polynomial of nonzero degree in (ξ,η). Since r_i is obtained from $\tilde{q}_i$ by the Cayley correspondence, r_i is $\mathfrak{U}$-invariant. Expanding $C_\alpha(q_{I(\alpha)}) = C_\alpha(r_{J(\alpha)}, s_{K(\alpha)})$ into its Taylor polynomial in $s_{K(\alpha)}$ about 0 gives

$$C_\alpha(r_{J(\alpha)}, s_{K(\alpha)}) = \sum_{\gamma} C_{\alpha\gamma}(r_{J(\alpha)}) s^{\gamma}_{K(\alpha)}. \tag{28}$$

Here $C_{\alpha\gamma}$ are polynomials and $s^{\gamma}_{K(\alpha)} = s^{\gamma_1}_{i_1}\ldots s^{\gamma_{k_\alpha}}_{i_{k_\alpha}}$ where $K(\alpha)=\{i_1,\ldots,i_{k_\alpha}\}\subseteq\{k+1,\ldots,\ell\}$. Writing $q^{\alpha} = r^{\beta_\alpha}s^{\delta_\alpha}$ and substituting (28) into ($*$) gives

$$C = \sum_{\alpha}\sum_{\gamma}\left(C_{\alpha\gamma}(r_{J(\alpha)}) r^{\beta_\alpha}\right) s^{\delta_\alpha} s^{\gamma}_{K(\alpha)}. \tag{$**$}$$

Clearly ($**$) is uniquely determined by (27). Moreover, (i) the coefficients $C_{\alpha\gamma}(r_{J(\alpha)})r^{\beta_\alpha}$ are $\mathfrak{U}$-invariant polynomials and (ii) the factor $s^{\delta_\alpha}s^{\gamma}_{K(\alpha)}$ is a homogeneous polynomial in (ξ,η) of degree n_i with coefficients which are polynomial in x. In other words, this factor may be written uniquely in the form

$$\sum_{j=0}^{n_i} p_j(x)\binom{n_i}{j}\xi^{n_i-j}\eta^{j} \tag{$\dagger$}$$

since C is a homogeneous polynomial of degree n_i in (ξ,η). Associate to ($\dagger$) the special equivariant $v\in\mathfrak{X}^{\mathfrak{U}}_{n+1,n_i+1}$, whose $j+1^{st}$ component is the polynomial $p_j(x)$. Thus we have shown that C can be written *uniquely* as a finite sum

$$\sum_{\beta\in\mathcal{G}} f_\beta v_\beta$$

where $f_\beta\in\mathcal{R}$, $v_\beta\in\mathfrak{X}^{\mathfrak{U}}_{n+1,n_i+1}$, and $\mathcal{G}$ is a finite collection of subsets of the form

$$H\times\{(m_1\ldots,m_{\ell-k+1})\in\mathbb{Z}^{\ell-k+1}_{\geq} \mid \sum_{j=1}^{\ell-k+1} m_j \leq n_i\}$$

where H is a subset of $\mathcal{F}$. This proves that $\mathfrak{X}^{\mathfrak{U}}_{n+1,n_i+1}$ has a Stanley decomposition.

We end this section by computing the Stanley decomposition of certain normal form modules. In part the calculation follows the existence argument given above.

Example 3. We treat the case when N is of type (2, 2, 2), that is, N is made up of three two by two Jordan blocks $\left(\begin{smallmatrix} 0 & 0 \\ 1 & 0 \end{smallmatrix}\right)$. To compute the Stanley decomposition of the normal form module $\mathcal{X}_6^{\mathcal{U}}$, it suffices to compute the Stanley decomposition of the special equivariant module $\mathcal{X}_{6,2}^{\mathcal{U}}$. The covariant algebra $\mathcal{C}$, in which $\mathcal{X}_{6,2}^{\mathcal{U}}$ is naturally embedded, is the algebra of polynomials on

$$\mathbf{R}^2 \oplus \mathbf{R}^2 \oplus (\mathbf{R}^2)^* \oplus (\mathbf{R}^2)^* \oplus \mathbf{R}^2$$

(with co-ordinates $((x_1, y_1), (x_2, y_2), (\xi, \eta), (X, Y), (x_3, y_3))$ which are invariant under the $S\ell_2(\mathbf{R})$ action induced by the representation $\rho_1 + \rho_1 + \tilde{\rho}_1 + \tilde{\rho}_1 + \rho_1$. Using the isomorphism $S : (\mathbf{R}^2)^* \longrightarrow \mathbf{R}^2 : (x, y) \longrightarrow \left(\begin{smallmatrix} y \\ -x \end{smallmatrix}\right)$ we have $S \circ \tilde{\rho}_1 = \tilde{\rho}_1 \circ S$, that is, the contragradient representation $\tilde{\rho}_1$ is equivalent to the standard representation ρ_1 under S. Thus $\mathcal{C}$ is the algebra of polynomials on

$$\mathbf{R}^2 \oplus \mathbf{R}^2 \oplus \mathbf{R}^2 \oplus \mathbf{R}^2 \oplus \mathbf{R}^2$$

(with co-ordinates $((x_1, y_1), (x_2, y_2), (\eta, -\xi), (Y, -X), (x_3, y_3)) = (w_1, w_2, w_3, w_4, w_5))$ which are invariant under the $S\ell_2(\mathbf{R})$ action induced by the representation $\rho_1 + \rho_1 + \rho_1 + \rho_1 + \rho_1$. By a theorem of Weyl [18], $\mathcal{C}$ is the algebra of polynomials in the brackets

$$[i, j] = w_i^1 w_j^2 - w_i^2 w_j^1 \quad \text{where } w_i = (w_i^1, w_i^2)$$

for $1 \le i < j \le 5$. Note that these brackets satisfy the Plücker relation

$$[i, j][k, \ell] - [i, k][i, \ell] + [i, \ell][j, k] = 0$$

for every distinct i, j, k, ℓ in $\{1, \ldots, 5\}$. Writing out the brackets $[i, j]$ gives

$$\begin{aligned}
[1,2] = x_1y_2 - x_2y_1, [1,3] = x_1\xi + y_1\eta, [1,4] = x_1X + y_1Y, [1,5] &= x_1y_3 - x_3y_1 \\
[2,3] = x_2\xi + y_2\eta, [2,4] = Xx_2 + y_2Y, [2,5] &= x_2y_3 - x_3y_2 \\
[3,4] = \xi Y - \eta X, \quad [3,5] &= x_3\xi + y_3\eta \\
[4,5] &= x_3X + y_3Y
\end{aligned}$$

Plot each bracket $[i, j]$ as a node (i, j) and consider all paths joining (1,2) with (4,5) which are made up of moves right or moves down and do not pass through any node of the form (i, i) $1 \le i \le 5$. Such paths are called *maximal monotone paths* and are given in figure 1.

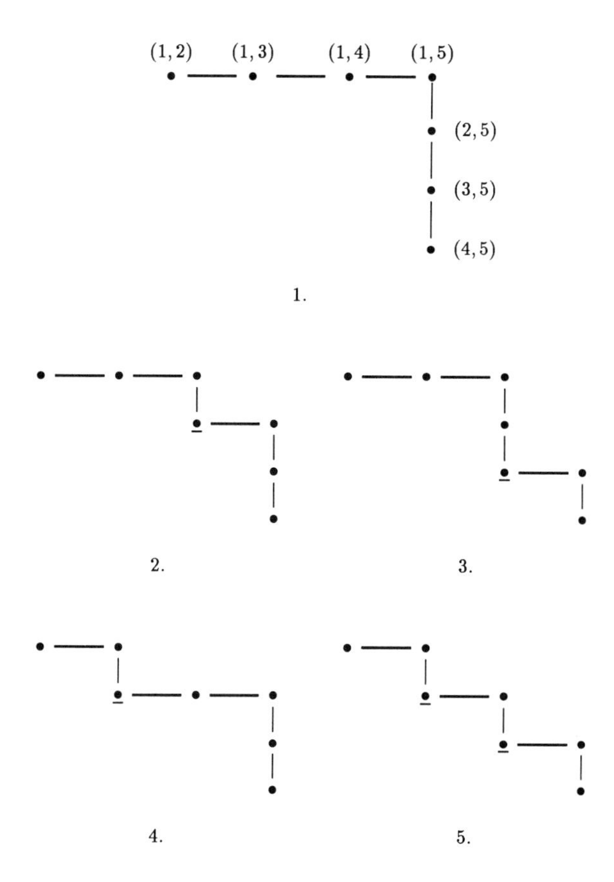

Figure 1. Maximal monotone paths joining (1,2) to (4,5) and not intersecting the diagonal.

A *corner* of a maximal monotone path is a node which comes from a move down followed by a move right. The corners are indicated by underlined dots in figure 1. The Stanley decomposition of $\mathcal{C}$ is

$$(*) \qquad \mathcal{C} = \mathbf{R}[\text{nodes of path 1}] \oplus [2,4]\mathbf{R}[\text{nodes of path 2}] \\ + \oplus [3,4]\mathbf{R}[\text{nodes of path 3}] \oplus [2,3]\mathbf{R}[\text{nodes of path 4}] \\ \oplus [2,3][3,4]\mathbf{R}[\text{nodes of path 5}]$$

The polynomial variables in the ith summand above are the brackets which appear as nodes in the ith maximal monotone path. The monomial factors in front of the ith summand are the product of the corners in the ith path. To prove this decomposition, it is clear that the polynomial algebras on the variables of two distinct maximal monotone paths have only 0 in common, because the paths are not the same. Also any proper collection of nodes not having a node which lies above and to the right of another node is contained in a unique maximal monotone path whose graph lies furthest above or to the right. For instance the underlined nodes in figure 2

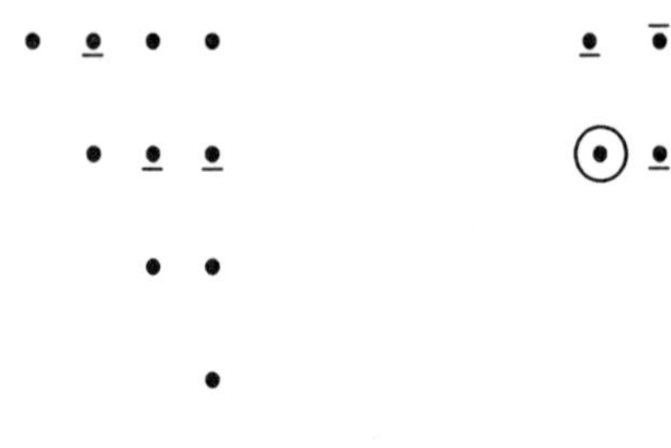

Figure 2 Figure 3

lie in the maximal monotone paths 4 and 2 of figure 1, but 2 lies further to the right than 4. All we now have to do is to explain the factors of the summands. Let P be a right most maximal monotone path containing a given proper set of nodes. Suppose the circled node in figure 3 is not contained in P. Then there is a larger monotone path Q (namely the one going through the barred node) containing the given nodes. This contradicts the maximality of P. Hence the circled node, which is a corner, is contained in P. This completes the proof that $(*)$ is the Stanley decomposition of $\mathfrak{C}$.

Using the expressions for the brackets in co-ordinates, and setting $(X,Y) = (0,1)$ and $\Delta_{ij} = x_i y_j - x_j y_i$ gives

$$\begin{aligned}
&\mathbf{R}[\Delta_{12}, x_1\xi + y_1\eta, y_1, \Delta_{13}, \Delta_{23}, x_3\xi + y_3\eta, y_3] \\
&\quad \oplus y_2\mathbf{R}[\Delta_{12}, x_1\xi + y_1\eta, y_1, y_2, \Delta_{23}, x_3\xi + y_3\eta, y_3] \\
&\quad \oplus \xi\mathbf{R}[\Delta_{12}, x_1\xi + y_1\eta, y_1, y_2, \xi, x_3\xi + y_3\eta, y_3] \\
&\quad \oplus (x_2\xi + y_2\eta)\mathbf{R}[\Delta_{12}, x_1\xi + y_1\eta, x_2\xi + y_2\eta, y_2, \Delta_{23}, x_3\xi + y_3\eta, y_3] \\
&\quad \oplus \xi(x_2\xi + y_2\eta)\mathbf{R}[\Delta_{12}, x_1\xi + y_1\eta, x_2\xi + y_2\eta, y_2, \xi, x_3\xi + y_3\eta, y_3].
\end{aligned}$$

Taking the coefficients of the polynomials which are of degree 1 and ξ and η (since we are interested in $\mathbf{R}^2$–valued special equivariants) gives

$$\begin{aligned}
&(x_1\xi + y_1\eta)\mathbf{R}[y_1, y_3, \Delta_{12}, \Delta_{13}, \Delta_{23}] \oplus (x_3\xi + y_3\eta)\mathbf{R}[y_1, y_3, \Delta_{12}, \Delta_{13}, \Delta_{23}] \\
&\oplus y_2(x_1\xi + y_1\eta)\mathbf{R}[y_1, y_2, y_3, \Delta_{12}, \Delta_{23}] \oplus y_2(x_3\xi + y_3\eta)\mathbf{R}[y_1, y_2, y_3, \Delta_{12}, \Delta_{23}] \\
&\oplus \xi\mathbf{R}[y_1, y_2, y_3, \Delta_{12}] \oplus (x_2\xi + y_2\eta)\mathbf{R}[y_2, y_3, \Delta_{12}, \Delta_{23}]
\end{aligned}$$

Identifying ξ with the vector $\binom{1}{0}$ and η with the vector $\binom{0}{1}$ gives

$$\mathsf{R}[y_1,y_3,\Delta_{12},\Delta_{13},\Delta_{23}]\begin{pmatrix}x_1\\ y_1\end{pmatrix}\oplus\mathsf{R}[y_1,y_3,\Delta_{12},\Delta_{13},\Delta_{23}]\begin{pmatrix}x_3\\ y_3\end{pmatrix}$$
$$\oplus\, y_2\mathsf{R}[y_1,y_2,y_3,\Delta_{12},\Delta_{23}]\begin{pmatrix}x_1\\ y_1\end{pmatrix}\oplus y_2\mathsf{R}[y_1,y_2,y_3,\Delta_{12},\Delta_{23}]\begin{pmatrix}x_3\\ y_3\end{pmatrix}$$
$$\oplus\,\mathsf{R}[y_2,y_3,\Delta_{12},\Delta_{23}]\begin{pmatrix}x_2\\ y_2\end{pmatrix}\oplus\mathsf{R}[y_1,y_2,y_3,\Delta_{12}]\begin{pmatrix}1\\ 0\end{pmatrix},$$

which is the Stanley decomposition of the special equivariant module $\mathcal{X}^{\mathcal{U}}_{6,2}$ over the ring $\mathcal{R}$, which has generators $\{y_1,y_2,y_3,\Delta_{12},\Delta_{13},\Delta_{23}\}$ subject to the relation $y_3\Delta_{12}-y_2\Delta_{13}+y_1\Delta_{23}=0$. This completes the computation of the Stanley decomposition of the normal form module $\mathcal{X}^{\mathcal{U}}_6$.

Example 4. We compute the Stanley decomposition of the normal form module of $\mathcal{X}^{\mathcal{U}}_3$ where N is a three by three Jordan block. From (20) the generating function $\mathcal{P}^3$ of the algebra of covariants $\mathcal{C}$ is

$$\begin{aligned}\mathcal{P}^3 &= \frac{1}{1-uv}\,\frac{1}{2\pi i}\int_{|z|=1}\frac{(1-z^2)}{(1-tz^2)(1-\frac{t}{z^2})(1-zu)(1-zv)}\,\frac{dz}{z}\\ &= \frac{1+uvt}{(1-uv)(1-u^2t)(1-v^2t)(1-t^2)}\,.\end{aligned}\tag{29}$$

Applying a fundamental result from the umbral calculus [19] it follows that $\mathcal{C}$ is generated by brackets in variables called symbols. In the case at hand consider the symbols

$$\alpha=(\alpha_1,\alpha_2),\;\beta=(\beta_1,\beta_2),\;\Xi=(\xi,\eta),\quad X=(X,Y)$$

and the following table

basic bracket expression	umbral evaluation	term in generating function
$[X,\Xi]$	$\pi_1=X\eta-Y\xi$	uv
$(\alpha X)^2$	$\sigma_3=x_1X^2+2x_2XY+x_3Y^2$	u^2t
$(\alpha\Xi)^2$	$\sigma_2=x_1\xi^2+2x_2\xi\eta+x_3\eta^2$	v^2t
$[\alpha\beta]^2$	$\pi_2=2(x_1x_3-{x_2}^2)$	t^2
$(\alpha X)(\alpha\Xi)$	$\sigma_1=(x_1\xi+x_2\eta)X+(x_2\xi+x_3\eta)Y$	uvt

Table 3

where we have used the traditional round and square bracket notation of the umbral calculus. Only the third column of the above table needs to be explained. We used the following rules: (1) The number of times X or Ξ appears in the basic bracket expression is the power of u or v respectively; (2) For each appearance of α or β add a factor of $t^{\frac{1}{2}}$. A straightforward calculation shows that the relation

$$g={\sigma_1}^2+\sigma_2\sigma_3-\frac{1}{2}{\pi_1}^2\pi_2=0$$

holds among the covariants. Using table 3, we see that the algebra

$$\mathcal{B} = \mathbb{R}[\sigma_1, \sigma_2, \sigma_3, \pi_1, \pi_2] \big/ _{I},$$

where I is the ideal generated by g, has the same Poincaré polynomial as $\mathcal{C}$, namely, $\mathcal{P}^3$. Hence $\mathcal{B}$ is a presentation of the algebra $\mathcal{C}$. We now apply the Sturmfels–White theorem to compute the Stanley decomposition of $\mathcal{B}$. Since I is not generated by a monomial, if we order the variables by

$$\sigma_1 > \sigma_2 > \sigma_3 > \pi_1 > \pi_2 > 1,$$

we see that g is a Gröbner basis for I with the ideal *init* I of initial terms in the Gröbner basis generated by the monomial σ_1^2. Thus $\mathcal{B}$ has the same Stanley decomposition as

$$\mathbb{R}[\sigma_1, \sigma_2, \sigma_3, \pi_1, \pi_2] \big/ _{init\ I}.$$

As is easy to see, this is

$$\mathbb{R}[\pi_1, \pi_2, \sigma_2, \sigma_3] \oplus \sigma_1 \mathbb{R}[\pi_1, \pi_2, \sigma_2, \sigma_3] \,. \tag{30}$$

Using table 3 and substituting $(X, Y) = (0, 1)$ into (30) gives

$$\begin{aligned} &\xi^2 \mathbb{R}[x_3, x_1 x_3 - x_2^2] + (x_1 \xi^2 + 2x_2 \xi\eta + x_3 \eta^2)\mathbb{R}[x_3, x_1 x_3 - x_2^2] \\ &\quad + 2\xi(x_2 \xi + x_3 \eta)\mathbb{R}[x_3, x_1 x_3 - x_2^2] \end{aligned} \tag{31}$$

Identifying ξ^2 with $(1,0,0), 2\xi\eta$ with $(0,1,0)$ and η^2 with $(0,0,1)$, (31) becomes

$$\mathbb{R}[x_3, x_1 x_3 - x_2^2] \begin{pmatrix} 1 \\ 0 \\ 0 \end{pmatrix} \oplus \mathbb{R}[x_3, x_1 x_3 - x_2^2] \begin{pmatrix} x_1 \\ x_2 \\ x_3 \end{pmatrix} \oplus \mathbb{R}[x_3, x_1 x_3 - x_2^2] \begin{pmatrix} 2x_2 \\ x_3 \\ 0 \end{pmatrix}$$

which is the desired Stanley decomposition.

§**4. Notes.** These notes try to pinpoint the first explicit statement of certain basic ideas in normal form theory for formal power series vectorfields $\mathbf{X} = X_1 + X_2 + \cdots$ with leading term which is a <u>nonsemisimple</u> linear term. We also treat the Hamiltonian case where $\mathbf{H} = H_2 + H_3 + \cdots$ is a formal power series with quadratic leading terms. A nice survey of the general history of normal form theory can be found in [20, p.42].

In [21, p.55] Takens states that to find the normal form of $\mathbf{X}$ one needs to find a complement to the image of ad_{X_1} in the space $\mathbf{P}_\ell$ of homogeneous polynomial vectorfields of degree ℓ for $\ell = 2, 3, \ldots$. He also computes the normal form of $\mathbf{X}$ in the case (2) when $X_1 = \left(\begin{smallmatrix} 0 & 0 \\ 1 & 0 \end{smallmatrix}\right)$. This idea Belitskii refines somewhat in [22, p.112 and 23] when he shows that, after a suitable choice of inner product on $\mathbf{P}_\ell$, a complement to *im* ad_{X_1} is the kernel of the adjoint of ad_{X_1}. This observation was made again by Elphick et al. in [25] for vectorfields and also by Dragt and Finn [24, p.2652] in the

Hamiltonian case. Let $X_1 = S + N$ be commuting semisimple S and nilpotent N linear maps in the vectorfield case and $H_2 = S+N$ be Poisson commuting quadratic polynomials in the Hamiltonian case. Then van der Meer [26, p.133] shows that a complement to $im\ ad_{H_2}$ is given by a complement to $ker\ ad_S \cap ker\ ad_N$ in $ker\ ad_S$. This observation was also made by Meyer [27] for the vectorfield case.

In [1] Cushman uses the theorem of Jacobson-Morosov to embed $H_2 = N$ into a subalgebra of the space of quadratic polynomials under Poisson bracket, which is isomorphic to $sl_2(\mathbf{R})$ and is generated by $\{N, M, H\}$. He also notes that because $\{ad_N, ad_M, ad_H\}$ defines a representation of $sl_2(\mathbf{R})$ on the space of homogeneous polynomials of degree ℓ for $\ell = 3, 4, \dots$, a complement to $im\ ad_N$ is given by $ker\ ad_M$ (see also [3]). In [2] Cushman uses these ideas to compute the normal form for the Hamiltonian 1:1 nonsemisimple resonance. The use of representation theory of $sl_2(\mathbf{R})$ in normal form theory for vectorfields was independently observed in a footnote in Bogaevskii and Povsner [28]. An elaboration of this remark can be found in [29, p.68]; however, no examples of normal forms are explicitly computed. In [27] the normal form for the (2) case for vectorfields is obtained using representation theory and a formula for the normal form of a vectorfield in nonsemisimple 1:1 resonance is given. Cushman and Sanders [4] use representation theory and generating function techniques to calculate and prove the correctness of the normal form for the (3) and (2,2) cases for vectorfields. They also calculate and prove the correctness of the normal form for the 1:1 nonsemisimple resonance, rectifying an error in [27]. These same normal forms and the (3,2) case were obtained in [25] using the Belitskii approach and invariant theory. The (3,2) and (4) cases are recomputed in [31] using the technique of [4]. In [5] Cushman and Sanders show that normal form theory for Hamiltonian functions with nilpotent quadratic terms is equivalent to classical invariant theory of pairs of binary forms. In [30] some of the normal forms given in [5] are recomputed using essentially the same techniques as [4].

§5. Appendix. Here we give a list of the Stanley decompositions of normal form modules for vectorfields with nilpotent linear part N. We say that N has a Jordan block structure $(n_1 + 1, n_2 + 1, ., n_m + 1)$ if it is made up of m Jordan blocks of size $n_i + 1$ of the form

$$N|\mathbf{R}^{n_i+1} = \begin{pmatrix} 0 & & & & & \\ 1 & 0 & & & & \\ & 2 & 0 & & & \\ & & \cdot & \cdot & & \\ & & & \cdot & \cdot & \\ & & & & n_i & 0 \end{pmatrix}.$$

The list consists of the cases

$$(2),\ (3), (2,2), (4), (2,3), (5), (2,2,2), (3,3).$$

We also give the generators and relations for the ring $\mathcal{R}$ of polynomial functions which are invariant under $\mathcal{U} = \exp sM$ where $\{M, N, H\}$ spans a subalgebra of the Lie algebra of linear vectorfields which is isomorphic to $s\ell_2(\mathbf{R})$.

A Table of Stanley decompositions of normal form modules

N	generators and relations for $\mathfrak{R}$	Stanley decomposition
(2)	$p_1 = x_1$	$f_1(p_1)\binom{1}{0} + f_2(p_1)\binom{x_1}{x_2}$ [4, p.36]
(3)	$p_1 = x_3, p_2 = x_1x_3 - x_2^2$	$f_1(p_1,p_2)\begin{pmatrix}1\\0\\0\end{pmatrix} + f_2(p_1,p_2)\begin{pmatrix}x_1\\x_2\\x_3\end{pmatrix} + f_3(p_1,p_2)\begin{pmatrix}2x_2\\x_3\\0\end{pmatrix}$ [4, p.45]
(2,2)	$p_1 = x_2, p_2 = x_4$ $p_3 = x_2x_3 - x_1x_4$	$\sum_{i=1}^{3}\Big[f_{1i}(p_1,p_2)\nu_1^{(i)} + f_{2i}(p_1,p_2,p_3)\nu_2^{(i)} + f_{3i}(p_1,p_2,p_3)\nu_3^{(i)}\Big]$ where $\nu_j^{(1)} = \binom{\nu_j}{0}$ and $\nu_j^{(2)} = \binom{0}{\nu_j}$ for $j = 1,2,3$ and $\nu_1 = \binom{1}{0}, \nu_2 = \binom{x_1}{x_2}, \nu_3 = \binom{x_3}{x_4}$. [4, p.46]
(4)	$p_1 = x_4, p_2 = x_2x_4 - x_3^2$ $p_3 = x_1x_4^2 + 2x_3^3 - 3x_2x_3x_4$ $p_4 = 6x_1x_2x_3x_4 - x_1^2x_4^2$ $-4x_1x_3^3 + 3x_2^2x_3^2 - 4x_2^3x_4$ relation: $p_3^2 + 4p_2^3 + p_1^2p_4 = 0$	$f_1(p_1,p_2,p_4)\nu_1 + f_2(p_1,p_2,p_4)\nu_2 + f_3(p_1,p_2,p_4)\nu_3$ $+f_4(p_2,p_4)\nu_4 + f_5(p_1,p_2,p_4)M\nu_1$ $+f_6(p_1,p_2,p_4)M\nu_2 + f_7(p_1,p_2,p_4)M\nu_3$ $+f_8(p_1,p_2,p_4)M^2\nu_1 + f_9(p_1,p_2,p_4)\nu_5$ where $M = \begin{pmatrix}0&3&&\\&0&2&\\&&0&1\\&&&0\end{pmatrix}$, and $\nu_1 = \begin{pmatrix}x_1\\x_2\\x_3\\x_4\end{pmatrix}, \nu_2 \begin{pmatrix}3x_1x_2x_3 - 2x_2^3 - x_1^2x_4\\2x_1x_3^2 - x_2^2x_3 - x_1x_2x_4\\x_1x_3x_4 - 2x_2^2x_4 + x_2x_3^2\\x_1x_4^2 + 2x_3^2 - 3x_2x_3x_4\end{pmatrix}$ $\nu_3 = \begin{pmatrix}3(x_1x_3 - x_2^2)\\x_1x_4 - x_2x_3\\x_2x_4 - x_3^2\\0\end{pmatrix}, \nu_4 = \begin{pmatrix}3(x_1x_3 - x_2^2)(x_1x_4 - x_2x_3)\\3x_1^2x_3^2 - 2x_1x_3^3 - 2x_2^3x_4 + x_1^2x_4^2\\3(x_1x_4 - x_2x_3)(x_2x_4 - x_3^2)\\6(x_2x_4 - x_3^2)^2\end{pmatrix}$ $\nu_5 = \begin{pmatrix}1\\0\\0\\0\end{pmatrix}$ [7, p.62]

N	generators and relations for $\mathfrak{R}$	Stanley decomposition
(3,2)	$p_1 = x_3, p_2 = x_5$ $p_3 = x_1x_3 - x_2^2$ $p_4 = x_3x_4 - x_2x_5$ $p_5 = 2x_2x_4x_5 - x_3x_4^2 - x_1x_5^2$ relation : $p_4^2 = p_3p_2^2 - p_1p_5$	$f_1(p_1,p_2,p_3,p_5)\binom{\nu_1}{0} + f_2(p_1,p_2,p_3,p_5)\binom{\nu_2}{0}+$ $f_3(p_1,p_2,p_3,p_5)\binom{\nu_3}{0} + f_4(p_1,p_2,p_3,p_5)\binom{\nu_4}{0} + f_5(p_1,p_2,p_3,p_5)\binom{\nu_5}{0}$ $+p_4f_6(p_1,p_2,p_3,p_5)\binom{\nu_6}{0} + f_7(p_2,p_3,p_5)\binom{\nu_7}{0}+$ $+f_8(p_2,p_3,p_5)\binom{\nu_8}{0} + g_1(p_1,p_2,p_3,p_5)\binom{0}{\nu_1}$ $+g_2(p_1,p_2,p_3,p_5)\binom{0}{\nu_2} + g_3(p_1,p_{2,3},p_5)\binom{0}{\nu_3}$ $g_4(p_1,p_2,p_3,p_5)\binom{0}{\nu_4}$ where $\nu_1 = \begin{pmatrix}1\\0\\0\end{pmatrix}, \nu_2 = \begin{pmatrix}2x_4\\x_5\\0\end{pmatrix}, \nu_3 \begin{pmatrix}x_1\\x_2\\x_3\end{pmatrix}, \nu_4 = \begin{pmatrix}2x_2\\x_3\\0\end{pmatrix}$ $\nu_5 = \begin{pmatrix}2x_2x_4\\x_3x_4+x_2x_5\\2x_3x_5\end{pmatrix}, \nu_6 = \begin{pmatrix}x_1\\x_2\\x_3\end{pmatrix}, \nu_7 = \begin{pmatrix}x_4^2\\x_4x_5\\x_5^2\end{pmatrix},$ $\nu_8 = \begin{pmatrix}2x_4(x_2x_4 - x_1x_5)\\x_3x_4^2 - x_1x_5^2\\2x_5(x_3x_4 - x_2x_5)\end{pmatrix}$ $\eta_1 = \binom{1}{0}, \eta_2 = \binom{x_4}{x_5}, \eta_3 = \binom{x_2}{x_3}, \eta_4 = \binom{x_2x_4 - x_1x_5}{x_3x_4 - x_2x_5}$ [25, p.125].
(5)	$p_1 = x_5, p_2 = x_3x_5 - x_4^2$ $p_3 = x_1x_5 - 4x_2x_4 + 3x_3^2$ $p_4 = 3x_3x_4x_5 - 2x_4^3 - x_2x_5^2$ $p_5 = x_1x_3x_5 - x_1x_4^2 - x_3^2 - x_2^2x_5$ $+2x_2x_3x_4$ relation: $0 = p_4^2 + p_1^3p_5 - p_1^2p_2p_3$ $+4p_2^3$	$f_1(p_1,p_2,p_3,p_5)\nu_1 + \sum_{i=0}^{3} f_{2+i}(p_1,p_2,p_3,p_5)M^i\nu_2$ $+ \sum_{i=0}^{3} f_{6+i}(p_1,p_2,p_3,p_5)M^i\nu_3 + f_{10}(p_1,p_2,p_3,p_5)\nu_4$ where $M = \begin{pmatrix}0 & 4 & & & \\ & 0 & 3 & & \\ & & 0 & 2 & \\ & & & 0 & 1\\ & & & & 0\end{pmatrix}$, and $\nu_1 = \begin{pmatrix}1\\0\\0\\0\\0\end{pmatrix}, \nu_2 = \begin{pmatrix}x_1\\x_2\\x_3\\x_4\\x_5\end{pmatrix} \nu_3 = \begin{pmatrix}x_1x_3 - x_2^2\\ \frac{1}{2}(x_1x_4 - x_2x_3)\\ \frac{1}{6}(x_1x_5 + 2x_2x_4 - 3x_2^2)\\ \frac{1}{2}(x_2x_5 - x_3x_4^2)\\ x_3x_5 - x_4^2\end{pmatrix}$ $\nu_4 = \begin{pmatrix}\frac{1}{3}(x_1x_2x_3 + 2x_2^2x_4 - 3x_1x_3x_4)\\ \frac{1}{2}(x_2^2x_5 - x_1x_4^2)\\ \frac{1}{3}(-x_1x_4x_5 + 3x_2x_3x_5 - 2x_2x_4^2)\\ \frac{1}{6}(-2x_2x_4x_5 + 9x_3^2x_5 - 6x_3x_4^2 - x_1x_5^2)\\ 3x_3x_4x_5 - 2x_4^3 - x_2x_5^2\end{pmatrix}$

N	generators and relations for $\mathcal{R}$	Stanley decomposition
(2,2,2)	y_1, y_2, y_3 $\Delta_{12} = x_1 y_2 - x_2 y_1$, $\Delta_{13} = x_2 y_3 - x_3 y_2$ $\Delta_{23} = x_2 y_3 - x_3 y_2$ relation: $y_3 \Delta_{12} - y_2 \Delta_{13} + y_1 \Delta_{23}$ $= 0$	$\sum_{i=1}^{3} [f_{1i}(y_1, y_2, \Delta_{12}, \Delta_{13}, \Delta_{23}) \nu_1^{(i)}$ $+ f_{2i}(y_1, y_2, \Delta_{12}, \Delta_{13}, \Delta_{23}) \nu_2^{(i)}$ $+ y_2 f_{3i}(y_1, y_2, \Delta_{12}, \Delta_{23}) \nu_1^{(i)}$ $+ y_2 f_{4i}(y_1, y_2, \Delta_{12}, \Delta_{23}) \nu_2^{(i)}$ $+ f_{5i}(y_2, y_3, \Delta_{12}, \Delta_{23}) \nu_3^{(9)}$ $+ f_{6i}(y_1, y_2, y_3, \Delta_{12}) \nu_4^{(i)}]$ where $\nu_j^{(1)} = \begin{pmatrix} \nu_j \\ 0 \\ 0 \end{pmatrix}, \nu_j^{(2)} = \begin{pmatrix} 0 \\ \nu_j \\ 0 \end{pmatrix}, \nu_j^{(3)} = \begin{pmatrix} 0 \\ 0 \\ \nu_j \end{pmatrix}$ for $j = 1, \ldots, 4$ and $0 = \binom{0}{0}, \nu_1 = \binom{x_1}{x_i}$, $\nu_2 = \binom{x_3}{y_3}, \nu_3 = \binom{x_2}{y_2}$, and $\nu_4 = \binom{1}{0}$.
(3,3)	$p_1 = x_3, p_2 = y_3$, $p_3 = x_2 y_3 - x_3 y_2$, $p_4 = x_1 x_3 - x_2^2$, $p_5 = y_1 y_3 - y_2^2$ $p_6 = x_1 y_3 + x_3 y_1 - 2 x_2 y_2$ relation: $p_3^2 = -p_1^2 p_5 + p_1 p_2 p_6 - p_2^2 p_4$	$\sum_{i=1}^{2} \Big[f_{1i}(p_1, p_2, p_4, p_5, p_6) \nu_i^{(i)}$ $+ f_{2i}(p_1, p_2, p_4, p_5, p_6) \nu_2^{(i)}$ $+ f_{3i}(p_1, p_2, p_4, p_5, p_6) \nu_3^{(i)}$ $+ f_{4i}(p_1, p_2, p_4, p_5, p_6) \nu_4^{(i)}$ $+ f_{5i}(p_1, p_2, p_4, p_5, p_6) \nu_5^{(i)}$ $+ f_{6i}(p_1, p_2, p_4, p_5, p_6) \nu_6^{(i)} \Big]$ where $\nu_j^{(i)} = \binom{\nu_j}{0}$ and $\nu_j^{(2)} = \binom{0}{\nu_j}$ for $j = 1, 2$ and $0 = \begin{pmatrix} 0 \\ 0 \\ 0 \end{pmatrix}, \nu_1 = \begin{pmatrix} 1 \\ 0 \\ 0 \end{pmatrix}$, $\nu_2 = \begin{pmatrix} x_1 \\ x_2 \\ x_3 \end{pmatrix}, \nu_3 = \begin{pmatrix} 2x_2 \\ x_3 \\ 0 \end{pmatrix}, \nu_4 = \begin{pmatrix} y_1 \\ y_2 \\ y_3 \end{pmatrix}$ $\nu_5 = \begin{pmatrix} 2y_2 \\ y_3 \\ 0 \end{pmatrix}, \nu_6 = \begin{pmatrix} 2(x_1 y_2 - x_2 y_1) \\ x_1 y_3 - x_3 y_1 \\ 2(x_2 y_3 - x_3 y_2) \end{pmatrix}$.

REFERENCES

[1] CUSHMAN, R., *Representation theory and normal form*, preprint # 180, Rijksuniversiteit Utrecht (1980).

[2] CUSHMAN, R., *Reduction of the 1:1 nonsemisimple resonance*, Hadronic J., 5 (1982), 2109–24.

[3] CUSHMAN, R., DEPRIT, A, AND MOSAK, R., *Normal form and representation theory*, J. Math. Phys. 24 (1983), 2103–17.

[4] CUSHMAN, R. AND SANDERS, J., *Nilpotent normal forms and representation theory of* $sl_2(\mathbf{R})$ *in: Multiparameter bifurcation theory*, eds. M. Golubitsky and J. Guckenheimer, Contemporary Mathematics, vol. 56 Amer. Math Soc., Providence (1986), 31–51.

[5] CUSHMAN, R. AND SANDERS, J., *Invariant theory and normal form of Hamiltonian vectorfields with nilpotent linear part, in: Oscillation, bifurcation and chaos*, CMS conference proceedings vol. 8, eds. F.V. Atkinson et al. , AMS, Providence (1987), 353–371.

[6] CUSHMAN, R., SANDERS, J. AND WHITE, N., *Normal form for the (2;n) nilpotent vectorfield, using invariant theory*, Physica D 30 (1988), 399–412.

[7] CUSHMAN, R. AND SANDERS, J., *Nilpotent normal form in dimension 4*, in: NATO ASI Series, vol. F37: Dynamics of infinite dimensional systems, eds. S.–N. Chow and J.K. Hale, Springer–Verlag, New York (1987), 61–66.

[8] CUSHMAN, R. AND SANDERS, J., *Splitting algorithm for nilpotent normal forms*, Dynamics and Stability of Systems 2 (1988), 235–246.

[9] BILLERA, L., CUSHMAN, R., AND SANDERS, J., *The Stanley decomposition of the harmonic oscillator*, Proc. Ned. Akad. Wet. series A 91 (1988), 375-393.

[10] STURMFELS, B. AND WHITE, N., *Computing Combinatorial decompositions of rings*, preprint, RISC, Johannes–Kepler Universität, Linz (1988).

[11] HOWE, R.E., *The classical groups and invariants of binary forms*, Proc. Sympos. in Pure Math. 46 (1988), 133–166.

[12] HELGASON, S., *Differential geometry, Lie groups, and symmetric spaces*, Academic Press, New York (1982).

[13] HUMPHREYS, J., *Introduction to Lie algebras and representation theory*, Springer–Verlag, New York (1972).

[14] FOGARTY, J., *Invariant theory*, Benjamin, New York (1963).

[15] SPRINGER, T., *Invariant theory*, Lecture Notes in Mathematics 583 (1977) Springer–Verlag, New York.

[16] BRÖCKER, T., AND TOM DIECK, T., *Representations of Compact Lie groups*, Springer–Verlag, New York (1985).

[17] SPRINGER, T., *On the invariant theory of* SU_2, Proc. Kon. Ned. Akad. Wet. Series A 83 (1980), 339–345.

[18] WEYL, H., *The classical groups*, Princeton Univ. Press, Princeton (1946).

[19] KUNG, J.P.S. AND ROTA, G–C., *Invariant theory of binary forms*, Bull. Amer. Math. Soc 10 (1984), 27–85.

[20] VAN DER MEER, J.–C., *The Hamiltonian Hopf bifurcation*, Lect. Notes in Math. 1160, Springer–Verlag, New York (1985).

[21] TAKENS, F., *Singularities of vectorfields*, Publ. Math. I.H.E.S. 43 (1974), 47–100.

[22] BELITSKII, G.R., *Equivalence and normal forms of germs of smooth mappings*, Russ. Math. Surveys 33 (1978), 107–177.

[23] BELITSKII, G.R., *Normal forms relative to a filtering action of a group*, Trans. Moscow Math. Soc. 40 (1981), 1–39.

[24] DRAGT, A.J. AND FINN, J.M., *Normal form for mirror machine Hamiltonians*, J. Math. Phys. 20 (1979), 375-393.

[25] ELPHICK, C. ET AL., *A simple global characterization for normal forms of singular vectorfields*, Physica D 29 (1987), 95–127.

[26] VAN DER MEER, J.–C., *Nonsemisimple 1:1 resonance at an equilibrium*, Celestial Mech. 27 (1982), 131–149.

[27] MEYER, K., *Normal forms for the general equilibrium*, Funcialaj ekvacioj 27 (1984), 261–271.

[28] BOGAEVSKI, V.N. AND POVSNER, A.YA., *A nonlinear generalization of a shearing transformation*, Funct. Anal. and Appl. 16 (1982), 45–46.

[29] *Encyclopedia of mathematical sciences*, vol 1, Dynamical systems I, eds. D.V. Anosov and V.I. Arnold, Springer–Verlag, New York, (1988).

[30] ELPHICK, C., *Global aspects of Hamiltonian normal forms*, Phys. Lett. A 127 (1988), 418–24.

[31] WANG, D., *Some new results on the method of representation theory of $s\ell_2(\mathbf{R})$ in normal form theory*, preprint, Institute of Math., Acad. Sinica, Beijing (1988).

OPERATORS COMMUTING WITH COXETER GROUP ACTIONS ON POLYNOMIALS

CHARLES F. DUNKL*

Abstract. The inverse of the operator $f \mapsto \sum_{i=1}^{N} x_i \frac{\partial}{\partial x_i} f(x) + \sum_{j=1}^{m} \alpha_j (f(x) - f(x\sigma_j))$ on smooth functions on $\mathbf{R}^N$ is constructed where $\{\sigma_i : 1 \leq i \leq m\}$ is the set of reflections in a Coxeter group G, and the values of α_i are constant on conjugacy classes. This operator arises in the theory of orthogonal polynomials on the sphere with G–invariant weight functions which are products of power of linear functions. The construction depends on some structure and character theory of Coxeter groups.

Key words. Coxeter groups, reflections, symmetric group, differential–difference operators.

AMS(MOS) subject classifications. 20C30, 33A75, 20C05.

The conjugacy classes of reflections in a Coxeter group G generate the subalgebra $\mathcal{R}$ of the center $\mathcal{Z}\mathbb{C}G$ of the group algebra. The complex homomorphisms of $\mathcal{R}$ come from the irreducible characters of G. For the infinite families of types A_N, B_N, D_N the characters and their values on these conjugacy classes can be described in terms of partitions and Ferrers diagrams. The purpose of this paper is to exhibit the use of the algebra $\mathcal{R}$ in the analysis of a commutative set $\{T_i : 1 \leq i \leq N\}$ of differential–difference operators associated to G (acting on $\mathbf{R}^N$), see [4]. These operators are the analogues of $\{\frac{\partial}{\partial x_i} : 1 \leq i \leq N\}$ in the theory of spherical harmonics when the usual rotation–invariant measure on the sphere is multiplied by the G–invariant function

$$h(x)^2 := \prod_{j=1}^{m} |\langle x, v_j \rangle|^{2\alpha_j},$$

where $\{v_j : 1 \leq j \leq m\}$ is a set of positive roots for $G, \alpha_j > 0$ for each j, and $\alpha_i = \alpha_j$ whenever the reflections corresponding to v_i and v_j are conjugate. The operators $\{T_i\}$ depend on $\{\alpha_j\}$. The "h–Laplacian" $\sum_{i=1}^{N} T_i^2$ is used to define "h–harmonic" polynomials (see [3]). When G is a dihedral group ($N = 2$), these h–harmonic polynomials are closely related to the classical Gegenbauer and Jacobi polynomials.

For each smooth function F on a spherical neighborhood of $0 \in \mathbf{R}^N$, the functions $f_i := T_i F$ satisfy the exactness conditions $T_i f_j = T_j f_i (1 \leq i,\ j \leq N)$. The algebra $\mathcal{R}$ will be used in this paper to solve the inverse problem: given an N–tuple $(f_i)_{i=1}^N$ of smooth functions satisfying the exactness conditions, construct a function F with $T_i F = f_i$, $(1 \leq i \leq N)$. Also, $\mathcal{R}$ has a simpler structure than the full center $\mathcal{Z}\mathbb{C}G$, as will be illustrated by examples, including non–crystallographic Coxeter groups. Integers suffice to express the various structure constants and values of complex homomorphisms.

---

*Department of Mathematics, Mathematics–Astronomy Building, University of Virginia, Charlottesville, VA 22903. The author was partially supported by NSF grants DMS–8601670 and DMS–8802400.

The sections of the paper are as follows:

(1) definitions and results from previous work;

(2) the algebra $\mathfrak{R}$, its complex homomorphisms, some examples (A_N, F_4, H_3, H_4);

(3) the inversion formula, applications, examples.

§1. Definitions and Previous Results. Suppose that G is a Coxeter (also called "finite reflection") group realized as a finite subgroup of the orthogonal group on $\mathbf{R}^N$ with the set $\{\sigma_i : 1 \le i \le m\}$ of reflections (see the text [1] by Benson and Grove for basic theorems, description, and classification). Choose a set of vectors $\{v_i : 1 \le i \le m\} \subset \mathbf{R}^N$ such that σ_i is the reflection along v_i (that is,

$$x\sigma_i = x - 2\langle x, v_i\rangle v_i/|v_i|^2, x \in \mathbf{R}^N), \quad (1 \le i \le m), \quad \text{and} \quad |v_i| = |v_j|$$

whenever $\sigma_i \sim \sigma_j$. The inner product and norm are $\langle x, y\rangle := \sum_{i=1}^N x_i y_i$ and $|x| := \langle x, x\rangle^{1/2}$, $(x, y \in \mathbf{R}^N)$. We choose positive parameters α_i, $1 \le i \le m$ such that $\alpha_i = \alpha_j$ whenever $\sigma_i \sim \sigma_j$.

Define $h(x) := \prod_{j=1}^m |\langle x, v_j\rangle|^{\alpha_j}$, a G–invariant positively homogeneous function on $\mathbf{R}^N$.

We will use the term "smooth function" to mean C^∞ functions defined on some spherical neighborhood $\{x : |x| < r\}$ of $0 \in \mathbf{R}^N$. By an operator homogeneous of degree k, we mean a linear operator which maps homogeneous polynomials of degree n to homogeneous polynomials of degree $n + k$.

The operators $\{T_i\}$ mentioned in the introduction are the components of the h–gradient.

1.1 Definition. *For a smooth function f, the h–gradient $\nabla_h f$ is defined by*

$$\nabla_h f(x) := \nabla f(x) + \sum_{i=1}^m \alpha_i \frac{f(x) - f(x\sigma_i)}{\langle x, v_i\rangle} v_i.$$

The components are $T_i f(x) := \langle \nabla_h f(x), e_i\rangle$, where e_i is the standard unit vector (with 1 as the ith coordinate), $1 \le i \le N$.

The operators $\nabla_h, T_i, (1 \le i \le N)$ are homogeneous of degree -1. The right regular representation R of G on smooth functions is defined by $R(w)f(x) := f(xw)$, and $\nabla_h(R(w)f)(x) = (\nabla_h f(xw))w^{-1}$; the proof of this depends on the fact that $\sigma_j w = w\sigma_i$ implies $\alpha_i = \alpha_j$.

The following was shown in [4] in the study of orthogonal polynomials of spherical harmonic type associated to the measure $h(x)^2 dm(x)$ on the unit sphere $S := \{x \in \mathbf{R}^N : |x| = 1\}$ (where $dm(x)$ is the rotation–invariant measure on S).

1.2 Theorem. *$\{T_i : 1 \le i \le N\}$ is a commutative set of operators. Furthermore, the h–Laplacian Δ_h satisfies*

$$\sum_{i=1}^N T_i^2 f(x) = \Delta_h f(x)$$

$$= \Delta f(x) + \sum_{j=1}^m \alpha_j \left[\frac{2\langle \nabla f(x), v_j\rangle}{\langle x, v_j\rangle} - |v_j|^2 \frac{f(x) - f(x\sigma_j)}{\langle x, v_j\rangle^2}\right]$$

for smooth functions f.

The main orthogonality result (Theorem 1.6 [3]) is:

1.3 THEOREM. *If p is a homogeneous polynomial on $\mathbf{R}^N$, then $\int_S pq\, h^2 dm = 0$ for all polynomials q of lower degree than p, if and only if $\Delta_h p = 0$.*

Let P_n denote the space of homogeneous polynomials of degree n on $\mathbf{R}^N$, and let $H_n^h = \{p \in P_n : \Delta_h p = 0\}$, the space of h–harmonic polynomials. Each $p \in P_n$ has a unique decomposition $p(x) = \sum\limits_{0 \leq j \leq n/2} |x|^{2j} p_j(x)$ with $p_j \in H_{n-2j}^h$, and there is an orthogonal decomposition $L^2(S, h^2 dm) = \sum\limits_{n=0}^{\infty} \oplus\, H_n^h$ (see Theorem 1.7 [3]).

The adjoint of T_i (as operator on $L^2(S, h^2 dm)$) is

$$T_i^* p(x) = (N + 2n + 2\gamma)(x_i p(x) - (N + 2n + 2\gamma - 2)^{-1} |x|^2 \cdot T_i p(x)) \quad \text{for } p \in H_n^h,$$

where $\gamma := \sum_{i=1}^m \alpha_i$ (see Theorem 2.1 [4]). Consequently, the self–adjoint operator $\sum_{i=1}^N T_i^* T_i$ satisfies

$$\sum_{i=1}^N T_i^* T_i p(x) = (N + 2n + 2\gamma - 2) \sum_{i=1}^N x_i T_i p(x)$$

$(p \in H_n^h,\ n = 0, 1, 2, \dots)$. We are thus led to consider

$$\begin{aligned}\sum_{i=1}^N x_i T_i f(x) &= \langle x, \nabla_h f(x) \rangle \\ &= \sum_{i=1}^N x_i \frac{\partial}{\partial x_i} f(x) + \sum_{j=1}^m \alpha_j (f(x) - f(x\sigma_j)),\end{aligned}$$

for any smooth function f. The first part of this expression is multiplication by the degree when $f \in P_n$, while the second part is the image under the regular representation R of $\sum_{j=1}^m \alpha_j (1 - \sigma_j) \in \mathbf{R}G$ (the group algebra).

§2. The Algebra Generated by Conjugacy Classes of Reflections.

Now $\sum_{j=1}^m \alpha_j (1 - \sigma_j)$ is a central element, and hence its eigenvalues (as an operator on $\mathbf{R}G$) are given by $\lambda(\tau) := \sum_{j=1}^m \alpha_j (1 - \tau(\sigma_j)/\tau(1))$, where τ is an irreducible character of G. These eigenvalues are, in fact, integer combinations of the distinct values of α_j. To state this properly, we label the conjugacy classes of reflections: suppose G has l such classes $\{\sigma_{i,1}, \sigma_{i,2} \dots, \sigma_{i,m_i}\}$ for $1 \leq i \leq l$, so that $m = \sum_{i=1}^l m_i$. Let β_i be the common value of the parameters α_j associated to the elements of class #i. In this notation, $\lambda(\tau) = \sum_{i=1}^l m_i \beta_i (1 - \tau(\alpha_{i,1})/\tau(1))$. For indecomposable Coxeter groups $l = 1$ or 2.

2.1 LEMMA. *For any irreducible character τ of G, and for $1 \leq i \leq l$, $m_i\tau(\sigma_{i,1})/\tau(1) \in Z$.*

Proof. By Theorem (2.17) in Feit [5, pp. 17–18], $m_i\tau(\alpha_{i,1})/\tau(1)$ is an algebraic integer. Since $\sigma_{i,1}^2 = 1$, the trace $\tau(\sigma_{i,1})$ is a sum of square roots of 1, hence is rational. □

2.2 COROLLARY. *$\lambda(\tau) = \sum_{i=1}^{l} \beta_i n_i$ with $n_i \in Z$ and $0 \leq n_i \leq 2m_i$ (n_i depends on τ). Further, $\lambda(\tau) = 0$ exactly when $\tau = 1$, and $\lambda(\tau) = 2\sum_{i=1}^{l} \beta_i m_i$ exactly when τ is the determinant.*

Proof. The general inequality $|\tau(w)| \leq \tau(1)$ for $w \in G$ implies $0 \leq n_i \leq 2m_i$. The value $\lambda(\tau) = 0$ (requiring each $n_i = 0$) is obtained when each $\tau(\sigma_{i,j}) = \tau(1)$, which implies that the image of $\sigma_{i,j}$ under the representation corresponding to τ is the identity; since $\{\sigma_{i,j}\}$ generates G, this shows $\tau = 1$. A similar argument applies to $\lambda(\tau) = 2\sum_{i=1}^{l} \beta_i m_i$. □

Note that the number of distinct values of $\lambda(\tau)$ (as elements of $Z[\beta_1, \ldots, \beta_l]$) is the dimension of $\mathfrak{R}$, the subalgebra of $\mathbb{C}_G$ generated by $\{\sum_{j=1}^{m_i} \sigma_{i,j} : 1 \leq i \leq l\}$, because each complex homomorphism of $\mathfrak{R}$ is the restriction of a complex homomorphism of $\mathfrak{Z}\mathbb{C}G$ (the center), and thus comes from an irreducible character of G. In general, $\mathfrak{R} \neq \mathfrak{Z}\mathbb{C}G$, in fact, asymptotically, $\mathfrak{R}$ is much smaller than $\mathfrak{Z}\mathbb{C}G$ for the cases $G = A_N, B_N$, or D_N as $N \to \infty$.

2.3 *Example.* (the symmetric groups S_N (Weyl groups of type A_{N-1})). Let $h(x) := |\prod_{1\leq i<j\leq N}(x_i - x_j)|^\alpha$, restricted to $\{x \in \mathbb{R}^N : \sum_{i=1}^{N} x_i = 0\}$. The irreducible characters of S_N are determined by partitions $\mu = (\mu_1, \mu_2, \ldots, \mu_N)$ of N (that is, $N = \mu_1 + \mu_2 + \cdots + \mu_N, \mu_1 \geq \mu_2 \geq \cdots \geq \mu_N \geq 0, \mu_i \in Z$). There is one class of reflections, namely the transpositions, and by a formula of Young [9],

$$\tau_\mu(\sigma_{1,1})/\tau_\mu(1) = \sum_i \mu_i(\mu_i + 1 - 2i)/(N(N-1)) = \left(\sum_i \binom{\mu_i}{2} - \sum_i \binom{\mu_i'}{2}\right) / \binom{N}{2},$$

where μ' is the conjugate partition of μ (transposed Ferrers diagram).

When μ is self–conjugate, this value is clearly zero; but this is not a necessary condition: the examples with the lowest values of N are $\mu = (6,3,2,2,2)$, and $\mu' = (5,5,2,1,1,1)$ for $N = 15$, $\mu = (7,2,2,2,2,1)$ and $\mu' = (6,5,1,1,1,1,1)$, $\mu = (6,3,3,2,2)$ and $\mu' = (5,5,3,1,1,1)$ for $N = 16$.

The eigenvalue $\lambda(\tau_\mu) = \alpha\left(\binom{N}{2} - \sum_i \binom{\mu_i}{2} + \sum_i \binom{\mu_i'}{2}\right)$. It is known that the dimension of $\mathfrak{Z}\mathbb{C}S_N$, namely the partition function, has an asymptotic expression $p(N) \sim (1/(4\sqrt{3}N)) \exp(\pi\sqrt{2N/3})$, as $N \to \infty$, while the dimension of $\mathfrak{R}$ is bounded by $N(N-1)+1$.

Similar formulas for $\lambda(\tau)$ hold for the hyperoctahedral group B_N, with

$$h(x) = |\prod_{i=1}^{N} x_i|^\alpha \; |\prod_{1\leq i<j\leq N} (x_i^2 - x_j^2)|^\beta,$$

see Proposition 2.7 in [4].

2.4 Example. (The groups F_4, H_3, H_4). The group F_4 has two conjugacy classes of 12 reflections each, and 25 characters (see the table computed by Kondo [7]). In terms of parameters β_1, β_2, the values of $\sum_{i=1}^{2} 12\beta_i\tau(\sigma_{i,1})/\tau(1)$ are $0, \pm 4\beta_1 \pm 4\beta_2, \pm 6\beta_1, \pm 6\beta_2, \pm 6\beta_1 \pm 6\beta_2, \pm 12\beta_1, \pm 12\beta_2, \pm 12\beta_1 \pm 12\beta_2$ (a total of 21).

The groups H_3 and H_4 are the symmetry groups of the icosahedron and the 120–cell (see Coxeter [2, p. 153]) respectively, and are not crystallographic. Each group has just one class of reflections. From Grove's [6] character table, we find the values $m_1\tau(\sigma_{1,1})/\tau(1)$ to be

$$0,\ \pm 3,\ \pm 5,\ \pm 15 \quad \text{for} \quad H_3, \quad \text{and}$$
$$0,\ \pm 10,\ \pm 12,\ \pm 15,\ \pm 20,\ \pm 30,\ \pm 60 \quad \text{for } H_4.$$

§3. The Inverse of $\sum_{i=1}^{N} x_i T_i$. By an h–exact 1–form, we mean an N–tuple $f = (f_i)_{i=1}^{N}$ of smooth functions, defined on $\{x : |x| < r\}$ for some $r > 0$, satisfying the conditions $T_i f_j = T_j f_i$ for $1 \le i, j \le N$. For any smooth function $F, \nabla_h F$ is an h–exact 1–form. In this section we will state and prove a formula which determines the unique smooth function F with $F(0) = 0$ so that $\nabla_h F = f$, for any given h–exact 1–form f.

For $0 < t \le 1$, let

$$\exp((\log t)\sum_{j=1}^{m} \alpha_j(1 - \sigma_j)) = \frac{1}{|G|}\sum_{w \in G} p_w(t)w \in \mathbb{R}G,$$

thus defining the coefficients $p_w(t)$. We will show that each $p_w(t)$ is a polynomial in $t^{\beta_1}, t^{\beta_2}, \ldots, t^{\beta_l}$ with coefficients in Z (notation from §2). Given an h–exact 1–form f, let

$$F(x) = \frac{1}{|G|}\sum_{w \in G} \int_0^1 p_w(t)\langle xw, f(txw)\rangle dt.$$

Then $\nabla_h F = f$, $F(0) = 0$, and F is the unique function satisfying these conditions. If $f = \nabla_h g$ for some smooth function g, then the formula produces $F(x) = g(x) - g(0)$.

We proceed to the proofs and a detailed discussion of $p_w(t)$. The Fourier transform and inversion formula for $\mathcal{Z}\mathbb{C}G$ can be stated as follows: Let $c := \sum_{w \in G} c_w w \in \mathcal{Z}\mathbb{C}G$ (thus $w \mapsto c_w$ is constant on conjugacy classes), then the complex homomorphisms on $\mathcal{Z}\mathbb{C}G$ are given by $\phi_\tau(c) = \sum_{w \in G} c_w \tau(w)/\tau(1)$, where $\tau \in \hat{G}$ (the set of irreducible characters); further $c_w = \frac{1}{|G|}\sum_\tau \tau(1)\phi_\tau(c)\overline{\tau}(w)$.

Applying this formula to $c = \exp((\log t)\sum_{j=1}^{m} \alpha_j(1 - \sigma_j))$ yields

$$\begin{aligned}\phi_\tau(c) &= \exp((\log\ t)\sum_{j=1}^{m} \alpha_j(\phi_\tau(1) - \phi_\tau(\sigma_j)) \\ &= \exp((\log\ t)\sum_{j=1}^{m} \alpha_j(1 - \tau(\sigma_j)/\tau(1)) = t^{\lambda(\tau)}\end{aligned}$$

(where $\lambda(\tau)$ is the eigenvalue of $\sum_{j=1}^{m} \alpha_j(1-\sigma_j)$ corresponding to $\tau \in \hat{G}$, and $\lambda(\tau) \in Z[\beta_1, \ldots, \beta_l]$ (see Corollary 2.2). Note that $t^{\lambda(\tau)}$ is a monomial in $t^{\beta_1}, \ldots, t^{\beta_l}$. Thus, $p_w(t) = \sum_\tau \tau(1)\overline{\tau}(w)t^{\lambda(\tau)}$. For Coxeter groups, the complex conjugate is not necessary; in fact, a stronger statement is possible. We restrict our attention to indecomposable Coxeter groups; for direct products one multiplies appropriate p_w factors (roughly $p_w(t) = p_{w_1}(t)p_{w_2}(t)$ if $w = (w_1, w_2) \in G_1 \times G_2$ with independent choices of parameters β_i for G_1 and G_2).

3.1 THEOREM. *For any Coxeter group G, and $w \in G$, the coefficients of $t^{\lambda(\tau)}$ in $p_w(t)$ are integers.*

Proof. If G is crystallographic (that is, a Weyl group), then each character takes only integer values (this is classical work of Frobenius, Young, and Specht for the types A_N, B_N, D_N, Frame for E_6, E_7, E_8, and Kondo for F_4). Springer [8] has a general proof using the theory of regular elements. For the other indecomposable Coxeter groups (dihedral $I_2(m), H_3, H_4$), it turns out (by inspection) that

$$\sum\{\tau(1)\tau(w) : \lambda(\tau) = \lambda\} \in Z \quad \text{for each possible } \lambda.$$

For the dihedral groups, $p_w(t)$ can be written out explicitly (see Example 3.4 for the even case). For the groups H_3 and H_4 we use Grove's [6] character tables; the only irrational values involve $A = 2\cos\frac{\pi}{5}$, and $B = 2\cos\frac{2\pi}{5}$, which always combine as $A - B = 1$ in the sums. □

3.2 Example. (the symmetric group S_4). Using the notation from 2.3, and with $u := t^\alpha, p_1(t) = 1 + 9u^4 + 4u^6 + 9u^8 + u^{12}$. It would be interesting if there were a generating function for p_1 for an arbitrary S_N; a lot of information is involved, but perhaps not quite as much as all the degrees of the irreducible characters.

3.3 Example. (The octahedral group B_3). Let

$$h(x) := |\prod_{i=1}^{3} x_i|^\alpha \ |\prod_{1 \le i < j \le 3} (x_i^2 - x_j^2)|^\beta$$

($\alpha, \beta > 0$ and $x \in \mathbf{R}^3$). Then

$$p_1(t) = (1 + u^6)(1 + 4v^6 + v^{12}) + 9u^2v^4(1 + u^2)(1 + v^4),$$

with $u := t^\alpha$, $v := t^\beta$.

3.4 Example. (The even dihedral groups $I_2(2k)$). In terms of complex coordinates $z = x_1 + ix_2$ for $\mathbf{R}^2$, let

$$h(z) = |(z^k - \overline{z}^k)/2i|^\alpha \ |(z^k + \overline{z}^k)/2|^\beta.$$

Let $\omega := e^{\pi i/k}$, then the group consists of the rotations $z \mapsto z\omega^j$, and the reflections corresponding to even and odd values of j. As symmetries of the regular $(2k)-gon\{\omega^j : j = 0, 1, \ldots, 2k-1\}$, the "even" reflections have vertex–vertex mirrors, while the "odd" reflections have edge–edge mirrors. Note that $z^k - \overline{z}^k = \prod_{j=0}^{k-1}(z - \overline{z}\omega^{2j})$, and $z^k + \overline{z}^k = \prod_{j=0}^{k-1}(z - \overline{z}\omega^{2j+1})$.

The eigenvalues $\lambda(\tau)$ are 0, $2k\alpha$, $2k\beta$, $2k(\alpha+\beta)$, $k(\alpha+\beta)$ (the first four come from the linear characters, and the $k-1$ characters of degree two all yield $k(\alpha+\beta)$). Let $u := t^{k\alpha}$, $v := t^{k\beta}$. There are five different expressions for $p_w(t)$:

$$\begin{aligned}
p_1(t) &= 1 + 4(k-1)uv + u^2 + v^2 + u^2v^2; \\
p_w(t) &= (1-u^2)(1+v^2), \quad \text{for} \quad zw = \overline{z}\omega^{2j} \quad (0 \le j \le k-1); \\
p_w(t) &= (1+u^2)(1-v^2) \quad \text{for} \quad zw = \overline{z}\omega^{2j+1} \quad (0 \le j \le k-1); \\
p_w(t) &= 1 - 4uv + u^2 + v^2 + u^2v^2 \quad \text{for} \quad zw = z\omega^{2j} \quad (1 \le j \le k-1); \\
p_w(t) &= (1-u^2)(1-v^2) \quad \text{for} \quad zw = z\omega^{2j+1} \quad (0 \le j \le k-1).
\end{aligned}$$

3.5 Lemma. *The functions $p_w(t)$ satisfy $t\,p'_w(t) = \sum_{j=1}^m \alpha_j(p_w(t) - p_{w\sigma_j}(t))$, $p_1(1) = |G|$, $p_w(1) = 0$ for $w \neq 1$, and $\sum_{w\in G} p_w(t) = |G|$, $0 < t \le 1$. Also, $p_w(t) \ge 0$ for $0 < t \le 1$.*

Proof. Indeed,

$$\begin{aligned}
\frac{d}{dt}\sum_w p_w(t)w &= |G|\exp((\log\ t)\sum_{j=1}^m \alpha_j(1-\sigma_j))(\sum_{j=1}^m \alpha_j(1-\sigma_j))/t \\
&= \sum_{j=1}^m \alpha_j \sum_w p_w(t)t^{-1}(w - w\sigma_j) \\
&= \sum_w \sum_{j=1}^m \alpha_j(p_w(t) - p_{w\sigma_j}(t))t^{-1}w.
\end{aligned}$$

Further, $(1/|G|)\sum_w p_w(1)w = \exp(0) = 1 \in \mathbb{R}G$. Let $\tau = 1$, the trivial character, then

$$\phi_1(\sum_w p_w(t)w) = \sum_w p_w(t) = |G|\exp((\log\ t)\cdot\sum_{j=1}^m \alpha_j(1-1)) = |G|.$$

Finally,

$$\sum_w p_w(t)w = |G|\left((t^{\Sigma\alpha_j})1\right)(\exp((-\log\ t)\sum_{j=1}^m \alpha_j\sigma_j);$$

the argument of the exponential has positive coefficients for $0 < t < 1$, hence $p_w(t) \ge 0$ for each $w \in G$. □

Lemma 3.6. *Suppose f is an h–exact 1–form, then*

$$\nabla_h(\langle x, f(tx)\rangle) = f(tx) + t\frac{\partial}{\partial t}f(tx) + \sum_{j=1}^m \alpha_j(f(tx) - f(tx\sigma_j)\sigma_j),$$

for $0 < t < 1$.

Proof. Fix l (with $1 \leq l \leq N$), then

$$
\begin{aligned}
&T_l\langle x, f(tx)\rangle \\
&= f_l(tx) + \sum_{i=1}^{N} x_i t \frac{\partial f_i}{\partial x_l}(tx) \\
&+ \sum_{j=1}^{m} \alpha_j \Big(\sum_{i=1}^{N} x_i f_i(tx) - \sum_{i=1}^{N} (x\sigma_j)_i f_i(tx\sigma_j)\Big) \cdot (v_j)_l / \langle x, v_j\rangle \\
&= f_l(tx) + 2\sum_{j=1}^{m} \alpha_j \langle f(tx\sigma_j), v_j\rangle (v_j)_l / |v_j|^2 \\
&+ \sum_{i=1}^{N} x_i t \left(\frac{\partial f_i}{\partial x_l}(tx) + \sum_{j=1}^{m} \alpha_j (f_i(tx) - f_i(tx\sigma_j)) \cdot (v_j)_l / \langle tx, v_j\rangle \right).
\end{aligned}
$$

Observe $x\sigma_j = x - 2\langle x, v_j\rangle v_j/|v_j|^2$, and $2\langle f(tx\sigma_j), v_j\rangle v_j/|v_j|^2 = f(tx\sigma_j) - f(tx\sigma_j)\sigma_j$; also the coefficient of $x_i t$ in the last term is $T_l f_i = T_i f_l$ by hypothesis. Hence,

$$
\begin{aligned}
&T_l(\langle x, f(tx)\rangle) \\
&= f_l(tx) + \sum_{i=1}^{N} x_i t \frac{\partial}{\partial x_i} f_l(tx) \\
&+ \sum_{j=1}^{m} \alpha_j (f_l(tx\sigma_j) - (f(tx\sigma_j)\sigma_j)_l) \\
&+ \sum_{j=1}^{m} \sum_{i=1}^{N} x_i t (v_j)_i \alpha_j (f_l(tx) - f_l(tx\sigma_j)) / \langle tx, v_j\rangle \\
&= f_l(tx) + t\frac{\partial}{\partial t} f_l(tx) + \sum_{j=1}^{m} \alpha_j (f_l(tx) - (f(tx\sigma_j)\sigma_j)_l);
\end{aligned}
$$

the lth component of the asserted identity. □

3.7 COROLLARY. *For $w \in G$,*

$$
\begin{aligned}
\nabla_h(\langle xw, f(txw)\rangle) = f(txw)w^{-1} &+ t\frac{\partial}{\partial t} f(txw)w^{-1} \\
&+ \sum_{j=1}^{m} \alpha_j (f(txw)w^{-1} - f(txw\sigma_j)\sigma_j w^{-1}).
\end{aligned}
$$

Proof. Apply the formula $\nabla_h(R(w)g)(x) = (\nabla_h g)(xw)w^{-1}$ to $g(x) := \langle x, f(x)\rangle$. □

3.8 THEOREM. *Suppose f is an h–exact 1–form on $\{x : |x| < r\}$ and define*

$$F(x) = \frac{1}{|G|} \sum_{w \in G} \int_0^1 p_w(t)\langle xw, f(txw)\rangle dt,$$

then $\nabla_h F(x) = f(x)$ for $|x| < r$, and $F(0) = 0$.

Proof. Apply ∇_h to F, interchange with the integral, and use Corollary 3.7 to obtain:

$$\begin{aligned}
\nabla_h F(x) &= (1/|G|) \sum_{w \in G} \int_0^1 p_w(t)[f(txw)w^{-1} \\
&\quad + t\frac{\partial}{\partial t} f(txw)w^{-1} + \sum_{j=1}^m \alpha_j (f(txw)w^{-1} - f(txw\sigma_j)\sigma_j w^{-1})]dt \\
&= (1/|G|) \sum_{w \in G} \int_0^1 \{p_w(t)(f(txw) + t\frac{\partial}{\partial t} f(txw) \\
&\quad + \sum_{j=1}^m \alpha_j (p_w(t) - p_{w\sigma_j}(t)) f(txw)\} w^{-1} dt \\
&= (1/|G|) \sum_{w \in G} \int_0^1 \{p_w(t)(f(txw) + t\frac{\partial}{\partial t} f(txw) + tp'_w(t) f(txw)\} w^{-1} dt \\
&= (1/|G|) \sum_{w \in G} \int_0^1 \frac{\partial}{\partial t}(tp_w(t) f(txw) w^{-1}) dt \\
&= (1/|G|) \sum_{w \in G} p_w(1) f(xw) w^{-1} = f(x)
\end{aligned}$$

(since $p_1(1) = |G|$ and $p_w(1) = 0$ for $w \neq 1$). The term $\sum_w p_w(t) f(txw\sigma_j)\sigma_j w^{-1}$ was rewritten as $\sum_w p_{w\sigma_j}(t) f(txw) w^{-1}$, by change of summation variable. □

3.9 COROLLARY. *Suppose f and g are h–exact 1–forms, and $\langle x, f(x)\rangle = \langle x, g(x)\rangle$ for $|x| < r$ (some $r > 0$), then $f = g$.*

Proof. Apply the formula from 3.8 to f(x)-g(x). □

3.10 THEOREM. *Suppose F is a smooth function on $\{x : |x| < r\}$, some $r > 0$, then*

$$F(x) - F(0) = (1/|G|) \sum_{w \in G} \int_0^1 p_w(t)\langle xw, (\nabla_h F)(txw)\rangle dt,$$

for $|x| < r$.

Proof. Evaluate $\sum_{i=1}^N x_i T_i F(x)$ at txw and divide by t to obtain $\sum_{i=1}^N (xw)_i T_i F(txw) = \frac{\partial}{\partial t} F(txw) + t^{-1} \sum_{j=1}^m \alpha_j (F(txw) - F(txw\sigma_j))$ (for $|x| < r, w \in G$). Use this identity in the integral, then

$$\sum_{w\in G}\int_0^1 p_w(t)\langle xw, (\nabla_h F)(txw)\rangle dt$$

$$= \sum_{w\in G}\int_0^1 p_w(t)\Big(\frac{\partial}{\partial t}F(txw) + t^{-1}\sum_{j=1}^m \alpha_j(F(txw) - F(txw\sigma_j))\Big)dt$$

$$= \sum_{w\in G}\int_0^1 [p_w(t)\frac{\partial}{\partial t}F(txw) + t^{-1}\sum_{j=1}^m \alpha_j F(txw)(p_w(t) - p_{w\sigma_j}(t))]dt$$

$$= \sum_{w\in G}\int_0^1 (p_w(t)\frac{\partial}{\partial t}F(txw) + p_w'(t)F(txw))dt$$

$$= \sum_{w\in G}(p_w(1)F(xw) - p_w(0)F(0))$$

$$= |G|(F(x) - F(0)),$$

since $\sum_w p_w(t) = |G|$ for $0 < t \le 1$, and p_w is continuous on $0 \le t \le 1$ (see Lemma 3.5). □

This result establishes the uniqueness of the integral of an h–exact 1–form.

These formulas can also be used to construct an operator V which intertwines $\{T_i : 1 \le i \le N\}$ and $\{\frac{\partial}{\partial x_i} : 1 \le i \le N\}$ and which is homogeneous of degree 0.

3.11 THEOREM. *There exists a unique operator V, homogeneous of degree 0, such that $V1 = 1$, and $T_i V f(x) = V(\frac{\partial}{\partial x_i} f)(x)$, for any polynomial $f, 1 \le i \le N$. Furthermore, V is nonsingular.*

Proof. Proceeding by induction, suppose that the operator V with the asserted properties has been constructed for all polynomials of degree $\le n$, some $n \ge 0$. Let f be homogeneous of degree $n+1$ (in p_{n+1}) and define

$$Vf(x) = \frac{1}{|G|}\sum_{w\in G}\int_0^1 p_w(t)\sum_{i=1}^N (xw)_i(V\frac{\partial}{\partial x_i} f)(txw)dt.$$

By the inductive hypothesis, $(V\dfrac{\partial}{\partial x_i} f)_{i=1}^N$ is an h–exact 1–form. It is easy to check that $Vf(x) \in P_{n+1}$. By Theorem 3.10, $T_i V f(x) = V\dfrac{\partial}{\partial x_i} f(x)$. The uniqueness and nonsingularity are proved similarly. □

This is admittedly not a very explicit definition. For $N = 1$ and $G = Z_2$, it turns out that V can be expressed as a fractional integral. The author conjectures that V can be expressed as an integral transform with positive kernel, in the general situation.

REFERENCES

[1] BENSON, C.T. AND GROVE, L. C., *Finite Reflection Groups*, 2nd ed'n., Springer–Verlag, Berlin, Heidelberg, New York, 1985.

[2] H.S.M. COXETER, *Regular Polytopes*, 3rd ed'n., Dover, New York, 1973.

[3] C.F. DUNKL, *Reflection groups and orthogonal polynomials on the sphere*, Math. Z. 197 (1988), 33–60.

[4] C.F. DUNKL, *Differential–difference operators associated to reflection groups*, (to appear) Trans. Amer. Math. Soc..

[5] W. FEIT, *Characters of Finite Groups*, W.A. Benjamin, New York, 1967.

[6] L.C. GROVE, *The characters of the hecatonicosahedroidal group*, J. Reine u. Angew. Math. 265 (1974), 160–169.

[7] T. KONDO, *The characters of the Weyl group* F_4, J. Fac. Sci. Univ. Tokyo 1 (1965), 145–153.

[8] T.A. SPRINGER, *Regular elements of finite reflection groups*, Inventiones math. 25 (1974), 159–198.

[9] A. YOUNG, *On quantitative substitutional analysis*, V, Proc. London Math. Soc. (2) 31 (1930), 273–288.

THE MÖBIUS FUNCTION OF SUBWORD ORDER

ANDERS BJÖRNER*

Abstract. The structure of intervals in the subword ordering of a free monoid is investigated. It is shown that such intervals are lexicographically shellable, and a formula for their Möbius function is derived. When applied to monotone words this formula specializes to the classical number-theoretic Möbius function.

1. Introduction. Let A^* denote the free monoid over an alphabet A. The elements of A^* are finite strings of elements from A called **words**. The **length** $|\alpha|$ of a word α is the number of letters. There is a unique word $\lambda \in A^*$ of length zero, the **empty word**.

Several partial orderings of various subsets of A^* have been considered [4,5,6,7,11]. Here we shall say that β is a **subword** of α if after deleting some letters from α, the string of remaining letters yields β. This relation, written $\beta \leq \alpha$, is a partial ordering of A^*. For instance, $ac < aabc$.

We assume familiarity with the concept of **Möbius function** μ of a locally finite poset [8,9]. What can be said about the Möbius function of the subword ordering of A^*? To gain intuition for the combinatorial structure of intervals in A^* we encourage the reader to draw diagrams of the intervals [a,abba], [b,abba] (both of size 8) and [ab,abcab] (with 14 elements), and directly from the recursive definition compute $\mu(a, abba) = 0$, $\mu(b, abba) = -1$ and $\mu(ab, abcab) = -3$.

In preparation for the general rule we need a few more definitions. Given a word $\alpha = a_1 a_2 \ldots a_n \in A^*$, its **repetition set** is $\mathcal{R}(\alpha) = \{i : a_{i-1} = a_i\}$. An **embedding** of β in α is a sequence $1 \leq i_1 < i_2 < \cdots < i_k \leq n$ such that $\beta = a_{i_1} a_{i_2} \ldots a_{i_k}$. It is a **normal embedding** if also $\mathcal{R}(\alpha) \subseteq \{i_1, i_2, \ldots, i_k\}$. For $\alpha, \beta \in A^*$ let

$$\binom{\alpha}{\beta} = \text{number of embeddings of } \beta \text{ in } \alpha,$$

$$\binom{\alpha}{\beta}_n = \text{number of normal embeddings of } \beta \text{ in } \alpha.$$

For instance, $\binom{babba}{ba} = 4$ and $\binom{babba}{ba}_n = 1$. Clearly, $\beta \leq \alpha$ if and only if $\binom{\alpha}{\beta} > 0$.

Notice that if $|A| = 1$, then A^* under subword order has an obvious identification with the set of natural numbers $\mathbb{N}$ under cardinality order, and $\binom{\alpha}{\beta}$ specializes to the usual binomial coefficient while $\binom{\alpha}{\beta}_n$ specializes to the Möbius function of the chain $\mathbb{N}$ (up to a sign factor).

The general rule for the Möbius function of subword order is as follows.

*Department of Mathematics, Royal Institute of Technology, S-10044 Stockholm, Sweden.

THEOREM 1. *For all* $\alpha, \beta \in A^*$,

$$\mu(\beta, \alpha) = (-1)^{|\alpha|-|\beta|} \binom{\alpha}{\beta}_n .$$

The special case for lower intervals $[\lambda, \alpha]$ follows from Farmer [4] ("claim" on p. 609) and was also discovered by Viennot [11].

COROLLARY 1. *(Farmer-Viennot).* $\mu(\lambda, \alpha) = \begin{cases} (-1)^{|\alpha|}, & \text{if } \mathcal{R}(\alpha) = \emptyset, \\ 0, & \text{otherwise}. \end{cases}$

The general rule lends itself to quick computation by hand, as e.g. in the following examples:

$$\mu\,(\text{comics, combinatorics}) = -2,$$
$$\mu\,(\text{nano, nonriemannianmanifold}) = -5.$$

Theorem 1 contains the number-theoretic Möbius function as a special case. To see this take the alphabet $A = \{2, 3, 5, 7, \ldots\}$ of prime numbers and consider the set A^*_{mon} of **monotone** words, meaning words $a_1 a_2 \ldots a_n \in A^*$ such that $a_1 \leq a_2 \leq \cdots \leq a_n$. Clearly, A^*_{mon} is an order ideal in A^*, and prime factorization gives a bijection $\mathbb{N} \leftrightarrow A^*_{\text{mon}}$ under which divisibility in $\mathbb{N}$ corresponds to subword order. There is at most one normal embedding of one monotone word into another, and such an embedding exists if an only if the number of occurrences of each letter in the two words differ by at most one. Hence, for $\alpha, \beta \in A^*_{\text{mon}}$ the formula in Theorem 1 specializes to the defining formula for the classical Möbius function.

For instance, letting $a = 2, b = 3, c = 5, \ldots$, we have

$$\mu(6, 108) = \mu(ab, aabbb) = 0,$$
$$\mu(30, 300) = \mu(abc, aabcc) = 1,$$

etc..

By counting normal embeddings one is led to the following rational expressions.

THEOREM 2. *Suppose that* $|A| = n$. *Then:*

(i) $\sum_{\alpha \in A^*} \mu(\beta, \alpha) t^{|\alpha|} = \frac{t^k (1-t)}{(1+(n-1)t)^{k+1}}, \quad$ *if* $|\beta| = k, \beta \in A^*$,

(ii) $\sum_{\alpha, \beta \in A^*} \mu(\beta, \alpha) t^{|\alpha|} q^{|\beta|} = \frac{1-t}{1-(nq-n+1)t}.$

A surprising feature of formula (i) is that the right-hand side depends only on the length of the word β and not on its combinatorial structure. See Remark 3.

The method of proof for Theorem 1 is lexicographic shellability. We assume from now on some familiarity with the notions of shellable and Cohen-Macaulay posets [1,2,3].

THEOREM 3. *Every interval $[\beta, \alpha]$ in A^* is dual CL-shellable (and hence homotopy Cohen-Macaulay).*

A considerable amount of information about the structure of intervals in A^* can be deduced from this. See the cited references for details.

Suppose, finally, that the alphabet A is finite. Since then the poset $A_k^* = \{\alpha \in A^* : 1 \leq |\alpha| \leq k\}$, $k \in \mathbb{N}$, can be obtained by rank-selection from a suitable interval in A^*, the following is implied using [3, Theorem 8.1].

COROLLARY 2. *The poset A_k^* of all nonempty words of length at most k is dual CL-shellable (and hence homotopy Cohen-Macaulay).*

The corollary strengthens some of the results of Farmer [4] concerning the topological structure of A_k^*.

2. Proofs. Let $[\beta, \alpha]$ be an interval in A^*, $|\beta| = k$, $\alpha = a_1 a_2 \ldots a_n$. Then every maximal chain $\alpha = \alpha_0 > \alpha_1 > \cdots > \alpha_{n-k} = \beta$ is of length $n - k$, so $[\beta, \alpha]$ is a graded poset. Let $\mathcal{M}$ denote the set of maximal chains in $[\beta, \alpha]$.

Each maximal chain $m = (\alpha = \alpha_0 > \alpha_1 > \cdots > \alpha_{n-k} = \beta)$ is assigned a label $\lambda(m) = (\lambda_1(m), \lambda_2(m), \ldots, \lambda_{n-k}(m))$ as follows. If i_1 is the least number such that the erasure of a_{i_1} in $a_1 a_2 \ldots a_n$ yields α_1, then $\lambda_1(m) = i_1$. If i_2 is the least number such that the erasure of a_{i_2} in $a_1 \ldots \hat{a}_{i_1} \ldots a_n$ yields α_2, then $\lambda_2(m) = i_2$. And so on. Thus, the idea is to label by the position in α of the leftmost letter whose removal gives the successively correct subword as we trace the chain m downward from α to β. This idea is similar to the one used in [2].

For instance, the chain $zyzygy > zzygy > zzyy > zyy > zy$ would be labelled (2,5,1,4).

We claim that the labeling $\lambda : \mathcal{M} \to \mathbb{N}^{n-k}$ is a (dual) CL-labeling. See [1,3] for the definition of this concept.

Clearly, if two chains m and m' coincide along their first f edges, then $\lambda(m)$ and $\lambda(m')$ coincide in their first f entries. Also, the construction shows that for every rooted subinterval of $[\beta, \alpha]$ the induced labeling of maximal chains is combinatorially equivalent to a labeling by positions like the one we are considering for $[\beta, \alpha]$.

What remains to show in order to prove that λ is a CL-labeling, and hence that $[\beta, \alpha]$ is CL-shellable, is therefore the following:

(i) there is a unique chain $m_0 \in \mathcal{M}$ such that $\lambda_1(m_0) \leq \lambda_2(m_0) \leq \cdots \leq \lambda_{n-k}(m_0)$,

(ii) $\lambda(m_0) \leq \lambda(m)$ in lexicographic order for all $m \in \mathcal{M}$.

Let $\mathcal{E}$ denote the set of embeddings of β in α. There is a mapping $\pi : \mathcal{M} \to \mathcal{E}$ given by $\pi(m) = \{1, 2, \ldots, n\} - \{\lambda_1(m), \lambda_2(m), \ldots, \lambda_{n-k}(m)\}$. Let $\mathcal{E}' = \{\epsilon \in \mathcal{E} : \pi^{-1}(\epsilon) \neq \emptyset\}$. Also, let $\epsilon_0 = \{j_1, j_2, \ldots, j_k\}_<$ be the **final embedding** of β in α determined by $i_e \leq j_e$, $1 \leq e \leq k$, for every other embedding $\epsilon = \{i_1, i_2, \ldots, i_k\}_<$.

We will prove the following:

(iii) there exists $m \in \pi^{-1}(\epsilon)$ such that $\lambda_1(m) \leq \lambda_2(m) \leq \cdots \leq \lambda_{n-k}(m)$ if and only if $\epsilon = \epsilon_0$,

(iv) there exists $m \in \pi^{-1}(\epsilon)$ such that $\lambda_1(m) > \lambda_2(m) > \cdots > \lambda_{n-k}(m)$ if and only if ϵ is a normal embedding.

Since each set $\pi^{-1}(\epsilon)$ can contain at most one chain m with increasing label $\lambda(m)$, (iii) implies (i). Also from ϵ_0's property of being final (ii) follows. Hence, (iii) will prove Theorem 3.

On the other hand, since each set $\pi^{-1}(\epsilon)$ can contain at most one chain m with decreasing label $\lambda(m)$, (iv) implies that the total number of such chains equals the number of normal embeddings. In view of the combinatorial interpretation of the Möbius function in CL-shellable posets ([1, Corollary 2.3], [2, Theorem 3.4]) this proves Theorem 1.

Suppose $\epsilon \in \mathcal{E}'$. The labels $\lambda(m)$ of chains $m \in \pi^{-1}(\epsilon)$ are permutations of the set $\hat{\epsilon} = \{1, 2, \ldots, n\} - \epsilon$. Hence, the only chain that could possibly be in $\pi^{-1}(\epsilon)$ and have an increasing label is the chain m_ϵ obtained by successively erasing each letter a_i with $i \in \hat{\epsilon}$ from left to right (i.e., in order of increasing i). So, there exists $m \in \pi^{-1}(\epsilon)$ with increasing label if and only if $\pi(m_\epsilon) = \epsilon$. If $\epsilon = \{i_1, i_2, \ldots, i_k\}_<$, then clearly $\pi(m_\epsilon) = \epsilon$ if and only if for all $j \in \hat{\epsilon}$ if $i_e < j < i_{e+1}$, $1 \leq e \leq k$ (letting $i_{k+1} = n+1$), then $a_j \neq a_{i_e}$. But this means precisely that $\epsilon = \epsilon_0$, so (iii) follows.

Again, let $\epsilon \in \mathcal{E}'$. The only chain that could possibly be in $\pi^{-1}(\epsilon)$ and have a decreasing label is the chain m^ϵ obtained by successively erasing each letter with position in $\hat{\epsilon}$ from right to left. So, there exists $m \in \pi^{-1}(\epsilon)$ with decreasing label if and only if the entries of $\lambda(m^\epsilon)$ are the elements of $\hat{\epsilon}$ listed decreasingly. This, in turn, happens precisely if for every $j \in \hat{\epsilon}$, $a_j \neq a_{j-1}$, which is equivalent to saying that the embedding ϵ is normal. With this, also (iv) has been proven.

We now turn to the computation of the generating functions in Theorem 2. Let $F_\beta(t) = \sum\limits_{\alpha} \binom{\alpha}{\beta}_n t^{|\alpha|} = \sum\limits_{m \geq 0} c_m t^m$, and $|\beta| = k$. Then

$$c_m = \sum_{\mathcal{A}} \sum_{\mathcal{B}} \chi(a_{i_1} a_{i_2} \ldots a_{i_k} = \beta \,\&\, \mathcal{R}(a_1 a_2 \ldots a_m) \subseteq \{i_1, i_2, \ldots, i_k\}),$$

where $\mathcal{A} = \{a_1 a_2 \ldots a_m \in A^*\}$, $\mathcal{B} = \{1 \leq i_1 < i_2 < \cdots < i_k \leq m\}$ and $\chi(P)$ has value 1 if the statement P is true and 0 otherwise. Changing the order of summation to $\sum\limits_{\mathcal{B}} \sum\limits_{\mathcal{A}}$ we get

$$c_m = \binom{m-1}{k-1}(n-1)^{m-k} + \binom{m-1}{k} n(n-1)^{m-k-1},$$

where the first term comes from summation over the $i_1 = 1$ case and the second from the $i_1 \neq 1$ case. Hence,

$$\begin{aligned} F_\beta(t) &= t^k \sum_m \binom{m-1}{k-1} \left((n-1)t\right)^{m-k} + nt^{k+1} \sum_m \binom{m-1}{k} \left((n-1)t\right)^{m-k-1} \\ &= \frac{t^k}{(1-(n-1)t)^k} + \frac{nt^{k+1}}{(1-(n-1)t)^{k+1}} = \frac{t^k(1+t)}{(1-(n-1)t)^{k+1}}. \end{aligned}$$

This expression for $F_\beta(t)$ in turn gives:

$$\sum_{\alpha,\beta}\binom{\alpha}{\beta}_n t^{|\alpha|}q^{|\beta|} = \sum_\beta F_\beta(t)q^{|\beta|} = \sum_{k\geq 0} n^k q^k \frac{t^k(1+t)}{(1-(n-1)t)^{k+1}} =$$

$$= \frac{1+t}{(1-(n-1)t)}\cdot\left(1-\frac{nqt}{(1-(n-1)t)}\right)^{-1} = \frac{1+t}{1-(nq+n-1)t}.$$

The substitutions $t\mapsto -t$ and $q\mapsto -q$ then lead to Theorem 2.

3. Remark. For $\beta\in A^*$ let $[\beta,\infty)=\{\alpha\in A^* : \alpha\geq\beta\}$. If $|\beta|=|\gamma|$ there exist length-preserving bijections $\varphi:[\beta,\infty)\to[\gamma,\infty)$. For instance, if $\beta = b_1b_2\dots b_k$, $\gamma=c_1c_2\dots c_k$ and if the final embedding of β in α yields the factorization $\alpha=\alpha_0b_1\alpha_1b_2\alpha_2\dots\alpha_{k-1}b_k\alpha_k$, then let $\varphi\alpha=\alpha_0c_1\widehat{\alpha}_1c_2\widehat{\alpha}_2\dots\widehat{\alpha}_{k-1}c_k\widehat{\alpha}_k$, where $\widehat{\alpha}_i$ is obtained from α_i via the substitution $c_i\mapsto b_i$, $1\leq i\leq k$. The existence of such bijections implies that the number of words in $[\beta,\infty)$ of length p depends only on $k=|\beta|$. Call this number $Z_{k,p}$. Counting the length p superwords of $bb\dots b$ we find that

$$Z_{k,p}=\sum_{i=0}^{p-k}\binom{p}{i}(n-1)^i.$$

Let $\zeta(\beta,\alpha)=1$ if $\beta\leq\alpha$ and $=0$ otherwise. (This defines the *zeta function* of subword order, i.e., the incidence algebra inverse of its Möbius function [8,9].) The formula for $Z_{k,p}$ makes it possible to compute the generating functions:

$$\text{(i)}\quad \sum_{\alpha\in A^*}\zeta(\beta,\alpha)t^{|\alpha|}=\frac{t^k}{(1-nt)(1-(n-1)t)^k},\qquad \text{if } |\beta|=k,$$

$$\text{(ii)}\quad \sum_{\alpha,\beta\in A^*}\zeta(\beta,\alpha)t^{|\alpha|}q^{|\beta|}=\frac{1-(n-1)t}{(1-nt)(1-(nq+n-1)t)}.$$

Formulas (i) here and in Theorem 2 show that from an enumerative point of view the posets $[\beta,\infty)$ and $[\gamma,\infty)$ behave as if they were combinatorially isomorphic, when $|\beta|=|\gamma|$. However, this is in general not the case. For instance, $[aa,\infty)\not\cong[ab,\infty)$ since $\mu(ab,abab)=3$ and $\mu(aa,\delta)\leq 2$ for all $\delta\in A^*$, $|\delta|=4$.

4. Remark. The symmetric group S_n acts on the poset A_k^* by permuting letters in words, $|A|=n$. The poset A_k^* is Cohen-Macaulay (by Corollary 2) and has nonvanishing homology only in dimension $k-1$. Following Stanley [10] one can then consider the induced action on the homology of A_k^* (and of various rank-selected subposets). This homology representation of S_n has degree $(n-1)^k$. This follows from the "Remark" on p. 611 of Farmer [4] and can easily be checked by computing the Möbius function of $A_k^*\cup\{\hat{0},\hat{1}\}$ using Corollary 1.

The following result was stated as a conjecture in a preliminary version of this paper. The proof appearing below was then communicated by R. Stanley in private correspondence (I am grateful for his permission to include it here).

THEOREM 4. (STANLEY). *The S_n - module $\tilde{H}_{k-1}(A_k^*, \mathbf{C})$ is the k-fold tensor product of the irreducible representation of shape $(n-1, 1)$.*

Proof. Since A_k^* is Cohen-Macaulay, the character χ of this action satisfies $\chi(w) = (-1)^{k-1}\mu_w(\hat{0}, \hat{1})$, where μ_w is the Möbius function of the subposet $(A_k^*)^w$ of elements in A_k^* fixed by $w \in S_n$, with a bottom element $\hat{0}$ and a top element $\hat{1}$ adjoined. But w fixes a word α if an only if every letter of α is a fixed point of w. Hence, $(A_k^*)^w = B_k^*$, where $B = \text{Fix}\,(w)$ is the set of fixed points of w. So, $\mu_w(\hat{0}, \hat{1}) = (-1)^{k-1}(|\text{ Fix }(w)| - 1)^k$. But the irreducible character $\chi^{(n-1,1)}$ of S_n satisfies $\chi^{(n-1,1)}(w) = |\text{ Fix }(w)| - 1$, so its k-th tensor power agrees with χ.

5. Remark. Theorem 1 implies that $\mu(\beta, \alpha)$ can be computed in polynomial time. This was pointed out by J. Håstad, who suggested the following "dynamic programming" algorithm for computing the number of normal embeddings of β in α, and hence $\mu(\beta, \alpha)$.

Assume that $\beta = b_1\, b_2 \ldots b_k$ and $\alpha = a_1\, a_2 \ldots a_n, k \le n$. For $1 \le i \le k$, $1 \le j \le n$, let m_{ij} be the number of normal embeddings of $b_1\, b_2 \ldots b_i$ in $a_1\, a_2 \ldots a_j$ such that $b_i \mapsto a_j$. Directly from the definition:

(1) $m_{ij} = 0$ if $i > j$ or if $b_i \ne a_j$, and

(2) $m_{1j} = 1$ if $b_1 = a_j$ and $\mathcal{R}(\alpha) \cap \{1, \ldots, j-1\} = \emptyset$ and $= 0$ otherwise.

Letting e_j denote the greatest element of $\mathcal{R}(\alpha) \cup \{1\}$ such that $e_j \le j$, $1 \le j \le n$, we have for $i \ge 2$:

$$m_{ij} = \begin{cases} \sum\limits_{d=e_{j-1}}^{j-1}, & \text{if } b_i = a_j, \\ 0, & \text{otherwise.} \end{cases}$$

Thus, all values m_{ij} can be recursively computed in at most $O(kn^2)$ steps, and $\binom{\alpha}{\beta}_n = \sum_{j=e_n}^{n} m_{kj}$.

6. Remark. The rationality of the formal power series appearing in Theorem 2 and in Remark 3 has been further explored in joint work by C. Reutenauer and the author. Corresponding series for powers of the Möbius and zeta functions as well as analogous series in noncommuting variables were shown to be rational. The results will appear in a forthcoming paper.

REFERENCES

[1] A. BJÖRNER, A.M. GARSIA AND R.P. STANLEY, *An introduction to Cohen-Macaulay partially ordered sets*, in *Ordered Sets*, (I. Rival, ed.), Reidel, Dordrecht/Boston, 1982, pp. 583–615.

[2] A. BJÖRNER AND M. WACHS, *Bruhat order of Coxeter groups and shellability*, Advances in Math., 43 (1982), pp. 87–100.

[3] ——————, *On lexicographically shellable posets*, Trans. Amer. Math. Soc., 277 (1983), pp. 323–341.

[4] F.D. FARMER, *Cellular homology for posets*, Math. Japonica, 23 (1979), pp. 607–613.

[5] ——————, *Homotopy spheres in formal language*, Studies in Appl. Math., 66 (1982), pp. 171–179.

[6] C. GREENE, *Posets of shuffles*, J. Combin. Theory Ser. A. (to appear).

[7] M. LOTHAIRE, *Combinatorics on words*, Addison-Wesley, Reading, Mass., 1983.

[8] G.-C. ROTA, *On the foundations of combinatorial theory I: Theory of Möbius functions*, Z. Wahrsch. Verw. Gebiete, 2 (1964), pp. 340–368.

[9] R.P. STANLEY, *Enumerative combinatorics, Vol. 1*, Wadworth, Monterey, Cal., 1986.

[10] ———, *Some aspects of groups acting on finite posets*, J. Combin. Theory Ser. A., 32 (1982), pp. 132–161.

[11] G. VIENNOT, *Maximal chains of subwords and up-down sequences of permutations*, J. Combin. Theory Ser. A, 34 (1983), pp. 1–14.

KEYS & STANDARD BASES

ALAIN LASCOUX AND MARCEL-PAUL SCHÜTZENBERGER*

1. Introduction. The irreducible characters of the linear group on $\mathbb{C}$ {*Schur Functions*} are combinatorially interpreted as sums of *Young tableaux*.

Demazure **[D1]** **[D2]** has given a "Formule des caractères" which interpolates between a dominant weight, corresponding to a partition I, and the Schur function of index I. For every permutation μ, he obtains a "partial" character which can be interpreted as the class of the space of section $\mathcal{V}_{I,\mu}$ of the line bundle associated to I over the Schubert variety of index μ, in an appropriate Grothendieck ring; identifying this ring with the ring of polynomials, we can view $\mathcal{V}_{I,\mu}$ as a polynomial $\mathcal{D}(\mu, I)$.

An independent study of the same spaces $\mathcal{V}_{I,\mu}$, and more precisely, of their "standard bases", is due to Lakshmibai-Musili-Seshadri **[L-M-S]**. Extending the work of Hodge, they interpret Young tableaux as products of *Plücker coordinates* of the flag variety and associate to them *chains* of permutations to describe the different bases (see also **[L-W]**).

The link between the two constructions is not immediate. Moreover, none of these two point of view furnishes the multiplicative structure of sections which is needed in geometry to describe the *postulation* of Schubert varieties. Indeed, the product $\mathcal{V}_{I,\mu} \otimes \cdots \otimes \mathcal{V}_{I,\mu}$ contains more than the sections corresponding to a multiple of the weight I and thus the products of standard bases are not standard bases.

The answer comes from working in the free algebra rather than the Grothendieck ring or the ring of coordinates. Young tableaux (2.1) are now *words* which are representatives of certain congruence classes (th.2.4). More general words (*frank words*, 2.7) obtained from tableaux by permutation allow to associate to each congruence class two special tableaux *right* and *left keys* (2.9). The set of keys (2.12) is in fact the image of the embedding of the symmetric group in the set of tableaux (embedding originally defined by Ehresmann [E] to describe how cells attach in a cellular decomposition of the flag variety).

Now, a standard basis is a set (or a sum) of tableaux having the same right key (th.3.6). To generate it, one uses *symmetrizing operators* (3.5) on the free algebra which lift the operators on the ring of polynomial used by Demazure and Bernstein-Gelfand-Gelfand. Thus, the polynomial $\mathcal{D}(I, \mu)$ is just the commutative image of a sum of standard bases (th.3.8).

For what concerns the multiplicative structure of sections, the answer is also given by keys: the product of two tableaux t, t' belongs to a standard basis iff the right key of t is less than the left key of t' (2.11 and 4.2) . This allows us in section 4 to give a combinatorial interpretation of the Hilbert function associated to a weight

*L.I.T.P., Université Paris 7, 2 Place Jussieu, 75251 PARIS Ced 05, France, supported by the P.R.C. Mathématiques-Informatique

as an enumeration of chains of tableaux (th.4.3 and 4.4). See **[L-S6]** for the related order on the symmetric group and its Eulerianity properties. The link between keys, reduced decompositions of permutations and the Schubert cycles (i.e. the classes of the Schubert varieties in the cohomology ring of the flag manifold) is given in **[L-S4]**.

In section 5, we explicit some different ways of describing the standard bases.

In appendix 6, we have isolated a property of actions of the symmetric group which is of independent interest.

Caution. As usual, operators operate *on their left.*

2. Frank words and keys. Let $\mathbb{A}^*$ be the free monoid generated by the alphabet $\mathbb{A} = \{a_1 < a_2 < ...\}$. A word $v = x_1...x_r (x_i \in \mathbb{A})$ is called a *column* iff $x_1 > ... > x_r$ and a *row* iff $x_1 \leq x_2 \leq \cdots \leq x_r$. Let $\mathbb{V}$ denote the set of all columns. Every word $w \in \mathbb{A}^*$ admits a unique factorisation as a product of a minimal number of columns : $w = v_1 v_2 \cdots v_k (v_i \in \mathbb{V})$. We shall call it the *column factorisation* of w and denote it occasionally by $w = v_1 \cdot v_2 \cdot \ldots \cdot v_k$, v_1 being the *left column* $w\pounds$ of w and v_k the *right column* $w\$$ of w . The *shape* of w is the sequence $\|w\| = (|v_1|, ..., |v_r|)$ of the degrees (or lengths) of the column factors of w.

To use a traditional term (see **[Mc]**), $\|w\|$ is a *composition* of the integer $|w|$ and the $|v_i|$ are the parts of $\|w\|$. On the set of compositions, one has the following preorder: $I \geq J$ iff for every k, the sum of the k biggest parts of I is bigger than the sum of the k biggest parts of J. It is clear that if $I \geq J$ and $J \geq I$, then I is a permutation of J , and that if H is any composition, $IH \geq JH \Leftrightarrow I \geq J$. One can imbed the set of compositions into the set of words : $I = (1I, 2I, ..., rI) \rightarrow (1I \cdots 1)((1I+2I) \cdots (1I+1)) \cdots ((1I+2I+\cdots+rI) \cdots (1I+2I+\cdots+(r-1)I))$. We note this word $I\mathbf{M}$ and call it a *composition word*. It can be looked as the maximal element (as a permutation) of the Young group $\mathbb{S}_I \hookrightarrow \mathbb{S}_{1I+\cdots+rI}$. For instance, the composition word $(2,4,1)\mathbf{M}$ is $(21)(6543)(7)$, which is the maximal element of the subgroup $\mathbb{S}_2 \times \mathbb{S}_4 \times \mathbb{S}_1$ of $\mathbb{S}_7$.

Taking the underlying set of a column defines a bijection $v \rightarrow \{v\}$ between the set $\mathbb{V}$ of the columns and the family $2^{\mathbb{A}}$ of the subsets of $\mathbb{A}$; one extends to $\mathbb{V}$ the order $\leq$ on $\mathbb{A}$ by letting $u \leq v$ iff there is an increasing injection of $\{u\}$ into $\{v\}$. Thus $u \leq v$ is the least order on $\mathbb{V}$ that contains both the inclusion order $\{u\} \subset \{v\}$ and the term to term order between equipotent subsets of $\mathbb{A}$.

DEFINITION 2.1. A *contretableau* is a word which is an increasing product of columns.

For instance, if $\mathbb{A} = \{1 < 2 < 3 < \cdots\}$, the word 2 3 41 421 is a contretableau because of $2 \leq 3 \leq 41 \leq 421$.

It is convenient to define another order $\triangleright$ on $\mathbb{V}$ by letting $u \triangleright v$ iff there is a decreasing injection of $\{v\}$ into $\{u\}$ and to call a *tableau* any product $u_1 u_2 \cdots u_k$ where the columns u_i are decreasing for $\triangleright$, i.e. $u_1 \triangleright u_2 \triangleright \cdots \triangleright u_k$. For instance, 321 31 2 4 is a tableau because $321 \triangleright 31 \triangleright 2 \triangleright 4$.

The two orders $\leq$ and $\triangleright$ are linked by the fact that $u \triangleright v$ iff $v' \leq u'$ where $v \to v'$ is the bijection on V induced by the "reversal" of the alphabet A assumed to be finite (i.e. by the morphism $a_i \to a_{n+1-i}$, n being the cardinal of A).

It follows from the definitions that the shape of any tableau t is a partition of its degree. This suggests writing the letters of t into the boxes of the Ferrers diagram of $\|t\|$. For instance, 321 31 2 4 can be represented by $\begin{smallmatrix}3\\23\\1124\end{smallmatrix}$. A similar remark holds for contretableaux and we can represent 2 3 41 421 by $\begin{smallmatrix}2344\\12\\1\end{smallmatrix}$. It is a direct consequence of the definition of the orders $\leq$ and $\triangleright$ that each row of the planar writings of tableaux or contretableaux is a weakly increasing sequence of letters.

There is essentially one natural congruence $\equiv$ on A^* that admits as a section the set of contretableaux, where *natural* means that it commutes with the order preserving injections of alphabets. It is called the *plactic congruence*. As shown by D. Knuth, it can be defined by the following identities where a, b, c are any three letters of A such that $a < b < c$:

$$(2.2) \qquad baa \equiv aba \; ; \; bba \equiv bab \; ; \; cab \equiv acb \; ; \; bca \equiv bac$$

As it has been observed by **[K-L]**, these generating congruences are exactly all the pairs of words of degree 3 that are not a column nor a line and that differ by the transposition of two adjacent letters. Thus one member of each pair is a tableau (e.g. $\begin{smallmatrix}b\\ac\end{smallmatrix}$ or $\begin{smallmatrix}b\\aa\end{smallmatrix}$) and the other one, a contretableau (e.g. $\begin{smallmatrix}bc\\a\end{smallmatrix}$ or $\begin{smallmatrix}ab\\a\end{smallmatrix}$) .

It turns out that the set of all tableaux is also a section of the plactic congruence and the defining relations (2.2) are simply the expression of this fact for the words of degree 3.

Another remarkable property of the plactic congruence is that every tableau (or contretableau) is congruent to the word obtained when reading by rows (from top to bottom) its planar representation. For instance, 321 31 2 4 $\equiv$ 3 23 1124 or 2 3 41 421 $\equiv$ 2344 12 1.

This phenomenon is closely tied with another definition of the plactic congruence $\equiv$ as one of the least congruences on A^* such that the subalgebra of $\mathbb{Z}(\mathsf{A}^*/\equiv)$ generated by the (non commutative) symmetric sums

$$\Lambda_p = \sum \{v : v \in V \; , |v| = p\} \; , \; p = 1, 2,$$

is a commutative algebra isomorphic to the usual algebra of symmetric polynomials in the letters of A .

To obtain the complete characterisation of the plactic congruence, one needs to add a further condition which follows immediately from 2.2 and which will be used later.

PROPOSITION 2.3. *The plactic congruence $\equiv$ is the least congruence on A^* for which $\Lambda_1\Lambda_2 \equiv \Lambda_2\Lambda_1$, and which moreover satisfies for any interval B of A the relation*

$$w \equiv w' \Rightarrow w \cap \mathsf{B}^* \equiv w' \cap \mathsf{B}^*$$

($w \cap \mathsf{B}^*$ *denotes the word obtained by erasing the letters not in* B).

Taking B equal to a single letter, 2.3 implies that the plactic congruence commutes with the natural morphism $w \to \underline{w}$ of A^* onto the free commutative monoid; this can also be directly checked on relations 2.2.

The plactic congruence is no other than the algebraic formalization of Schensted's construction, whose main result can be summarized in the following theorem (**[Sche]**, **[L-S1]**).

THEOREM 2.4. 1) *Each plactic class contains a unique tableau* t *and a unique contretableau.*

2) *The elements of the class of* t *are in bijection with the set of permutation tableaux (called* insertion tableaux*) of the same shape as* t .

By a permutation tableau, we mean, of course, a permutation (of any alphabet) which is at the same time a tableau. Given any word w, we denote $w\mathbf{R}$ the tableau congruent to it and $w@$ its insertion tableau. It is well known (see **[Schu]**) that the involution $w \to w^{-1}$ on permutation words corresponds to the exchange of $w\mathbf{R}$ and $w@$; we shall not use this fact.

More explicitely, the insertion tableau (which is the Q-symbol of Schensted) of a word $w = x_1 x_2 \ldots$ describes the increasing sequence of the shapes of the tableaux $x_1\mathbf{R}$, $x_1x_2\mathbf{R}$, $x_1x_2x_3\mathbf{R}$,... . The particular choice of the alphabet being irrelevant, $\substack{2\\13}$, $\substack{6\\28}$ and $\substack{\beta\\ \alpha\gamma}$, with $\alpha < \beta < \gamma$, must be considered as the same insertion tableau representing the sequence of shapes $\varnothing \to \diamond \to \substack{\diamond\\ \diamond} \to \substack{\diamond\\ \diamond\diamond}$.

More generally, any word congruent to $w@$ will be called an *insertion word* for w. Insertion words are compatible with restriction of alphabets (see **[L-S1]**):

LEMMA 2.5. *Given any word* $w = x_1 \ldots x_{m-1} x_m \ldots x_{m+r} x_{m+r+1} \ldots$, *then the word* $w@ \cap \{m, \ldots, m+r\}$ *is an insertion word for the factor* $x_m \cdots x_{m+r}$.

In particular, as pointed out by Schensted, $w@$ contains the subword $m\ m+1$ iff $x_m \le x_{m+1}$ and the subword $m+1\ m$ iff $x_m > x_{m+1}$. Call *file* of a permutation of $\{1, 2, 3 \ldots\}$ any maximal subword of the type $(m+k) \cdots (m+1)m$. The shape $\|w\|$ corresponds to the files of any insertion word for w. More precisely, one has the following lemma:

LEMMA 2.6. *Let* $w = v_1 \cdot \ldots \cdot v_k$ *be a word,* μ *an insertion word for* w . *Then*

1) *The files of* μ *are the same as those of the composition word* $\|w\|\mathbf{M}$.
2) $\|w\| \le \|w\mathbf{R}\|$; *equality happens iff* $\|w\|\mathbf{M}$ *is an insertion word for* w.
3) *For each permutation* J *of the shape* $\|w\mathbf{R}\|$, *there exists one and only one word of shape* J *congruent to* w.

Proof. Assertion *1)* is a direct corollary of 2.5: the files of μ are the same as the files of $w@$ and they encode exactly the inequalities $x_i \le x_{i+1}$ or $x_i > x_{i+1}$ for all the pairs of adjacent letters in w . For what concerns *2)*, it is easy to check that the tableau $\mu\mathbf{R}$ has shape greater than the composition corresponding to the files of μ ; this composition being $\|w\|$ and the shape of $\mu\mathbf{R} = w@$ being equal to that of $w\mathbf{R}$, we get the required inequality. In the case of equality, the tableau $\mu\mathbf{R}$ is determined by its files : consecutive entries in a file must be in consecutive rows of $\mu\mathbf{R}$. A mild intimacy with the jeu de taquin shows that this last condition is equivalent to requiring that $\|w\|\mathbf{M} \equiv \mu\mathbf{R}$. Finally, condition *3)* is a rewriting of the case where $\|w\|$ is a permutation of $\|w\mathbf{R}\|$; we just saw that in this case the insertion tableau $w@$ is uniquely determined, which means, thanks to the bijection 2.4.*2* that w is uniquely determined. □

For example, $w = 53 \cdot 61 \cdot 2 \cdot 4$ has shape $2211 \leq 321 = \|w\mathbf{R}\|$; the sequence of tableaux congruent to the left factors of w : $\emptyset \longrightarrow 5 \longrightarrow \begin{smallmatrix} 5 \\ 3 \end{smallmatrix} \longrightarrow \begin{smallmatrix} 5 \\ 3\,6 \end{smallmatrix} \longrightarrow \begin{smallmatrix} 5 \\ 3 \\ 1\,6 \end{smallmatrix} \longrightarrow \begin{smallmatrix} 5 \\ 3\,6 \\ 1\,2 \end{smallmatrix} \longrightarrow \begin{smallmatrix} 5 \\ 3\,6 \\ 1\,2\,4 \end{smallmatrix}$ shows that $w@ = \begin{smallmatrix} 4 \\ 2\,5 \\ 1\,3\,6 \end{smallmatrix}$; w admits the insertion word $\mu = 452361$, since $\mu \equiv w@$; the files of μ are 21, 43, 5, 6 and are identical to those of the composition word $\|w\|\mathbf{M} = 21\ 43\ 5\ 6$. On the other hand, in the same congruence class, we have a unique word w' of shape 213; it is determined by its insertion tableau congruent to the composition word $213\mathbf{M} = 21\ 3\ 654 \equiv \begin{smallmatrix} 6 \\ 2\,5 \\ 1\,3\,4 \end{smallmatrix}$. Indeed, $w' = 51\ 3\ 642$ as we can check from the sequence of tableaux congruent to its left factors : $\emptyset \longrightarrow 5 \longrightarrow \begin{smallmatrix} 5 \\ 1 \end{smallmatrix} \longrightarrow \begin{smallmatrix} 5 \\ 1\,3 \end{smallmatrix} \longrightarrow \begin{smallmatrix} 5 \\ 1\,3\,6 \end{smallmatrix} \longrightarrow \begin{smallmatrix} 5\,6 \\ 1\,3\,4 \end{smallmatrix} \longrightarrow \begin{smallmatrix} 5 \\ 3\,6 \\ 1\,2\,4 \end{smallmatrix}$. The words corresponding to the other permutations of 321 are given next page.

The preceding lemma has detached in the congruence class of a tableau t, the set of those words w (among which the tableau and the contretableau) for which $\|w\|$ is a permutation of $\|t\|$:

DEFINITION 2.7. A word w is *frank* iff $\|w\|$ is a permutation of $\|w\mathbf{R}\|$.

Equivalently, thanks to 2.6.*2*, a word w is frank iff it admits the composition word $\|w\|\mathbf{M}$ as an insertion word.

For a two-columns tableau t, finding its congruent contretableau t' can be considered as using the generator $\flat$ of the symmetric group $\mathbb{S}(2)$ to transpose the two columns of t . This is best done with the jeu de taquin ([**L-S1**]) : $\boxed{\begin{smallmatrix} 4 \\ 3 \\ 1 \end{smallmatrix}}\ \boxed{\begin{smallmatrix} 6 \\ 2 \end{smallmatrix}} \Rightarrow \boxed{\begin{smallmatrix} 4 \\ 1 \end{smallmatrix}}\ \boxed{\begin{smallmatrix} 6 \\ 3 \\ 2 \end{smallmatrix}}$. We shall write $t' = t^\flat$ and $t = t'^\flat$. Notice that $t\$ = 62$ is a subword of $t'\$ = 632$ and that $t'\pounds = 41$ is a subword of $t\pounds = 431$.

More generally, on the set of k-columns words, one has an action (not everywhere defined; we use the symbol $\emptyset$ when it is not defined) of the symmetric group $\mathbb{S}(k)$. First, if the factor $v_r v_{r+1}$ of $w = v_1 \cdot \ldots \cdot v_k$, $v_i \in \mathrm{V}$, is a tableau or a contretableau, then the image of w by the simple transposition σ_r , $1 \leq r < k$, is set equal to $v_1 \cdots v_{r-1}(v_r v_{r+1})^\flat v_{r+2} \cdots v_k$ if moreover this last word has still k columns. In all other cases, the image of w by σ_r is set equal to $\emptyset$. It is checked in section 6 that this extends to an action of the symmetric group for which frank words play a special rôle that we summarize in the following theorem (*1* and *2* being a rewriting of 2.6.*2* and 2.6.*3*):

THEOREM 2.8.

1) *For each word* w , *one has* $\|w\| \geq \|w\mathbf{R}\|$, *with equality iff* w *is frank.*
2) *The set of frank words in the plactic class of a tableau* t *is in bijection with the set of permutations of the shape of* t .
3) *The product of two frank words* w , w' *is frank iff* $u\$.u'\pounds$ *is frank for any pair of frank words* u, u', *with* $u \equiv w$ *and* $u' \equiv w'$.

For example, the class of 531 62 4 contains the six frank words (read vertically!) which correspond to the six permutations of the shape 321 :

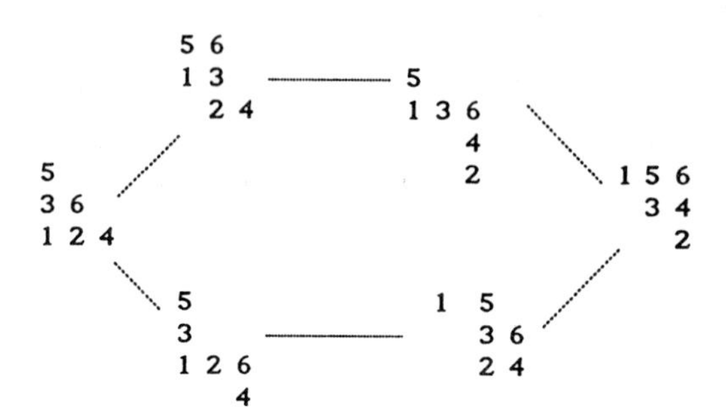

On the other hand, the product of the two frank words 31 42 and 4 51 is not frank: the insertion tableau of 31 42 4 51 is 721 43 5 6, which is not congruent to 21 43 5 76. Indeed, 4 51 $\equiv$ 41 5, and condition 3) is violated, because $42 \cdot 41$ is not frank.

We now come to the study of keys.

By definition, a *key* is a tableau such that its columns are pairwise comparable for the inclusion order. This condition implies that the action of the symmetric group giving the frank words is simply the permutation of the columns because this is true (and easily verified) in the special case of two-columns keys, where the operation $\flat$ reduces to just commutation. For example, 531 53 3 is a key and the frank words in its congruence class are

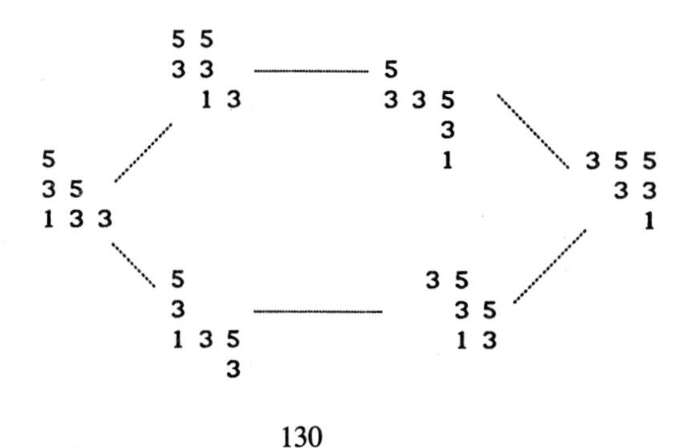

DEFINITION 2.9. The *right key* $t\mathbf{K}_+$ of a tableau t (or of any word congruent to t) is the tableau of the same shape as t whose columns belong to the set of columns $\{u\$\ ,\ u \equiv t \text{ and } u \text{ frank}\}$. The *left key* $t\mathbf{K}_-$ of t is the tableau of the same shape as t whose columns belong to the set $\{u\pounds,\ u \equiv t \text{ and } u \text{ frank}\}$.

In other words, the left key (resp. right key) of t is made of the left (resp. right) columns, repeated the appropriate number of times so as to fill the shape of t, of the frank words in the class of t . For instance, the above hexagon for the tableau 531 62 4 give the keys $t\mathbf{K}_- = \begin{smallmatrix}5\\3\,5\\1\,1\,1\end{smallmatrix}$ and $t\mathbf{K}_+ = \begin{smallmatrix}6\\4\,6\\2\,4\,4\end{smallmatrix}$. Notice that a tableau is a key iff it is equal to its right (resp. left) key. In other case, the keys of a tableau belong to different plactic classes.

Since the test that the product of two frank words w, w' is frank involves exactly the columns composing $w\mathbf{K}_+$ and $w'\mathbf{K}_-$, we can reformulate th.2.8:

THEOREM 2.10. *1) A word w is frank iff $\|w\mathbf{R}\|$ is a permutation of $\|w\|$.*
2) A product ww' of two frank words w, w' is frank iff the shape of $ww'\mathbf{R}$ is the union of the shapes $\|w\mathbf{R}\|$ and $\|w'\mathbf{R}\|$.
3) A product ww' of two frank words w, w' is frank iff $(w\mathbf{K}_+)(w'\mathbf{K}_-)$ is frank.

If a pair of columns satisfies $u \le v$, then $u' \le v'$ for any other pair of columns u', v' such that $\{u'\} \subseteq \{u\}, \{v'\} \supseteq \{v\}$; similarly, $u \triangleright v$ implies $u' \triangleright v'$ for any pair such that $\{u'\} \supseteq \{u\}$, $\{v'\} \subseteq \{v\}$. Thus, in the special case of two frank words w, w' having the same shape up to a reordering, condition *3)* can be restricted to the comparison of columns in $w\mathbf{K}_+$ and $w'\mathbf{K}_-$ of the same length instead of all pairs of columns (as required by 2.8.*3*). Recall that for columns of the same length, the order $\le$ is the componentwise order on words of the same degree (that we can denote by the same symbol $\le$). In short, one can replace in that case 2.10.*3* by:

THEOREM 2.11. *Assume that w, w' are two frank words such that $\|w\mathbf{R}\| = \|w'\mathbf{R}\|$, then ww' is frank iff $w\mathbf{K}_+ \le w'\mathbf{K}_-$.*

For example, the product of the two tableaux $w =$ 421 41 3 and $w' =$ 432 32 4 is not frank ; condition *2)* is violated since (421 41 3)(432 32 4)$\mathbf{R}$ = 421 431 432 3 4 is a tableau of shape (= 33311) different from 332211. In fact, 421 41 3 $\cong$ 421 1 43 and 432 32 4 $\cong$ 32 432 4, but $(43\ 32)^\flat = \emptyset$, and thus condition *3)* of the theorem is violated. Condition 2.11 has the same fate, since 431 43 3 $\big($ = (421 41 3)$\mathbf{K}_+\big)$ is not smaller than 432 32 3 $\big($ = (432 43 4)$\mathbf{K}_-\big)$.

On the other hand, 421 31 3 432 32 4 $\mathbf{R}$ = 421 431 32 32 3 4 has shape 332211, as is insured by the inequality 431 31 3 $\big($ = (421 31 3)$\mathbf{K}_+\big) \le$ 432 32 3 $\big($ = (432 43 4)$\mathbf{K}_-\big)$ required by 2.11.

Definition 2.12. *Key of a permutation*: to each pair consisting of a permutation $\zeta \in \mathcal{S}(n)$, and a partition $I = (1I, 2I, \dots)$, Ehresmann [**E**] has associated a key, noted $\mathbf{K}(\zeta, I)$, by taking the sequence of left reordered factors of ζ (considered as a word) of successive degrees $1I, 2I, \dots$.

For example, $\zeta = 316452$ and $I = 532$ give the key 65431 631 31.

In case that $I = n...21$, we shall simply write $\mathbf{K}(\zeta)$ instead of $\mathbf{K}(\zeta, n...21)$; thus $\zeta \to \mathbf{K}(\zeta)$ is an embedding of $\mathfrak{S}(n)$ into the set of tableaux of shape $n...21$. The reader may notice that the so-called "strong", "Bruhat order" on permutations (see **[Bj]**) is a special case of the pervading order on words (componentwise) which we have been repeatedly using, by way of the equivalence due to Ehresmann :

$$\eta \le \zeta \quad \Leftrightarrow \quad \mathbf{K}(\eta) \le \mathbf{K}(\zeta) \tag{2.13}$$

For example, the keys associated to $\zeta = 3241, \eta = 2143$ and $\mu = 1423$ are $\mathbf{K}(\zeta) = 4321\ 432\ 32\ 3$, $\mathbf{K}(\eta) = 4321\ 421\ 21\ 1$, $\mathbf{K}(\mu) = 4321\ 421\ 41\ 1$; thus $\zeta \ge \eta$, but ζ and μ are not comparable since the two columns 32 and 41 are not comparable.

3. Symmetrizations. The definition of tableaux is strictly dependent upon a chosen total order on $\mathbb{A}$. It is remarkable that nonetheless the commutative image of the sum S_I of all tableaux of a given shape I be a symmetrical function: this is the most constructive definition of the *Schur Function* of index I. To understand this phenomenon (see nevertheless Knuth's proof **[B-K]**), one must define an action of the symmetric group on the free algebra such that the S_I are invariant under this action. Further, this action must induce the usual action of the symmetric group when projected by $w \to \underline{w}$ on the commutative algebra. By the duality $w \to w^{-1}$ for permutation words, the new action we shall define now can be specialized to give the action that we have been using in our study of frank words.

Consider first the case of a two-letters alphabet $\mathbb{A} = \{a, b\}$. It is clear that the image by the transposition $\sigma = \sigma_{ab}$ of the tableau $t = (ba)^h a^k b^m$ must be $t^\sigma = (ba)^h a^m b^k$, since t^σ is the only tableau of the same shape as t whose commutative image is the monomial $a^{m+h} b^{h+k}$.

More generally, because words w in a, b are determined by their insertion tableau $w@$ and their commutative image $\underline{w}$ (we recover $w\mathbf{R}$ from its content $\underline{w}$ and its shape, equal to that of $w@$), one defines w^σ to be the word:

$$(w^\sigma)@ = w@ \quad \& \quad (\underline{w})^\sigma = \underline{(w^\sigma)} \tag{3.1}$$

In other terms σ , as it has been defined, preserves the insertion tableau and commutes with the projection $\mathbb{Z} < a, b > \to \mathbb{Z}[a, b]$.

For example, the image by σ of the word baa **a** $bbaa$ **aa** b is baa **b** $bbaa$ **bb** b (we have marked the letters which change) because these two words have the same insertion tableau, equal to $\begin{smallmatrix}1\,6\,7\\0\,2\,3\,4\,5\,8\,9\,X\end{smallmatrix}$, and they project onto $a^7 b^4$ and $a^4 b^7$ in $\mathbb{Z}[a, b]$.

Since the column ba commutes (plactically) with a and b , shifting the factors ba of a word w in a, b generates the congruence class of w. This remark implies the following easy algorithm to compute $w \to w^\sigma$:

$$\left\{ \begin{array}{l} \text{fix the successive factors } ba \text{ of } w, \text{ then change the remaining} \\ \text{subword } a^k b^m \text{ into } a^m b^k. \end{array} \right. \tag{3.2}$$

For instance the preceding word gives (ba) aa $\big(b(ba)a\big)$ aab and we have to change the remaining word $aa\ aab$ into $ab\ bbb$ to get $(ba)\ ab\ \big(b(ba)a\big)\ bbb\ =\ w^{\sigma}$.

Consider now the more general case of a simple transposition σ_i of consecutive letters a_i, a_{i+1} . One defines w^{σ_i} to be the word in which the subword $w \cap \{a_i, a_{i+1}\}$ has been modified according to 3.1 or 3.2, the other letters being left unchanged. For example, denoting by $x...x$ any word in letters different from a and b, the above computation shows that the image of $bxaxaxa\mathbf{a}xbxbxaxaxa\mathbf{a}xa\mathbf{a}xb$ is $bxaxax\mathbf{b}xbxbxaxax\mathbf{b}x\mathbf{b}xb$.

It is proven in **[L-S1]** that $w \to w^{\sigma_i}$ extends to an action of the symmetric group on $\mathbb{Z}\langle\mathbb{A}\rangle$, i.e. that given a permutation μ and a word w, all factorizations of $\mu = \sigma\ \sigma' \ldots\ \sigma''$ into simple transpositions produce the same word $\big((w^{\sigma})^{\sigma'}\ldots\big)^{\sigma''}$ denoted w^{μ}.

One can note in reference to a previous remark that in section 2, we have acted on the insertion words to generate from a tableau the frank words which are congruent to it, thus preserving $w\mathbf{R}$, and that the action described here preserves $w@$.

At the commutative level, on $\mathbb{Z}[\mathbb{A}]$, we have at our disposal other actions of the symmetric group $\mathfrak{S}(\mathbb{A})$ (see **[L-S 2]**).

In particular, two operators $\overline{\pi}_{\mu}$ and π_{μ} on $\mathbb{Z}[\mathbb{A}]$ are associated to each permutation μ. For a simple transposition σ_i the operator $\overline{\pi}_{\sigma_i}$ (abbreviated $\overline{\pi}_i$, and acting as always on its left) is

$$(3.3) \qquad f \longrightarrow \left(f^{\sigma_i} - f\right) \Big/ \left(1 - a_i/a_{i+1}\right) = f\overline{\pi}_i$$

and the operator π_i is just the sum of $\overline{\pi}_i$ and the identity :

$$(3.4) \qquad \pi_i = \overline{\pi}_i + 1$$

Let $\underline{w} = \underline{v}a_i^k \in \mathbb{Z}[\mathbb{A}]$, with $\underline{v}$ symmetrical in a_i and a_{i+1}.

Then direct computation gives

$$\underline{w}\pi_i \ = \ \underline{v}a_i^k + \underline{v}a_i^{k-1}a_{i+1} + \cdots + \underline{v}a_{i+1}^k$$

i.e. $\underline{w}\pi_i$ is the sum of all monomials between $\underline{w}$ and $\underline{w}^{\sigma_i}$, and $\underline{w}\overline{\pi}_i$ is the same sum apart from the first term ($=w$) missing.

This indicates how we can lift $\overline{\pi}_i$ into an operator, denoted θ_i, on the free algebra. Given i and a word w, let its degree in a_{i+1} be m and its degree in a_i be $m+k$. Then w and w^{σ_i} differ by the exchange of a subword $\mathbf{a}_i^k$ into $\mathbf{a}_{i+1}^k$ if $k \geq 0$, or of $\mathbf{a}_{i+1}^{-k}$ into $\mathbf{a}_i^{-k}$ if $k \leq 0$.

In the first case, we define $w\theta_i$ to be the sum of all words in which the subword $\mathbf{a}_i^k$ has been changed respectively into $\mathbf{a}_i^{k-1}\mathbf{a}_{i+1}$, $\mathbf{a}_i^{k-2}\mathbf{a}_{i+1}^2, \cdots, \mathbf{a}_{i+1}^k$; in the second, we put $w\theta_i = -\big(w^{\sigma_i}\big)\theta_i$ as in the commutative case. In other terms, θ_i interpolates between the identity and σ_i for the words having more occurences of a_i than of a_{i+1}. The corresponding algorithm is in this case ($k \geq 0$), putting $a_i = a, a_{i+1} = b, \theta = \theta_i$:

$$
(3.5) \quad \begin{cases} \text{fix the successive factors } ba \text{ of } w \text{ , then change the remaining} \\ \text{subword } a^{m+k}b^m \text{ into successively } a^{m+k-1}b^{m+1}, a^{m+k-2}b^{m+2}, \dots, \\ a^m b^{m+k} \text{ and take the sum of all the words so obtained.} \end{cases}
$$

For instance, for the word studied in 3.2, we have $(ba)aa(b(ba)a)aab\theta = (ba)aa(b(ba)a)abb + (ba)aa(b(ba)a)bbb + (ba)ab(b(ba)a)bbb$.

More generally, we can transform w by changing its subword $a^{m+k}b^m$, $k \in \mathbb{Z}$, into any row $a^r b^{2m+k-r}$ of the same degree. This operation will preserve the insertion tableau, as does σ_i (which is a special case). In particular, we shall need the projection of $a^{m+k}b^m$ onto a^{2m+k}, $k \in \mathbb{Z}$, that we shall denote λ (and λ_i for the pair of letters a_i, a_{i+1}):

$$
(3.6) \quad \begin{cases} \text{fix the successive factors } ba \text{ of } w, \text{ then change the remaining} \\ a^{m+k}b^m \text{ into } a^{2m+k} \text{ to obtain } w\lambda. \end{cases}
$$

Since $\sigma_i = \sigma, \theta_i = \theta, \lambda_i = \lambda$ preserve the insertion tableau, they are also compatible with the right and left keys : if w is a frank word congruent to t, then w^σ and $w\lambda$ are also frank, and $w\theta$ is a sum of frank words; $w^\sigma\$$ and $w\lambda\$$ are equal to $w\$^\sigma$ or $w\$$. Thus $t^\sigma \mathbf{K}_+$, $t\lambda\mathbf{K}_+$ and $t'\mathbf{K}_+$, with any t' in the sum $t\theta$, are equal to $t\mathbf{K}_+$ or $(t\mathbf{K}_+)^\sigma$. We shall give a more precise statement in theorem 3.8.

The operators θ_i do not satisfy the Coxeter relations $\theta_i\theta_{i+1}\theta_i = \theta_{i+1}\theta_i\theta_{i+1}$, contrary to the operators $\overline{\pi}_i$, π_i and λ_i; thus, if $\sigma_i \cdots \sigma_j$ and $\sigma_h \cdots \sigma_k$ are two reduced decompositions of the same permutation, the operators $\theta_i \cdots \theta_j$ and $\theta_h \cdots \theta_k$ will in general be different and there is no canonical way of defining operators θ_μ by products of operators θ_i.

Nevertheless, we recover this lost Coxeter relation when acting on dominant monomials, as we shall see in 3.8.

DEFINITION 3.7. The *standard basis* $\mathfrak{U}(\mu, I)$ associated to the pair μ, I (μ permutation, I partition) is the sum in the free algebra of all tableaux having right key $\mathbf{K}(\mu, I)$. The *costandard basis* $\mathfrak{B}(\mu, I)$ is the sum of all contretableaux having right key $\mathbf{K}(\mu, I)$.

Since by definition all the elements in a plactic class have the same right key, it is clear that $\mathfrak{U}(\mu, I) \equiv \mathfrak{B}(\mu, I)$, and more precisely, that $\mathfrak{B}(\mu, I)\mathbf{R} = \mathfrak{U}(\mu, I)$.

To any partition $I = (1I, 2I, ...)$ on associates the *dominant* monomial $a^I = (a_{1I} \dots a_2 a_1)(a_{2I} \dots a_2 a_1)(a_{3I} \dots a_2 a_1) \dots$.

THEOREM 3.8. *Let* a^I be a dominant monomial and $\sigma_i\sigma_j \dots \sigma_k$ *be any reduced decomposition of a permutation* μ. Then

$$
\mathfrak{U}(\mu, I) = a^I \theta_i \theta_j \cdots \theta_k.
$$

Proof. Let μ and i be such that $\ell(\mu\sigma) > \ell(\mu)$, with $\sigma = \sigma_i$, $a_i = a, a_{i+1} = b$. If w is a frank word such that $w\mathbf{K}_+ = \mathbf{K}(\mu, I)$ or $\mathbf{K}(\mu\sigma, I)$, then $w\lambda\mathbf{K}_+ = \mathbf{K}(\mu, I)$. Let t be a tableau such that $t\mathbf{K}_+ = \mathbf{K}(\mu, I)$, $t^\sigma\mathbf{K}_+ \neq \mathbf{K}(\mu, I)$ (this implies that $t^\sigma\mathbf{K}_+ = \mathbf{K}(\mu\sigma, I)$). Then there exists a frank word $w \equiv t$ such that the right factor $w^\sigma\$$ of w^σ contains the letter b and not the letter a; thus w contains a and not b; this implies that $t\lambda = t$. One checks moreover that all the tableaux (not only t^σ) in the sum $t\theta$ have the same right key $\mathbf{K}(\mu, I)$.

Conversely, if t is such that $t\mathbf{K}_+ = t^\sigma\mathbf{K}_+ = \mathbf{K}(\mu, I)$, then $(t + t^\sigma)\theta = 0$. Supposing the theorem true for μ, it is also true for $\mu\sigma$. □

For instance, suppose that we already know $\mathfrak{B}(426135, 321)$; we compute $\mathfrak{B}(436125, 321)$ by using the operator $\theta = \theta_2$, the contretableaux t such that t^σ also belong to $\mathfrak{B}(426135, 321)$ give a zero contribution; the others are of the type $t = t\lambda_2$:

4 42 642	$\longrightarrow$	4 42 643 + 4 43 643
4 41 642	$\longrightarrow$	4 41 643
3 42 642	$\longrightarrow$	3 42 643
2 41 642	$\longrightarrow$	2 41 643 + 3 41 643
3 41 642	$\longrightarrow$	0
3 32 642	$\longrightarrow$	0
3 31 642 + 2 31 642	$\longrightarrow$	0

All the contretableaux belonging to a costandard basis $\mathfrak{B}(\mu, I)$ having the same right column (since it is the reordering of the factor of μ of length $1I$), we have a faster way to compute the costandard bases, by induction on the number of parts of I :

LEMMA 3.9. *Let p be a positive integer, $I = (1I, ..., rI)$ be a partition with $r \geq p$, I' the resulting partition after deletion of the part pI, μ a permutation , v the column such that $\{v\} = \{1\mu, ..., (pI)\mu\}$. Then there exist permutations $\nu, \eta \ldots$ such that*

$$\mathfrak{B}(\mu, I) = \left[\mathfrak{B}(\nu, I') + \mathfrak{B}(\eta, I') + \cdots\right]v$$

Proof. Two congruent frank words w, w' have the same right column $w\$ = w'\$$ iff $|w\$| = |w'\$|$. Thus, to compute the right key of a tableau, we need only to generate a set of frank words $w^{(1)}, w^{(2)}, \ldots$ such that $\{|w^{(1)}\$|,\ |w^{(2)}\$|, \ldots\} = \{1I, 2I, \ldots\}$. We can require that the shapes of these frank words be $(rI, \ldots, 2I, pI, 1I)$, $(rI, ..., 1I, pI, 2I), \ldots, ((r-1)I, ..., 1I, pI, rI)$. The images of $w^{(1)}, \ldots, w^{(r)}$ by the transposition (of columns) σ_{r-1} will be frank words with right column of degree pI. Thus the right key of any frank word $w = v_1 \cdot \ldots \cdot v_r$ is equal to that of the frank word $\left(v_1 \cdot \ldots \cdot v_{r-1}\mathbf{K}_+\right) \cdot v_r$. To describe a standard basis, we need only to look for frank words of the type $w = w' \cdot v$, w' being a key of shape I' and v the column: $\{v_r\} = \{1\mu, ..., (pI)\mu\}$, such that $w\mathbf{K}_+ = \mathbf{K}(\mu, I)$. □

This lemma gives a fast induction when we take $p = 1$ to factorize the column of maximal length. For example, let $\mu = 32514$, $I = 4321$. Then $v = 5321$, $I' = 321$; $\mathfrak{B}(32514, 4321) = \left(32\ 532\ v\ +\ 3\ 31\ 532\ v\ +\ 2\ 31\ 532\ v\right) + \left(3\ 32\ 432\ v\ +\ 3\ 31\ 432\ v\ +\ 2\ 31\ 432\ v\right)$ decomposes into $\left[\mathfrak{B}(32514, 321) + \mathfrak{B}(32415, 321)\right]\ v$.

4 . Postulation. Let $\mathcal{A}$ be a vector bundle on any variety $\mathcal{M}$, $\mathcal{F}(\mathcal{A}) \to \mathcal{M}$ the relative flag manifold of complete flags of quotient bundles of $\mathcal{A}$. If $\mathcal{A}$ is of rank n, one has from definition (see **[Gr]**) n tautological line bundles $L_1, \dots, L_n$ on $\mathcal{F}(\mathcal{A})$. The Grothendieck ring $\mathcal{K}(\mathcal{F}(\mathcal{A}))$ of classes of vector bundles is a quotient of the ring of polynomials $\mathcal{K}(\mathcal{M})[\mathbb{A}]$, $\mathbb{A}$ being an alphabet of cardinal n, by a certain ideal $\mathcal{I}$, the images of $a_1, \dots, a_n$ being respectively the classes of $L_1, \dots, L_n$.

Since all constructions given here are compatible with $\mathcal{I}$, we can replace $\mathcal{K}(\mathcal{F}(\mathcal{A}))$ by $\mathbf{Z}[\mathbb{A}]$ and $K(\mathcal{M})$ by the ring of symmetric polynomials $\mathbf{Z}[\mathbb{A}]^{\mathcal{S}(\mathbb{A})}$. The projection $p : \mathcal{F}(\mathbb{A}) \to \mathcal{M}$ induces a morphism $p_* : \mathcal{K}(\mathcal{F}(\mathbb{A})) \to \mathcal{K}(\mathcal{M})$ which corresponds in fact to the operator $\pi_\omega : \mathbf{Z}[\mathbb{A}] \to \mathbf{Z}[\mathbb{A}]^{\mathcal{S}(\mathbb{A})}$ associated to the maximal permutation of $\mathcal{S}(\mathbb{A})$. We can express π_ω as a product of simple operators 3.4, but it can be directly defined by the following global expression (see **[L-S2]**):

$$\mathbf{Z}[\mathbb{A}] \ni f \longrightarrow \sum_{\mu \in \mathcal{S}(\mathbb{A})} \Big[f / \prod_{i<j} (1 - a_j/a_i)\Big]^{\mu} \tag{4.1}$$

In case that $\mathcal{M}$ is a point, the morphism p_* associates to any vector bundle $\mathcal{B}$ the Euler-Poincaré characteristics : $\sum_i (-1)^i \dim \mathcal{H}^i(\mathcal{B})$; in terms of polynomials, this should be interpreted as $\mathbf{Z}[\mathbb{A}] \ni f \;\to\; f \pi_\omega \varepsilon_{\mathbb{A}}$, f being any polynomial lifting the class of $\mathcal{B}$ and $\varepsilon_{\mathbb{A}}$ being the specialisation $a_1 \to 1, \dots, a_n \to 1$.

Let J be a partition, I its conjugate, L the line bundle $L = L_1^{1J} \otimes L_2^{2J} \otimes \cdots$. From Demazure's construction, **[D1]** **[D2]** **[L−S5]** we have that the number $a^I \pi_\mu \varepsilon_A$ is the postulation (that is to say, the dimension of the cohomology $\mathcal{H}^0$; the other spaces $\mathcal{H}^i$ being null, the postulation coincide in that case with the Euler-Poincaré characteristics) of the line bundle L on the Schubert variety of index $\omega\mu^{-1}$.

More generally, considering simultaneously all the powers of L together, we have the *Hilbert series* $\mathcal{H}_{I,\mu}(z) = (1 - z a^I)^{-1} \pi_\mu \varepsilon$ relative to L of the Schubert variety $Schub_{\omega\mu^{-1}}$ (L defines an embedding of the flag variety into a projective space if $1I > 2I > \dots$).

From considerations of dimension, we know that the series $\mathcal{H}_{I,\mu}(z)$ is rational of the type $\mathcal{N}_{I,\mu}(z)/(1-z)^{\ell(\mu)+1}$, $\mathcal{N}_{I,\mu}(z)$ being a polynomial of degree $\leq \ell(\mu)$. However $(1 - z a^I)^{-1} \pi_\mu$ has in general a denominator of degree greater than $\ell(\mu)+1$. Raising up to the free algebra, we shall get a combinatorial interpretation (4.4) of the Hilbert series and clarify in particular this drop in the degrees.

From 3.8 , given any reduced decomposition $\sigma_i ... \sigma_j$ of μ , then $\underbrace{a^I \dots a^I}_{k}(\theta_i + 1) \cdots (\theta_j + 1)$ is a sum of words having the same insertion tableau as $a^I \dots \dots a^I$, thus it is a product of k tableaux of shape I . On the other hand, again according to 3.8, $(a_{1I} ... a_1)^k (a_{2I} ... a_1)^k \cdots (\theta_i + 1) ... (\theta_j + 1)$ is the sum of tableaux $\mathcal{T}(\mu, I^k)$, I^k denoting the partition $\underbrace{1I \dots 1I}_{k}\ \underbrace{1I \dots 1I}_{k} \cdots$.

Since the operators θ_i are compatible with the plactic congruences, comparing the two sums gives that each tableau t in $\mathcal{T}(\mu, I^k)$ is congruent to a frank word which is a product $t_1 \cdots t_k$ of tableaux of shape I.

Conversely, from 2.11, we see that a product $t^{(1)} \cdots t^{(k)}$ of tableaux belonging to $\mathcal{T}(\mu, I)$ is congruent to a tableau $t \in \mathcal{T}(\mu, I^k)$ iff the following inequalities are satisfied:

$$(4.2) \qquad t^{(1)}\mathbf{K}_+ \leq t^{(2)}\mathbf{K}_- \ ; \ t^{(2)}\mathbf{K}_+ \leq t^{(3)}\mathbf{K}_- \ ; \ldots ; \ t^{(k-1)}\mathbf{K}_+ \leq t^{(k)}\mathbf{K}_-$$

Moreover, in such a case, if $v_1 \cdots v_r$ is the right key of $t^{(k)}$, then $v_1^k \cdots v_r^k$ is the right key of $t^{(1)} \ldots t^{(k)}$ because each frank word in the class of $t^{(1)} \cdots t^{(k)}$ has a right column which is one the columns $v_1, \ldots, v_r$.

Let us call *I-chain of length k* a product of tableaux of the same shape I satisfying the inequalities 4.2; the *right key* of a chain will be the right key of its last tableau, *the left key* of a chain being the left key of its first tableau.

The preceding results may be summarized in the following theorem:

THEOREM 4.3. *Let I be a partition, μ a permutation in $\mathcal{S}(\mathcal{A})$, $\sigma_i \ldots \sigma_j$ any reduced decomposition of μ . Then*

$$(1 - a^I)^{-1}\theta_i \cdots \theta_j \ = \ \sum_{\Gamma} \{\Gamma : \mathbf{K}_+(\Gamma) < \mathbf{K}(I, \mu)\}$$

sum of all I- chains Γ of right key $\mathbf{K}(I, \mu)$ and

$$(1 - a^I)^{-1}(\theta_i + 1) \cdots (\theta_j + 1) \ = \ \sum_{\Gamma} \{\Gamma : \mathbf{K}_+(\Gamma) \leq \mathbf{K}(I, \mu)\}$$

sum of all I- chains Γ of right key less or equal to $\mathbf{K}(I, \mu)$.

For instance, the 21 chains of length 2 for $\mathcal{S}(3)$ are all the 27 products $\neq \varnothing$ of two tableaux of shape 21 described below, and correspond bijectively to the 27 tableaux of shape 42. There are 8 tableaux of shape 21, only two being not keys; for them, one has $\left(\begin{smallmatrix}2\\1\,3\end{smallmatrix}\right)\mathbf{K}_- = \begin{smallmatrix}2\\1\,2\end{smallmatrix}$ and $\left(\begin{smallmatrix}2\\1\,3\end{smallmatrix}\right)\mathbf{K}_+ = \begin{smallmatrix}3\\1\,3\end{smallmatrix}$, $\left(\begin{smallmatrix}3\\1\,2\end{smallmatrix}\right)\mathbf{K}_- = \begin{smallmatrix}3\\1\,1\end{smallmatrix}$ and $\left(\begin{smallmatrix}3\\1\,2\end{smallmatrix}\right)\mathbf{K}_+ = \begin{smallmatrix}3\\2\,2\end{smallmatrix}$. On the second row, for example, one reads that the chain $\begin{smallmatrix}2\\1\,2\end{smallmatrix} \cdot \begin{smallmatrix}2\\1\,3\end{smallmatrix}$ factorizes the tableau $\begin{smallmatrix}2\,2\\1\,1\,2\,3\end{smallmatrix}$, its left key being $\left(\begin{smallmatrix}2\\1\,2\end{smallmatrix}\right)\mathbf{K}_-$ and its right key being $\left(\begin{smallmatrix}2\\1\,3\end{smallmatrix}\right)\mathbf{K}_+$.

	2 11	2 12	3 11	2 13	3 13	3 12	3 22	3 23
2 11	22 1111	22 1112	23 1111	22 1113	23 1113	23 1112	23 1122	23 1123
2 12	∅	22 1122	∅	22 1123	22 1133	∅	23 1222	23 1223
3 11	∅	∅	33 1111	∅	33 1113	33 1112	33 1122	33 1123
2 13	∅	∅	∅	∅	23 1133	∅	∅	23 1233
3 13	∅	∅	∅	∅	33 1133	∅	∅	33 1233
3 12	∅	∅	∅	∅	∅	∅	33 1222	33 1223
3 22	∅	∅	∅	∅	∅	∅	33 2222	33 2223
3 23	∅	∅	∅	∅	∅	∅	∅	33 2233

Reintroducing a parameter z , projecting to $\mathbb{Z}[\mathbb{A}]$ and using the specialization $\varepsilon_{\mathbb{A}} : \mathbb{A} \to \{1, \ldots, 1\}$, we get:

COROLLARY 4.4. *Let I be a partition, μ a permutation in $\mathfrak{S}(\mathbb{A})$. Then the postulation* $(1 - za^I)^{-1}\overline{\pi_\mu}\varepsilon_{\mathbb{A}}$ *(resp.* $(1 - za^I)^{-1}\pi_\mu\varepsilon_{\mathbb{A}}$ *) is equal to the generating function of the number of I-chains Γ having right key* $\mathbf{K}(I, \mu)$ *(resp. having right key less or equal to* $\mathbf{K}(I, \mu)$ *) , i.e.*

$$(1 - za^I)^{-1}\ \overline{\pi_\mu}\ \varepsilon \ = \ \sum_{\Gamma} z^{\text{length}\Gamma}\ \Gamma\ \varepsilon_A$$

sum on all I-chains Γ having right key $\mathbf{K}(I, \mu)$.

5. Avatars of standard bases. According to theorem 3.8, if μ and σ_k are such that $\ell(\mu\sigma_k) > \ell(\mu)$, then for any partition I, $\mathfrak{U}(\mu, I)\theta_k = \mathfrak{U}(\mu\sigma_k, I)$. Thus the operators θ_k allow to connect the standard bases corresponding to different permutations. Using the same induction $\mu \to \mu\sigma_k$, it is not too difficult, but we shall abstain from doing it, to check that standard bases can also be defined in the following two other manners 5.2 and 5.8.

First, according to **[L-M-S]**, a tableau can be considered as an increasing chain of permutations (with respect to the Ehresmann order 2.13). One says that a chain of permutations $\mu^{(1)} \le \mu^{(2)} \le \ldots \le \mu^{(r)}$ *lifts* a tableau $t = v_1 \cdot \ldots \cdot v_r$ if $\widetilde{v}_1, \ldots, \widetilde{v}_r$ are respective left factors of $\mu^{(1)}, \ldots, \mu^{(r)}$, where, for a word $v = x_1 \cdots x_m$, the notation $\widetilde{v}$ stands for the reverse word $x_m \cdots x_1$.

It is clear that given a tableau, there exists a unique minimal lift of it. Indeed, putting $\mu^{(0)} = \textit{identity}$ and having found the minimal chain $\mu^{(0)} \le \mu^{(1)} \le \ldots \le$

$\mu^{(p-1)}$ with respective left factors $\widetilde{v}_1, \dots, \widetilde{v}_{p-1}$, given moreover $v_p = x_1 \cdots x_m$, we see that the set of permutations μ such that $\mu \geq \mu^{(p-1)}$ and $1\mu = x_m, \dots, m\mu = x_1$ admits a unique minimal element $\mu^{(p)}$. An induction on p thus gives a lift $\mu^{(1)}(t) \leq \cdots \mu^{(r)}(t)$ that we shall call the *canonical lift of* t. From the construction, for any other lift $\zeta^{(1)} \leq \cdots \leq \zeta^{(r)}$, one has $\mu^{(1)} \leq \zeta^{(1)}, \dots, \mu^{(r)} \leq \zeta^{(r)}$, i.e. the canonical lift is minimal with respect to the Ehresmann order.

For example, the canonical lift of the tableau 531 62 4 is **135** 246 $\leq$ **26** 3145 $\leq$ **4** 62135. Let us illustrate on this example how to pass from $\mu^{(p-1)}$ to $\mu^{(p)}$, say for $p = 2$. The left reordered factors of $\mu^{(1)}$ are $1, 13, 135, 1235, 12345$; 236 is the minimum word having subword 26 bigger than 135, 1236 is the minimum word containing 236 bigger than 1235, and finally, 12346 is the minimum word containing 1235 bigger than 12345. These minimum words are the left reoredered factors of $\mu^{(2)} = 263145$ which therefore is the minimum permutation bigger than $\mu^{(1)}$ and beginning by **26**.

DEFINITION 5.1. *Given a partition I and a permutation μ , the L-M-S standard basis $\mathfrak{U}'(\mu, I)$ is the set of tableaux t such that the last permutation of their canonical lift is equal to μ.*

When $\mu = identity$, the set $\mathfrak{U}(\mu, I)$ reduces to the tableau $(1I \cdots 1)(2I \cdots 1) \times \cdots (rI \cdots 1)$ as well as $\mathfrak{U}(\mu, I)$; the induction $\mu \to \mu\sigma$ proves, as claimed in the beginning of this section, that $\mathfrak{U}(\mu, I)$ is the sum of the tableaux belonging to $\mathfrak{U}'(\mu, I)$. In other words, one has the following property showing that the L-M-S standard bases coincide with the one defined in 3.5, up to the change of the alphabet $\mathbb{A}$ with $\mathbb{N}$.

PROPOSITION 5.2. *A key* $\mathbf{K} = \mathbf{K}(\mu, I)$ *is the right key of a tableau t iff t has shape I and μ is the last permutation in the canonical lift of t.*

For example, the last permutation 462135 in the canonical lift of the tableau $t = 531\ 62\ 4$ gives the key 642 64 4, which is the right key of t as seen in 2.9.

One may favor horizontals rather than verticals. Reading the successive horizontals of a tableau t , one gets a word which is a product of rows (as defined in sect.2) and which is congruent to t; we shall call this word the *row-word* of t.

For example, the row-word of $\begin{smallmatrix} 5 & & \\ 3 & 6 & \\ 1 & 2 & 4 \end{smallmatrix}$ is 5 36 124. Row-words are characterized by their insertion tableau, as seen from property 2.5 :

LEMMA 5.3. *A word w is the row-word of a tableau t iff there exists a partition* $J = 1J \geq 2J \geq \cdots \geq pJ$ *such that* $\big[(pJ + ... + 2J + 1) \cdots (pJ + ... + 1J)\big] \cdots \big[(pJ + 1) \cdots (pJ + (p-1)J)\big] \big[1 \cdots pJ\big]$ *is an insertion word for w. In that case, J is the partition conjugate to the shape of t.*

For example, $(5\ 36\ 124)@ = \begin{smallmatrix} 4 & & \\ 2 & 5 & \\ 1 & 3 & 6 \end{smallmatrix}$, and this tableau is congruent to the word [456] [23] [1].

Apart from the symmetry between rows and columns, which means taking instead of composition words their reverse, the same property as 2.8.*2* holds : in the class of any tableau t of shape conjugate to a partition $J = (1J, \ldots, pJ)$, for any permutation H of J, there exists a unique word w congruent to t, which admits $\big[(pH + ... + 2H + 1) \cdots (pH + ... + 1H)\big] \cdots \big[1 \cdots pH\big]$ as an insertion word. This forces w to be a product $w_p \cdots w_1$ of rows of respective degrees $pH, \ldots, 2H, 1H$. For standard tableaux, transposition (i.e. exchange of the two axes of coordinates) commutes this construction with the one given in section 2 The hexagon generated by the action of the symmetric group on the row-word 5 36 124 is now

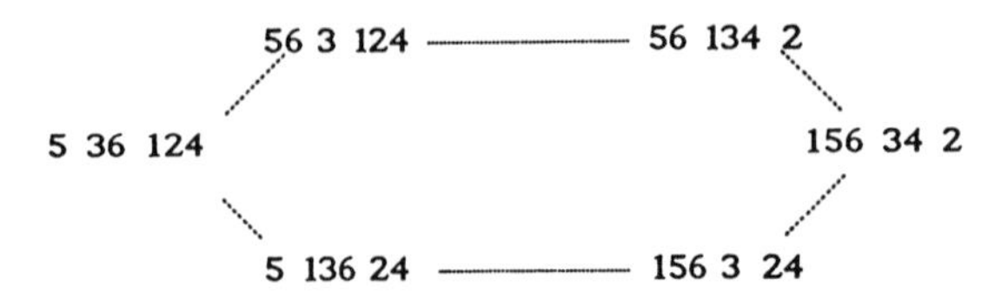

The corresponding insertion words are respectively

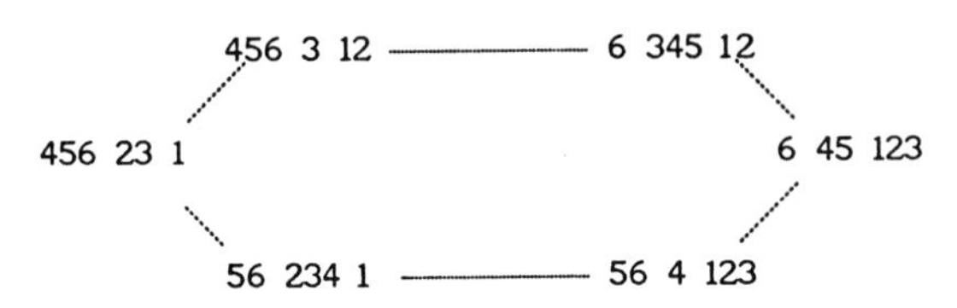

Given $H \in \mathbb{N}^p$, let T_H be the sum of words w such that:

(5.4) w has the insertion word

$$\varphi = \big[(pH + ... + 2H + 1) \cdots (pH + ... + 1H)\big] \cdots \big[1 \cdots pH\big]$$

(5.5) For the factorization $w = w_p \cdots w_1$ corresponding to φ , every w_j, $1 \leq j \leq p$, belongs to the monoid generated by $\mathfrak{A}_j = \{a_1, \ldots, a_j\}$, i.e. $w_1 \in \mathfrak{A}_1^*, w_2 \in \mathfrak{A}_2^*, \ldots, w_p \in \mathfrak{A}_p^*$.

Because of the explicit value of φ, the above factorization is the row-factorization of w, apart from *void* factors that we must specify in order to fix the flag conditions 5.5.

For example, $T_{1302} = (44\,222\,1 + 34\,222\,1 + 44\,122\,1 + 34\,122\,1) + (44\,112\,1 + 34\,112\,1) + (24\,122\,1) + (33\,222\,1 + 33\,122\,1) + (24\,112\,1) + (33\,112\,1) + (23\,122\,1) + (23\,112\,1)$ is the sum of words $w = w_3 w_2 w_1$such that $w@ \equiv 6\,345\,12$ $\left(w@ = \begin{smallmatrix} 6 & & \\ 3 & 4 & \\ 1 & 2 & 5 \end{smallmatrix}\right)$ and $w_1 \in \{1\}^*, w_2 \in \{1,2\}^*, w_3 \in \{1,2,3\}^*$.

As we already said, the induction $\mu \to \mu\sigma$, starting from the case T_J with J partition (T_J is the single word $\cdots 2^{2J} 1^{1J}$), allows to obtain the general case, summarized in the following proposition:

PROPOSITION 5.6. *Let J be a partition, μ a permutation, $H = J^\mu$, I the partition conjugate to J . Then T_H is congruent to $\sum_{\nu \leq \mu} \mathfrak{U}(I, \mu)$.*

In the preceding example, $J = 321$, $\mu = 2413$. One has 8 permutations in the interval $[1234, 2413]$. According to the proposition, $T_{1302} \equiv \mathfrak{U}(321, 2413) +$

$\mathfrak{U}(321,1423)+\mathfrak{U}(321,2143)+\mathfrak{U}(321,2314)+\mathfrak{U}(321,1243)+\mathfrak{U}(321,1324)+\mathfrak{U}(321,2134)+\mathfrak{U}(321,1234) =$

$\{421422+321422+421412+321412\}+\{421411+321411\}+\{421212\}+\{321322+321312\} + \{421211\} + \{321311\} + \{321212\} + \{321211\}$.

These tableaux are respectively congruent to the words enumerated in the same order above.

Flags of alphabets or of modules naturally occur in the study of Schubert polynomials **[L-S2]** or of Schubert subvarieties of a flag manifold.

One can restrict the sum T_H to its component T'_H congruent to $\mathfrak{U}(\mu, I)$. Indeed, one has the following property, which is also proved through the induction $\mu \to \mu\sigma$:

LEMMA 5.7. *Let t be a tableau of shape I conjugate to J, $\mathbf{K} = \mathbf{K}(\mu, I)$ its right key, ζ a permutation, $H = J^\zeta$. Then there exists a word in T_H congruent to t iff $\zeta \geq \mu$.*

In other terms, for any word w, the set of H such that w is congruent to a word in T_H is either void or admits a unique minimum element (i.e. an $H = J^\mu$ such that μ is minimal for the Ehresmann order, J being the partition conjugate to the shape of $w\mathbf{R}$). One can now define T'_H to be the restriction of T_H to such words. For example, the tableau 4321 321 31 41 3 is congruent to the words 3344 11233 2 11 $\in T_{2154}$, 3344 11233 12 1 $\in T_{1254}$, 13344 1233 2 11 $\in T_{2145}$, 13344 1233 12 1 $\in T_{1245}$ which correspond to all the permutations above 3412 ; it is also congruent to the words 3344 23 2 11113 , 34 12334 2 1113, 3344 3 11223 11, 4 13334 1223 11 but these words do not belong to respectively $T_{5124}, T_{4152}, T_{2514}, T_{2451}$ (which are just below in the Ehresmann order) because the flag condition 5.5 is violated. Thus t is congruent to a word in T'_{2154}. Proposition 5.6 can now be reformulated:

PROPOSITION 5.8. *Let J be a partition, μ a permutation, $H = J^\mu$, I the partition conjugate to J. Then T'_H is congruent $\mathfrak{U}(I, \mu)$.*

The key of the preceding tableau is $\begin{array}{l} 4 \\ 3\ 4 \\ 2\ 3\ 4\ 4 \\ 1\ 1\ 3\ 3\ 3 \end{array}$, i.e. is equal to $\mathbf{K}(3412, 5421)$, in accordance with the fact that $(5421)^{3412}$ is equal to 2154.

6. Appendix. Let U, Ξ be two sets, Ξ^* the free monoid generated by Ξ. An *action* of Ξ^* on U is a *function* (not everywhere defined; we use the symbol $\varnothing$ for the points of indeterminacy) : $U \times \Xi^* \to U \cup \{\varnothing\}$ such that $u(\xi\xi') = (u\xi)\xi'$ and $u\xi = \varnothing \Rightarrow u\xi\xi' = \varnothing$ for any $u \in U, \xi, \xi' \in \Xi$.

Let Ξ be finite and totally ordered: $\Xi = \{\xi_1, \ldots, \xi_{p+1}\}$. Suppose that "Moore-Coxeter" relations hold, i.e. that for any pair $\xi_i = \sigma, \xi_j = \tau$ and any u in U, one has identically:

(6.1) $\quad u\xi \neq \varnothing \;\Rightarrow\; u\xi\xi = u$

(6.2) $\quad$ if $|i-j| \geq 2$, $u\sigma\tau = u\tau\sigma$

(6.3) $\quad$ if $|i-j| = 1$, $u\sigma\tau\sigma = u\tau\sigma\tau$.

Remark 6.4. 1) Let $|i-j|=1$ and $u\sigma, u\tau, u\sigma\tau \neq \emptyset$, then $u\sigma\tau\sigma = u\tau\sigma\tau \neq \emptyset$. 2) Let $|i-j| \geq 2$ and $u\sigma, u\tau \neq \emptyset$. Then $u\sigma\tau = u\tau\sigma \neq \emptyset$.

Proof of 1): $u\tau \neq \emptyset$ implies $(u\tau)\tau = u$ according to 6.1; the hypothesis becomes $(u\tau)\tau, (u\tau)\sigma\tau, (u\tau)\tau\sigma\tau \neq \emptyset$ showing that $(u\tau)\sigma\tau\sigma = (u\tau)\tau\sigma\tau \neq \emptyset$ by 6.3. Therefore, $\emptyset \neq (u\tau)\sigma\tau\sigma\sigma = (u\tau)\tau\sigma\tau\sigma = u\sigma\tau\sigma$ as required.
Proof of 2): As above, we use 6.1 to write $u\sigma = (u\tau)\tau\sigma$; according to 6.2, $(u\tau)\tau\sigma = (u\tau)\sigma\tau$; since $u\sigma \neq \emptyset$, $u\tau\sigma\tau$ is different from $\emptyset$ as well as its factor $u\tau\sigma$, and $u\sigma\tau$ by symmetry. □

Choose any $u = u_0$ in U. The three preceding axioms allow to identify the orbit $\Omega = \{u\xi : u\xi \neq \emptyset\}$ to a quotient (the $u\xi$ need not to be all different) of a subset of $\mathcal{S}(p+1)$, u being sent to the identity element of $S(p+1)$. The following proposition gives a necessary and sufficient condition for the orbit to be a quotient of the full symmetric group.

PROPOSITION 6.5. *Let $n, m \geq 1$, $p = n+m$ and set $\rho = \xi_n, \Xi_1 = \{\xi_1, \dots, \xi_{n-1}\}$, $\Xi_2 = \{\xi_{n+1}, \dots, \xi_{p-1}\}$. Assume that both $u\Xi_1^*$ and $u\Xi_2^*$ do not contain $\emptyset$ and that $\xi_1 \in \Xi_1^*, \xi_2 \in \Xi_2^* \Rightarrow u\xi_1\xi_2\rho \neq \emptyset$.*

Then $u\Xi^*$ does not contain $\emptyset$.

Proof. We can suppose $n \leq m$ by symmetry, and deduce the general case from the case where all the points $\neq \emptyset$ in $u\Xi^*$ are different. Thus the orbit Ω is a subset of the symmetric group and we write its elements as permutations. If $n = m = 1$, there is nothing to prove. Consider the case where $n = 1, m = 2$. Then Ξ_1 is void, $\Xi_2 = \{\xi_2\}, \rho = \xi_1$. By hypothesis, $u = 123, u\rho = 213$, $u\xi_2 = 312$ and $u\xi_2\rho = 312$ are all different from $\emptyset$. Thus taking $u' = 213, \sigma = \rho, \xi_2 = \tau$ in 6.3, we get that $u'\tau = u\rho\xi_2 = 231$, $u'\tau\sigma = u\rho\xi_2\rho = 321$ are different from $\emptyset$; this proves the proposition in this case.

Let again $n = 1$ and $m \geq 3$. As above, $\rho = \xi_1$ and Ξ_1 is void. Using induction on m, we have that Ω contains all the permutations such that their rightmost letter is $\neq 1$. In particular, for any $i, j > 1$, Ω contains all the permutations such that their restriction to the third rightmost letters is $1ij, i1j, 1ji$ or $j1i$. Repeating the same argument with $\sigma = \xi_{p-2}$ and $\tau = \xi_{p-1}$, we conclude that Ω contains all the permutations such that their right factor of length 3 is $ij1$ or $ji1$, concluding the proof of the proposition for $n = 1$.

Consider now the general case where $n \geq 2, m \geq 1$. For any $k \leq n$, we can find some ξ in Ξ_1^* such that the first (ie. left) letter of $u\xi$ is k. Thus by induction on n, i.e. by considering the restriction of $u\xi$ to all its letters except the first, we have that Ω contains all the permutations such that their first letter is $h \leq k$. Considering now the first three letters on the left and applying the same argument as for the case of $n = 1$, we conclude that Ω is the full symmetric group. □

The action of "commutation" of columns seen in section 2 satisfy the axioms 6.1, 6.2 and 6.3 . Only 6.3 is not straightforward. Since it involves only triples of consecutive columns in a word, it needs to be checked only for 3-columns words. This we do just now.

LEMMA 6.6. *Let w be a 3-columns word, σ and τ be the two generators of $\mathcal{S}(3)$. Then* $\{w\sigma, w\sigma\tau, w\sigma\tau\sigma \neq \emptyset\} \Rightarrow \{w\tau, w\tau\sigma \neq \emptyset \ \&\ w\sigma\tau\sigma = w\tau\sigma\tau\}$.

Proof. One of the four words w , $w\sigma$, $w\sigma\tau$, $w\sigma\tau\sigma$ has its shape decreasing or increasing. Let it be $w\sigma = v_1 \cdot v_2 \cdot v_3$. Recall that a 2-columns word w is a tableau or a contretableau iff $w^\flat \neq \emptyset$, i.e. iff w is frank, and that a word is a tableau (resp. a contretableau) if each factor made of two consecutive columns is such. The two factors v_1v_2 and v_2v_3 being frank, $w\sigma$ is a tableau or a contretableau. According to 2.6.3 , the action of permutation of columns on a tableau or a contretableau generate the frank words in its class: thus $w, w\sigma, w\sigma\tau, w\sigma\tau\sigma = w\tau\sigma\tau, w\tau\sigma, w\tau$ are the frank words in the class of $w\sigma$.

Suppose now that this is $w = v_1 \cdot v_2 \cdot v_3$ which has a decreasing shape $\|w\| = ijh$, and let t be the insertion tableau of w and σ be the first generator of $\mathcal{S}(3)$. Since $v_1v_2\sigma \neq \emptyset$, the word $v_1 \cdot v_2$ is a tableau and this determines $t \cap \{1, \ldots, i+j\}$. Since $v_1 \cdot v_2 \cdot v_3\sigma\tau \neq \emptyset$, the digit $i+j+h$ cannot be in the first column of t; since $v_1 \cdot v_2 \cdot v_3\sigma\tau\sigma \neq \emptyset$, it cannot be either in the second column of t. It must be in the third, which means that t is equal to $[i \cdots 1]\,[(i+j)\cdots(i+1)] \times [(i+j+h)\cdots(i+j+1)] = ijh\mathrm{M}$. Thus w is a tableau and we conclude as before. This reasoning also applies to the case where $\|w\|$ is increasing, since then $\|w\sigma\tau\sigma\|$ is decreasing and we can exchange the rôle of w and $w\sigma\tau\sigma$. □

Pictorially, hypothesis 6.6 is that if the four consecutive words $w \to w\sigma \to w\sigma\tau \to w\sigma\tau\sigma$ are different from $\emptyset$, then we can "close the hexagon":

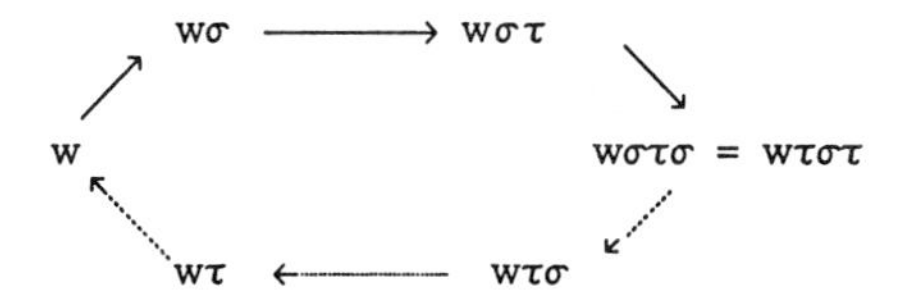

Let us finish with an example of a word whose orbit (under commutation of columns) is not a quotient of the full symmetric group.

Let $w = 31 \cdot 42 \cdot 4 \cdot 51$, and σ, ρ, τ be the three generators of $\mathcal{S}(4)$ We get four double points: $w = w\sigma, w\rho = 31\ 4\ 42\ 51 = w\sigma\rho$, $w\rho\sigma = \omega\rho\sigma\rho = 3 \cdot 41 \cdot 42 \cdot 51, w\tau = w\tau\sigma = 31 \cdot 42 \cdot 41 \cdot 5$. Since the words $42 \cdot 41$ and $42 \cdot 51$ are not frank, all the neighbours $w\rho\tau, w\tau\rho, w\sigma\tau\rho, w\sigma\rho\tau, w\rho\sigma\tau$ and $w\sigma\rho\sigma\tau$ are $\emptyset$ and thus the orbit Ω of w is restricted to the enumerated four double points. Indeed, condition 2.8.3 to ensure that w be frank is exactly that $42 \cdot 4$ (central factor of w) and $42 \cdot 41$ (central factor of $w\sigma$, $w\tau$ and $w\tau\sigma$) should be frank. Since this not the case for the last word, we already knew from th.2.8 that Ω could not be a quotient of the full symmetric group.

REFERENCES

[B-G-G] BERNSTEIN I.N., GELFAND I.M. GELFAND S.I., UMN, 28 (1973), pp. 1-26.

[B-K] A.BENDER, D.E.KNUTH, *Enumeration of plane partitions*, J. Comb. Th. A, 13 (1972), pp. 40–54.

[B] A.Björner, *Ordering of Coxeter Groups*, Contemp. M., 34 (1984), pp. 175–195.

[Bo] N.Bourbaki, *Eléments de Mathématiques modernes, Algèbre ch.1*, Paris, New York, C.C.L.S., 1971.

[D1] M.Demazure, *Désingularisation des variétés de Schubert*, Ann. E. N. S., 6 (1974), pp. 163–172.

[D2] M.Demazure, *Une formule des caractères*, Bull. Sc. M., 98 (1974), pp. 163–172.

[E] C.Ehresmann, *Sur la topologie de certains espaces homogènes*, Ann. M. (2), 35 (1934), pp. 396–443.

[Gr] A.Grothendieck, *Sur quelques propriétés fondamentales en théorie des intersections*, Séminaire Chevalley, Paris, 1958.

[K-L] D.Kazhdan and G.Lusztig, *Representations of Coxeter Groups and Hecke Algebras*, Invent., 53 (1979), pp. 165-184.

[K] D.E.Knuth, *Permutations matrices and generalized Young tableaux*, Pac. J. Math., 34 (1970), pp. 709–727.

[L-M-S] V.Lakshmibai, C.Musili and C.S.Seshadri, *Geometry of G/P IV*, Proc. Indian Acad. Sc., 88A (1979), pp. 280-362.

[L-S1] A.Lascoux & M.P.Schützenberger, *Le Monoide Plaxique*, Atti del C.N.R., Roma, 1981.

[L-S2] A. Lascoux & M.P.Schützenberger, *Symmetry and Flag Manifolds*, Lect. Notes in Math., 996 (1983).

[L-S3] A.Lascoux & M.P.Schützenberger, *Symmetrization operators on polynomial rings*, Funk. Anal., 21 (1987), pp. 77-78.

[L-S4] A.Lascoux & M.P.Schützenberger, *Tableaux and non commutative Schubert polynomials*, Funk. Anal (to appear).

[L-S5] A.Lascoux, *Anneau de Grothendieck de la variété de drapeaux*, submitted to the Comité d'évaluation des hommages à Grothendieck (1988).

[L-S6] A.Lascoux & M.P.Schützenberger, *Arêtes et tableaux*, Séminaire Lotharingien de Combinatoire Cagliari, 1988.

[L-W] V.Lakshmibai and J.Weyman, *Bases standards*, C. R. Acad. Sc. Paris, 1988.

[M] I.G.Macdonald, *Symmetric Functions and Hall Polynomials*, Oxford Mono., Oxford, 1979.

[Mc] Major P.MacMahon, *Combinatory Analysis*, Chelsea reprints, New York, 1960.

[Sche] C.Schensted, *Longest increasing and decreasing sequences*, Canad. J. Math., 13 (1961), pp. 179–191.

[Schu] M.P.Schützenberger, *Quelques remarques sur une construction de Schensted*, Math. Scand., 12 (1963), pp. 117–128.

VARIATIONS ON DIFFERENTIAL POSETS*

RICHARD P. STANLEY†

1. Introduction. Differential posets were introduced in [Sta$_3$]. They are partially ordered sets with many remarkable algebraic and combinatorial properties. In this paper we will consider ways to modify or extend the definition of differential posets and still retain some of their basic properties. This paper is essentially a sequel to [Sta$_3$], and familiarity with [Sta$_3$] will be useful but not essential for understanding this paper. In particular, if P is a poset and K a field, then KP denotes the K-vector space with basis P, while $\hat{K}P$ denotes the K-vector space of arbitrary (i.e., infinite) linear combinations of elements of P. If P is locally finite and $x \in P$, then define

$$\begin{aligned} C^+(x) &= \{y \in P : y \text{ covers } x\}, \\ C^-(x) &= \{y \in P : x \text{ covers } y\}. \end{aligned}$$

Furthermore, define continuous (i.e., infinite linear combinations are preserved) linear transformations $U, D : \hat{K}P \to \hat{K}P$ by

$$Ux = \sum_{y \in C^+(x)} y, \quad Dx = \sum_{y \in C^-(x)} y,$$

for all $x \in P$. If $S \subseteq P$ then we write

$$\mathbf{S} = \sum_{x \in S} x \in \hat{K}P.$$

If P is r-differential then we have

$$DU - UD = rI, \tag{1}$$

$$D\mathbf{P} = (U + r)\mathbf{P}. \tag{2}$$

(See Theorems 2.2 and 2.3 of [Sta$_3$].) We will consider three main variations of differential posets, all involving either modifications of the definition of U and D or modifications of equations (1) and (2).

*Partially supported by NSF grant # DMS-8401376. I am grateful also for the support and hospitality of the IMA, where some of the research for this paper was carried out.

†Department of Mathematics, MIT, Cambridge, MA 02139

2. Sequentially differential posets. Let $\mathbf{r} = (r_0, r_1, \ldots)$ be an infinite sequence of integers.

2.1 Definition. A poset P is called **r**-*differential* if it satisfies the following three conditions:

(S1) P is locally finite and graded, and has a $\hat{0}$ element.

(S2) If $x \neq y$ in P and there are exactly k elements of P which are covered by both x and y, then there are exactly k elements of P which cover both x and y.

(S3) If $x \in P_i$ and x covers exactly k elements of P, then x is covered by exactly $k + r_i$ elements of P. (Here P_i denotes the set of elements of P of rank i.)

If P is an **r**-differential poset for some sequence **r**, then we call P a *sequentially differential poset.* □

Properties (S1) and (S2) coincide with (D1) and (D2) of [Sta$_3$, Def. 1.1], while (S3) is a weakening of (D3). Thus Proposition 1.2 of [Sta$_3$] remains true for sequentially differential posets, i.e., the integer k of (S2) must be 0 or 1. (Thus given (S1), condition (S2) coincides with what Proctor calls *uniquely modular* [P_1, p. 270].) Moreover, the next three results are proved in exactly the same way as the corresponding results of [Sta$_3$] (viz., Proposition 1.3 and Theorems 2.2 and 2.3).

2.2 PROPOSITION. *Let L be a lattice satisfying (S1) and (S3). Then L is* **r**-*differential if and only if L is modular.* □

As in [Sta$_3$], if $A : \hat{K}P \to \hat{K}P$ is a linear transformation then A_j denotes the restriction of A to KP_j; and we can unambiguously use notation such as AB_j, since $A(B_j)$ and $(AB)_j$ have the same meaning. In particular I_j denotes the identity transformation $I : \hat{K}P \to \hat{K}P$ restricted to KP_j. We often omit subscripts if they are clear from context, e.g., in equation (18) it is clear that U^iD^j means $U^iD^j_k$.

2.3 PROPOSITION. *Let P be a locally finite graded poset with $\hat{0}$, with finitely many elements of each rank. Let $\mathbf{r} = (r_0, r_1, \ldots)$ be a sequence of integers. The following two conditions are equivalent:*

$$(3) \qquad \begin{array}{ll} (a) & P \text{ is } \mathbf{r}\text{–differential}, \\ (b) & DU_j - UD_j = r_j I_j, \quad \text{for all } j \geq 0. \end{array} \qquad \square$$

If P is finite in Proposition 2.3 and if j is greater than the rank of P (so that $P_j = \varnothing$), then (b) is regarded as vacuously true (whatever the value of r_j). For instance, a single point is **r**-differential for any sequence **r** with $r_0 = 0$.

2.4 PROPOSITION. *If P is an* **r**-*differential poset, then*

$$(4) \qquad D_{j+1}\mathbf{P_{j+1}} = U_{j-1}\mathbf{P_{j-1}} + r_j\mathbf{P_j}. \quad \square$$

Before discussing properties of **r**-differential posets, let us list some finite examples.

2.5 Example. The following finite posets are **r**-differential:

(a) An n-element chain ($r_0 = 1, r_i = 0$ for $1 \le i \le n-2, r_{n-1} = -1$).

(b) The boolean algebra B_n of rank n ($r_i = n - 2i$, $0 \le i \le n$).

(c) A product C_3^n of n 3-element chains C_3 ($r_i = n - i$, $0 \le i \le 2n$).

(d) The lattice $L_n(q)$ of subspaces of an n-dimensional vector space over the finite field $\mathbb{F}_q$ ($r_i = 1 + q + \cdots + q^{n-i-1} - (1 + q + \cdots + q^{i-1})$, $\quad 0 \le i \le n$).

(e) If P is $\mathbf{r}$-differential and finite of rank n (in which case it must have a $\hat{1}$) and Q is $\mathbf{s}$-differential, then the poset $P * Q$ obtained by identifying $\hat{1} \in P$ with $\hat{0} \in Q$ is $(r_0, \ldots, r_{n-1}, r_n + s_0, s_1, s_2, \ldots)$-differential.

(f) If P is $\mathbf{r}$-differential and finite of rank n, then the dual P^* is $(-r_n, -r_{n-1}, \ldots, -r_0)$-differential.

There are many other finite $\mathbf{r}$-differential posets, and it is probably hopeless to attempt to classify them all. It may be more tractable to find the finite $\mathbf{r}$-differential distributive lattices $L = J(P)$ (where $J(P)$ has the meaning of [Sta$_2$, Thm. 3.4.1]). Example 2.5(b, c, e) gives some examples. A further class of finite $\mathbf{r}$-differential distributive lattice may be constructed as follows. Suppose P and Q are finite posets of cardinalities m and n, respectively, such that $J(P)$ is $\mathbf{r}$-differential for $r = (r_0, \ldots, r_m)$ and $J(Q)$ is $\mathbf{s}$-differential for $\mathbf{s} = (s_0, \ldots, s_n)$. Suppose also that $r_m = s_0 = t$, say; and let $M = \{x_1, \ldots, x_t\}$ be the set of maximal elements of P and $N = \{y_1, \ldots, y_t\}$ the set of minimal elements of Q. Define a new poset $P\#Q$ on the disjoint union $P + Q$ by imposing the additional relations $x_i < y_j$ whenever $i \ne j$ (and all relations then implied by transitivity). Then $J(P\#Q)$ is $(r_0, \ldots, r_{m-2}, r_{m-1} + 1, 0, s_1 - 1, s_2, \ldots, s_n)$-differential. If we take P and Q to be 2-element antichains, then $J(P\#Q) = C_3 \times C_3$, where C_3 denotes a 3-element chain. If we take $P = Q = B_3$ (the boolean algebra of rank 3), then $J(P\#Q) = FD(3)$, the free distributive lattice on 3 generators [Sta$_2$, Exer. 3.24].

In general, one sees as in [Sta$_3$, Prop. 5.5] that for fixed $\mathbf{r}$ there is at most one $\mathbf{r}$-differential distributive lattice $L(\mathbf{r})$ (up to isomorphism). It is probably hopeless to determine for which $\mathbf{r}$ the lattice $L(\mathbf{r})$ exists; if the sequence $r_0, r_1, r_2, \ldots$ increases "sufficiently fast" then $L(\mathbf{r})$ exists, but it seems difficult to describe the necessary rate of growth precisely. Conceivably there are no other finite $\mathbf{r}$-differential distributive lattices besides those described above. A straightforward generalization of the construction of [Sta$_3$, Prop. 6.1] shows that if all $r_i \ge 0$, then there exists an $\mathbf{r}$-differential (modular) lattice.

Let us turn to the enumerative properties of sequentially differential posets. The basic principle here is that all the enumerative results of [Sta$_3$] can be extended to $\mathbf{r}$-differential posets, but their statements no longer involve generating functions and thus become more complicated. Consider, for instance, the number $\alpha(0 \to n)$ of saturated chains from $\hat{0}$ to P_n. According to [Sta$_3$, eqn. (12)], in an r-differential poset we have

$$\sum_{n \ge 0} \alpha(0 \to n) \frac{t^n}{n!} = \exp(rt + \frac{1}{2} rt^2).$$

Equivalently,

$$\alpha(0 \to n) = \sum_w r^{c(w)}, \tag{5}$$

where w ranges over all involutions in S_n and $c(w)$ is the number of cycles of w. For $\mathbf{r}$-differential posets, the term $r^{c(w)}$ in (5) is replaced by a certain monomial $r_0^{c_0}r_1^{c_1}\ldots$ (where $\sum c_i = c(w)$), as follows:

2.1 THEOREM. *Let P be an $\mathbf{r}$-differential poset. Then*

$$\alpha(0 \to n) = \sum_w \prod_m r_{\eta(w,m)}, \tag{6}$$

where (a) w ranges over all involutions $w_1w_2\cdots w_n$ in S_n, (b) m ranges over all weak excedances of w (i.e., $w_m \geq m$), and (c) $\eta(w,m)$ is the number of integers j satisfying $j < m$ and $w_j < w_m$.

For example, when $n = 3$ we have the following table, where the positions of the weak excedances are underlined:

involution w	values of $\eta(w,m)$
$\underline{1}\,\underline{2}\,\underline{3}$	$0,1,2$
$\underline{2}\,1\,\underline{3}$	$0,2$
$\underline{1}\,\underline{3}\,2$	$0,1$
$\underline{3}\,\underline{2}\,1$	$0,0$

Hence $\alpha(0\to 3) = r_0^2 + r_0r_1 + r_0r_2 + r_0r_1r_2$. Let us note that the number of weak excedances of an involution w is equal to $c(w)$, so that Theorem 2.1 reduces to (5) when each $r_i = r$.

Proof of Theorem 2.1. Consider the element

$$D^k\mathbf{P_{i+k}} = D_{i+1}D_{i+2}\cdots D_{i+k}\mathbf{P_{i+k}}$$

of KP_i. It is evident that repeated uses of Propositions 2.3 and 2.4 will express this element as a linear combination of elements of the form $U^j\mathbf{P_{i-j}}$. For instance,

$$\begin{aligned}
D^2\mathbf{P_{i+2}} &= D(U\mathbf{P_i} + r_{i+1}\mathbf{P_{i+1}})\\
&= (UD + r_i)\mathbf{P_i} + r_{i+1}(U\mathbf{P_{i-1}} + r_i\mathbf{P_i})\\
&= U(U\mathbf{P_{i-2}} + r_{i-1}\mathbf{P_{i-1}}) + r_{i+1}U\mathbf{P_{i-1}} + r_i(1+r_{i+1})\mathbf{P_i}\\
&= U^2\mathbf{P_{i-2}} + (r_{i-1} + r_{i+1})U\mathbf{P_{i-1}} + r_i(1+r_{i+1})\mathbf{P_i}.
\end{aligned}$$

We claim that in general,

$$D^k\mathbf{P_{i+k}} = \sum_{j=0}^{k}\left[\sum_w \prod_m r_{i+\eta(w,m)-\nu(w,m)}\right]U^j\mathbf{P_{i-j}}, \tag{7}$$

where (a) for j fixed, w runs over all involutions in S_k with j of the fixed points in w circled (so that w has $\geq j$ fixed points), (*b*) m ranges over all *uncircled* weak excedances of w, (c) $\eta(w,m)$ has the same meaning as above, and (d) $\nu(w,m)$ is the number of *circled* fixed points $d = w_d$ such that $d > w_m$. For example, if $w = 16③482⑦5$, then $\eta(1) = 0, \nu(1) = 2$, $\eta(2) = 1, \nu(2) = 1, \eta(4) = 2, \nu(4) = 1, \eta(5) = 4, \nu(5) = 1$, yielding $r_{i-2}r_ir_{i+1}r_{i+3}$.

We prove (7) by induction on k. For $k = 0$ it asserts that $\mathbf{P_i} = \mathbf{P_i}$, which is clear. Assume for k. Then

$$D^{k+1}\mathbf{P_{i+k+1}} = D\sum_{j=0}^{k}\left[\sum_w \prod_m r_{i+1+\eta(w,m)-\nu(w,m)}\right] U^j\mathbf{P_{i+1-j}}. \tag{8}$$

Now an easy induction argument shows that

$$DU^j_{i+1-j} = U^j D_{i+1-j} + (r_{i+1-j} + r_{i+2-j} + \cdots + r_i)U^{j-1}_{i+1-j}, \tag{9}$$

so that by Proposition 2.4,

$$DU^j\mathbf{P_{i+1-j}} = U^j(U\mathbf{P_{i-j-1}} + r_{i-j}\mathbf{P_{i-j}}) + (r_{i+1-j} + \cdots + r_i)U^{j-1}\mathbf{P_{i+1-j}}.$$

Hence (8) becomes

$$D^{k+1}\mathbf{P_{i+k+1}} = \sum_{j=0}^{k}\left[\sum_w \prod_m r_{i+1+\eta(w,m)-\nu(w,m)}\right] \cdot (U^{j+1}\mathbf{P_{i-j-1}} + r_{i-j}U^j\mathbf{P_{i-j}} + (r_{i+1-j} + \cdots + r_i)U^{j-1}\mathbf{P_{i-j+1}}).$$

Let $I(k,j)$ denote the set of involutions in S_k with j circled fixed points. Then we need to show that for $0 \leq j \leq k+1$,

(10)

$$\sum_{w\in I(k+1,j)} \prod_m r_{i+\eta(w,m)-\nu(w,m)} = \left(\sum_{w\in I(k,j-1)} + r_{i-j}\sum_{w\in I(k,j)} + (r_{i-j} + \cdots + r_i)\sum_{w\in I(k,j+1)}\right)\prod_m r_{i+1+\eta(w,m)-\nu(w,m)}.$$

Define a bijection

$$\varphi : I(k,j-1) \cup I(k,j) \cup [I(k,j+1) \times \{1,2,\cdots,j+1\}] \to I(k+1,j)$$

as follows. If $w = w_1w_2\cdots w_k \in I(k,j-1)$ (where w_i is an uncircled or circled integer) then $\varphi(w) = ①, w_1+1, w_2+1, \cdots, w_k+1$ (where here and below $w_i + 1$ is circled if and only if w_i is circled). If $w = w_1w_2\cdots w_k \in I(k,j)$, then $\varphi(w) = 1, w_1+1, \cdots, w_k+1$. If $w = w_1w_2\cdots w_k \in I(k,j+1)$ and $1 \leq t \leq j+1$, then $\varphi(w,t)$ is obtained from $w_1+1, w_2+1, \ldots, w_k+1$ by replacing the t-th circled term from the left, say $w_\ell + 1$, by an uncircled 1 and placing an uncircled $w_\ell + 1$ at the beginning.

Example. Let $k = 7$, $j = 2$. Then

$$\varphi(1\ 5\ 3\ 7\ 2\ \textcircled{6}\ 4) = \textcircled{1}\ 2\ 6\ 4\ 8\ 3\ \textcircled{7}\ 5$$
$$\varphi(\textcircled{1}\ 5\ 3\ 7\ 2\ \textcircled{6}\ 4) = 1\ \textcircled{2}\ 6\ 4\ 8\ 3\ \textcircled{7}\ 5$$
$$\varphi(\textcircled{1}\ 5\ \textcircled{3}\ 7\ 2\ \textcircled{6}\ 4,\ 1) = 2\ 1\ 6\ \textcircled{4}\ 8\ 3\ \textcircled{7}\ 5$$
$$\varphi(\textcircled{1}\ 5\ \textcircled{3}\ 7\ 2\ \textcircled{6}\ 4,\ 2) = 4\ \textcircled{2}\ 6\ 1\ 8\ 3\ \textcircled{7}\ 5$$
$$\varphi(\textcircled{1}\ 5\ \textcircled{3}\ 7\ 2\ \textcircled{6}\ 4,\ 3) = 7\ \textcircled{2}\ 6\ \textcircled{4}\ 8\ 3\ 1\ 5.$$

It is easily seen that φ is a bijection. Moreover, if $w \in I(k, j-1)$ then m is an uncircled weak excedance of $\varphi(m)$ if and only if $m-1$ is an uncircled weak excedance of w; and $\eta(\varphi(w), m) = 1 + \eta(w, m-1)$, $\nu(\varphi(w), m) = \nu(\varphi(w), m-1)$. If $w \in I(k, j)$ then m is an uncircled weak excedance of $\varphi(w)$ if and only if $m = 1$ or $m-1$ is an uncircled weak excedance of w; and $\eta(\varphi(w), 1) = 0$, $\nu(\varphi(w), 1) = j$, $\eta(\varphi(w), m) = 1 + \eta(w, m-1)$ for $m > 1$, $\nu(\varphi(w), m) = \eta(w, m-1)$ for $m > 1$. Finally if $w \in I(k, j+1)$ and $1 \leq t \leq j+1$, then m is an uncircled weak excedance of $\varphi(w, t)$ if and only if $m = 1$ or $m-1$ is an uncircled weak excedance of w; and $\eta(\varphi(w,t), 1) = 0, \nu(\varphi(w,t), 1) = t-1$, while for $m > 1$ it is not difficult to check that

$$\eta(\varphi(w,t), m) - \nu(\varphi(w,t), m) = 1 + \eta(w, m-1) - \nu(w, m-1).$$

From these observations (10) follows, and hence also (7) for $k+1$ by induction.

Now let $i = 0$ in (7). The left-hand side becomes $\alpha(0 \to k)\hat{0}$, while the right-hand side becomes (since $\mathbf{P_{i-j}} = 0$ for $j > 0$)

$$\Big(\sum_{w} \prod_{m} r_{\eta(w,m)}\Big)\hat{0},$$

where w and m are as in (6). This completes the proof. $\square$

Consider now the problem of evaluating $\alpha(n \to n+k)$, i.e., the number of saturated chains from P_n to P_{n+k}. In [Sta$_3$, Thm. 3.2] a certain polynomial $A_k(q)$ (whose coefficients are polynomials in r) was defined for which

$$\sum_{n \geq 0} \alpha(n \to n+k) q^n = A_k(q) F(P, q),$$

where $F(P,q) := \sum (\#P_n) q^n$ is the rank-generating function of P. An analogous but more complicated result holds for $\mathbf{r}$-differential posets. By (7) there is a polynomial $T_{jk}(y_0, y_{\pm 1}, \ldots)$ in the variables $y_0, y_{-1}, y_1, \ldots$ such that

$$D^k \mathbf{P_{i+k}} = \sum_{j=0}^{k} T_{jk}(r_i, r_{i\pm 1}, \ldots) U^j \mathbf{P_{i-j}}. \tag{11}$$

2.2 Theorem. *Let P be an $\mathbf{r}$-differential poset. Then the numbers $\alpha(n \to n+k)$ are given recursively (in k) by the formula*

$$\alpha(n \to n+k) = \sum_{\substack{0 \leq i \leq n \\ i \equiv n \ (\mathrm{mod}\ k)}} \sum_{j=0}^{k-1} T_{jk}(r_i, r_{i\pm 1}, \ldots) \alpha(i-j \to i). \tag{12}$$

Proof. Put $i = n$ in (11) and apply the linear transformation $\sigma : KP \to K$ defined by $\sigma(x) = 1$ for all $x \in P$. We obtain

$$\begin{aligned}\alpha(n \to n+k) &= \sum_{j=0}^{k} T_{jk}(r_n, r_{n\pm1}, \dots)\alpha(n-j \to n) \\ &= \alpha(n-k \to n) + \sum_{j=0}^{k-1} T_{jk}(r_n, r_{n\pm1}, \dots)\alpha(n-j \to n),\end{aligned}$$

since $T_{kk} = 1$. The solution to this recurrence (with the initial condition $\alpha(i \to i+k) = 0$ if $i < 0$) is clearly given by (12). □

By repeated applications of (12) we will eventually express $\alpha(n \to n+k)$ in the form

$$\alpha(n \to n+k) = \sum_{i=0}^{n} R_{ik}(r_{n-i}, r_{n-i+1}, \dots, r_{n+k-1})p_{n-i},$$

where $p_{n-i} = \#P_{n-i}$ and $R_{ik}(y_1, y_2, \dots, y_{k+i})$ is a polynomial in $y_1, y_2, \dots, y_{k+i}$ (independent of n). For instance,

$$\begin{aligned}\alpha(n \to n) &= p_n \\ \alpha(n \to n+1) &= \sum_{i=0}^{n} r_i p_i \\ \alpha(n \to n+2) &= \sum_{\substack{i \le n \\ i \equiv n \pmod 2}} [(r_{i-1} + r_{i+1})\sum_{j=0}^{i-1} r_j p_j + r_i(r_{i+1} + 1)p_i].\end{aligned}$$

It would be interesting to find a more explicit formula for $\alpha(n \to n+k)$, along the lines of (6) (the case $n = 0$).

Consider now a word $w = w(U, D) = w_1 w_2 \cdots w_\ell$ in the letters U and D. Let $x \in P$. We wish to compute the quantity $\langle w\hat{0}, x\rangle$ (using the scalar product defined in [Sta$_3$, §2]), i.e., the number of Hasse walks $\hat{0} = x_0, x_1, \dots, x_\ell = x$ with the cover relation $x_{i-1} < x_i$ or $x_{i-1} > x_i$ specified by w. For instance, $\langle UDDUDUU\hat{0}, x\rangle$ is the number of Hasse walks $\hat{0} = x_0 < x_1 < x_2 > x_3 < x_4 > x_5 > x_6 < x_7 = x$. Clearly $\langle w\hat{0}, x\rangle = 0$ unless (a) for all $1 \le i \le \ell$, the number of D's among $w_i, w_{i+1}, \dots, w_\ell$ does not exceed the number of U's, and (b) the difference between the number of U's and number of D's in w is the rank $\rho(x)$ of x. Let us call such a word w a *valid x-word.* As in [Sta$_3$], denote by $e(x)$ the number of saturated chains from $\hat{0}$ to x.

2.3 THEOREM. *Let P be an* $\mathbf{r}$*-differential poset, and let $x \in P$. Let $w = w_1 w_2 \cdots w_\ell$ be a valid x-word. Let $S = \{i : w_i = D\}$. For each $i \in S$, let a_i be the number of D's in w to the right of or including w_i, and let b_i be the number of U's in w to the right of w_i. Set $f_i = b_i - a_i$. Then*

$$\langle w\hat{0}, x\rangle = e(x)\prod_{i \in S}(r_0 + r_1 + \cdots + r_{f_i}). \tag{13}$$

Example. Let $w = DUDDUUUU$. Then $S = \{1, 3, 4\}$, $a_1 = 3, b_1 = 5, f_1 = 2, a_2 = 2, b_2 = 4, f_2 = 2, a_3 = 1, b_3 = 4, f_3 = 3$, so

$$\langle w\hat{0}, x\rangle = e(x)(r_0 + r_1 + r_2)^2(r_0 + r_1 + r_2 + r_3).$$

Proof of Theorem 2.3. Fix $n \geq 0$. Let $w_{(n)}$ denote the linear transformation w restricted to KP_n. By successive uses of Proposition 2.3(b) we can put $w_{(n)}$ in the form

$$w_{(n)} = \sum_{i,j} c_{ij}(w)U^i D^j, \tag{14}$$

where $c_{ij}(w)$ is a polynomial in $r_0, r_1, \ldots$. (depending on n), and where if $c_{ij} \neq 0$ then $i - j = \rho(x)$. Moreover, this representation is easily seen to be unique. Now

$$Uw_{(n)} = \sum_{i,j} c_{ij}(w)U^{i+1}D^j$$
$$\Rightarrow c_{ij}(Uw) = c_{i-1,j}(w).$$

Moreover, by (9) we have

$$\begin{aligned}
Dw_{(n)} &= \sum_{i,j} c_{ij}(w)DU^i_{n-j}D^j_n \\
&= \sum_{i,j} c_{ij}(w)(U^iD + (r_{n-j} + r_{n-j+1} + \cdots + r_{n-j+i-1})U^{i-1})D^j \\
&= \sum_{i,j} c_{ij}(w)U^iD^{j+1} \\
&\quad + \sum c_{ij}(w)(r_{n-j} + \cdots + r_{n-j+i-1})U^{i-1}D^j.
\end{aligned}$$

It follows that

$$c_{ij}(Dw) = c_{i,j-1}(w) + c_{i+1,j}(w)(r_{n-j} + \cdots + r_{n-j+i}).$$

In particular, when $j = n = 0$ we have

$$c_{i0}(Uw) = c_{i-1,0}(w) \tag{15}$$

$$c_{i0}(Dw) = c_{i+1,0}(w)(r_0 + \cdots + r_i). \tag{16}$$

Now put $n = 0$ in (14) and operate on $\hat{0}$. Since $D^j\hat{0} = 0$ for $j > 0$, we get (setting $\rho = \rho(x)$)

$$w\hat{0} = c_{\rho 0}(w)U^\rho\hat{0}.$$

Thus

$$\langle w\hat{0}, x\rangle = c_{\rho 0}(w)e(x).$$

It is easy to see from (15) and (16) that

$$c_{\rho 0}(w) = \prod_{i \in S}(r_0 + r_1 + \cdots + r_{f_i}),$$

so the proof follows. □

The previous theorem generalizes [Sta$_3$, Thm. 3.7]. When we put $w = D^nU^n$ in Theorem 2.3 (so $x = \hat{0}$), we obtain the following generalization of [Sta$_3$, Cor. 3.9]:

2.4 COROLLARY. *Let P be an* **r**-*differential poset. Then*

$$\alpha(0 \to n \to 0) = \sum_{x \in P_n} e(x)^2$$
$$= \prod_{i=0}^{n-1} (r_0 + r_1 + \cdots + r_i). \qquad \square$$

As a variation of Theorem 2.3, let us replace the word w by $(D+U)^n$. Thus $\langle (D+U)^n \hat{0}, x\rangle$ is the number of Hasse walks $\hat{0} = x_0, x_1, \ldots, x_n = x$ of length n from $\hat{0}$ to x.

2.5 THEOREM. *Let P be an* **r**-*differential poset, and let $x \in P_i$. Then*

$$\langle (D+U)^n \hat{0}, x\rangle = e(x) \sum_w \prod_s r_{\gamma(w,s)+\delta(w,s)} \tag{17}$$

where (a) $w = w_1 w_2 \cdots w_n$ ranges over all involutions in S_n with exactly i fixed points, (b) s ranges over the excedance set $\{s : w_s > s\}$, (c) $\gamma(w,s)$ is the number of fixed points $t = w_t$ such that $s < t < w_s$, and (d) $\delta(w,s) = \#\{t : s < t < w_s < w_t\}$.

Example. Let $n = 4$ and $i = 0$. For each involution $w \in S_4$ with no fixed points, let a and b denote the two indices s for which $w_s > s$. Then we have:

w	a	$\gamma(w,a)$	$\delta(w,a)$	b	$\gamma(w,b)$	$\delta(w,b)$
2143	1	0	0	3	0	0
3412	1	0	1	2	0	0
4321	1	0	0	2	0	0

Hence

$$\langle (D+U)^4 \hat{0}, \hat{0}\rangle = 2r_0^2 + r_0 r_1.$$

Proof of Theorem 2.5 (sketch). The proof is analogous to that of Theorem 2.1. Instead of (7), one proves by induction on n that

$$(D+U)_k^n = \sum_{i,j} b_{ij}(n) U^i D^j, \tag{18}$$

where

$$b_{ij}(n) = \sum_w \prod_s r_{k+\gamma(w,s)+\delta(w,s)-\epsilon(w,s)} \tag{19}$$

where (a) $w = w_1 w_2 \cdots w_n$ ranges over all involutions in S_n with i uncircled fixed points and j circled fixed points, (b) s ranges over the set $\{s : w_s > s\}$ (the number of such s is $\frac{1}{2}(n-i-j)$), (c) $\gamma(w,s)$ is the number of *uncircled* fixed points $t = w_t$ such that $s < t < w_s$ (d) $\delta(s) = \#\{t : s < t < w_s < w_t\}$, and (e) $\epsilon(s)$ is the number of *circled* fixed points $t = w_t$ such that $t < w_s$. The proof of (17) then follows by applying (18) to $\hat{0}$ (so $k = 0$) and taking the coefficient of x. □

When all $r_i = r$ (i.e, P is r-differential) then (17) simplifies considerably. For each w, the product over s is just $r^{\frac{1}{2}(n-i)}$, so that

$$\begin{aligned}\langle (D+U)^n \hat{0}, x\rangle &= e(x) r^{\frac{1}{2}(n-i)} I_i(n)\\ &= e(x) r^{\frac{1}{2}(n-i)} \binom{n}{i} (1\cdot 3\cdot 5 \cdots (n-i-1)),\end{aligned}$$

where $I_i(n)$ is the number of involutions in S_n with i fixed points. This is essentially the result appearing after Proposition 3.17 in [Sta$_3$].

In the special case when P is the infinite chain $0 < 1 < 2 < \cdots$ (so $\mathbf{r} = (1,0,0,\cdots)$), $\langle (D+U)^{2n}\hat{0}, \hat{0}\rangle$ is well-known to equal the Catalan number $C_n = \frac{1}{n+1}\binom{2n}{n}$. On the other hand, (17) yields that $\langle (D+U)^{2n}\hat{0},\hat{0}\rangle$ is the number of fixed-point free involutions $w = w_1 w_2 \cdots w_{2n}$ in S_{2n} such that we never have $i < j < w_i < w_j$. This is another well-known combinatorial interpretation of the Catalan number C_n.

A second major class of results in [Sta$_3$] dealt with the evaluation of eigenvalues and eigenvectors of certain linear transformations associated with differential posets. Analogous results hold for sequentially differential posets, and moreover most of the proofs are exactly the same. We therefore will simply state most results without proof. As in [Sta$_3$], Ch A or Ch(A,λ) denotes the characteristic polynomial $\det(\lambda I - A)$ (normalized to be monic) of the linear transformation $A : V \to V$ on a finite-dimensional vector space V. Moreover, we write $p_j = \#P_j$ and $\Delta p_j = p_j - p_{j-1}$.

2.6 THEOREM (see [Sta$_3$, Thm. 4.1]). *Let P be an $\mathbf{r}$-differential poset, and let $j \geq 0$. Then*

$$\text{(20)}\qquad \mathrm{Ch}(UD_j) = \prod_{i=0}^{j} (\lambda - (r_i + r_{i+1} + \cdots + r_{j-1}))^{\Delta p_i}. \qquad \square$$

2.7 COROLLARY (see [Sta$_3$, Cor. 4.2-4.4]). *Let $j \geq 0$. Suppose that for all $0 \leq i \leq j$, we have $r_i + r_{i+1} + \cdots + r_j \neq 0$. Then U_j is injective and D_{j+1} is surjective. Hence $p_j \leq p_{j+1}$, and there is an order-matching $\mu : P_j \to P_{j+1}$ (i.e., μ is injective and $\mu(x) > x$ for all $x \in P_j$).* □

We will omit here the (easy) extension to sequentially differential posets of the discussion of balanced endomorphisms in [Sta$_3$]. However, there is a special case which is worthy of mention here.

2.8 LEMMA (see [Sta$_3$, Prop. 4.7]). *Let P be an $\mathbf{r}$-differential poset. Then for all $j \geq n \geq 0$,*

$$U^n D_j^n = \prod_{i=1}^{n} (UD_j - r_{j-i+1} - r_{j-i+2} - \cdots - r_{j-1}).$$

Proof (sketch). First show by induction on n that

$$U^n D_{j-n+1} = (UD_j - r_{j-n+1} - \cdots - r_{j-1})U^{n-1}_{j-n+1}.$$

Then use the formula

$$U^n D^n_j = (U^n D_{j-n+1})D^{n-1}_j$$

to deduce the lemma by induction on n. □

2.9 PROPOSITION (see [Sta$_3$, Ex. 4.11]). *Preserve the conditions of Lemma 2.8. Then*

$$\text{(21)} \qquad \mathrm{Ch}(U^n D^n_j) = \lambda^{p_j - p_{j-n}} \prod_{i=0}^{j-n} (\lambda - \prod_{k=1}^{n} (r_i + r_{i+1} + \cdots + r_{j-k}))^{\Delta p_i}$$

$$\text{(22)} \qquad = \lambda^{p_j - p_{j-n}} \mathrm{Ch}(D^n U^n_{j-n}).$$

Proof. It follows from Theorem 2.5 and Proposition 2.7 (as in the proof of [Sta$_3$, Prop 4.12]) that
(23)

$$\mathrm{Ch}(U^n D^n_j) = \prod_{i=0}^{j} (\lambda - \prod_{k=1}^{n} (r_i + r_{i+1} + \cdots + r_{j-1} - r_{j-k+1} - r_{j-k+2} - \cdots - r_{j-1}))^{\Delta p_i}.$$

If $j - n < i \leq j$ and $k = j - i + 1$, then $r_i + \cdots + r_{j-1} - r_{j-k+1} - \cdots - r_{j-1} = 0$. Hence the factors in (22) for $i > j - n$ contribute $\lambda^{\Delta p_{j-n+1} + \cdots + \Delta p_j} = \lambda^{p_j - p_{j-n}}$. If $1 \leq i \leq j - n$ then $r_i + \cdots + r_{j-1} - r_{j-k+1} - \cdots - r_{j-1} = r_i + r_{i+1} + \cdots + r_{j-k}$, so the formula (21) for $\mathrm{Ch}(U^n D^n_j)$ follows. To obtain (22), use the fact (mentioned in the proof of [Sta$_3$, Thm. 4.1]) that if $A : V \to W$ and $B : W \to V$ are linear transformations on finite dimensional vector spaces V and W of dimensions v and w, respectively, then

$$\mathrm{Ch}(BA) = \lambda^{v-w} \mathrm{Ch}(AB). \quad □$$

Note. It may happen that $\Delta p_i < 0$ in (20) and (21), so the corresponding factor of $\mathrm{Ch}(UD_j)$ or $\mathrm{Ch}(U^n D^n_j)$ actually appears in the denominator. Hence (since $\mathrm{Ch}(A)$ is always a polynomial) it must be cancelled by some factor in the numerator. This puts constraints on the possible values of $\mathbf{r}$ and $F(P, q)$ which may be interesting to investigate further.

Recall now from [Sta$_1$] that a finite graded poset P of rank n is *unitary Peck* if for $0 \leq j \leq [n/2]$ the linear transformation

$$U^{n-2j} : KP_j \to KP_{n-j}$$

is a bijection. In particular, this condition implies that P is rank-symmetric ($p_i = p_{n-i}$) and rank unimodal (which in the presence of rank-symmetry means $p_0 \leq p_1 \leq \cdots \leq p_{[n/2]}$).

2.10 PROPOSITION. *Let P be a finite, rank-symmetric, rank-unimodal* **r**-*differential poset of rank n. Then the following two conditions are equivalent:*

(i) *P is unitary Peck,*

(ii) *if $0 \leq i < [n/2]$, $\Delta p_i > 0$, and $i \leq k \leq n-i-1$, then*

$$r_i + r_{i+1} + \cdots + r_k \neq 0.$$

Proof. Clearly P is unitary Peck if and only if for all $0 \leq j < [n/2]$, the linear transformations $D^{n-2j}U_j^{n-2j}$ and $U^{n-2j}D_{n-j}^{n-2j}$ have no zero eigenvalues. By Corollary 2.9, we have

$$\begin{aligned} \mathrm{Ch}(U^{n-2j}D_{n-j}^{n-2j}) &= \lambda^{p_{n-j}-p_j} \prod_{i=0}^{j} (\lambda - \prod_{k=1}^{n-2j} (r_i + \cdots + r_{n-j-k}))^{\Delta p_i} \\ &= \lambda^{p_{n-j}-p_j} \, \mathrm{Ch}(D^{n-2j}U_j^{n-2j}). \end{aligned}$$

Since P is rank-symmetric, we have $p_{n-j} = p_j$. Since P is rank-unimodal, we have $\Delta p_i \geq 0$ for $0 \leq i \leq j$. Hence the eigenvalues of $U^{n-2j}D_{n-j}^{n-2j}$ and $D^{n-2j}U_j^{n-2j}$ are given by

$$\prod_{k=1}^{n-2j} (r_i + \cdots + r_{n-j-k}), \tag{24}$$

for those i with $0 \leq i \leq j$ and $\Delta p_i > 0$. One easily checks that the non-vanishing of (24) for $0 \leq i \leq j$ and $\Delta p_i > 0$ is equivalent to (ii). □

Note: We do not know whether every finite, rank-symmetric, rank-unimodal **r**-differential poset is unitary Peck.

2.11 COROLLARY. *The boolean algebra B_n and subspace lattice $L_n(q)$ of Example 2.5(b,d) are unitary Peck.*

Proof. Using the values of r_i given by Example 2.5(b,d), it is easy to check that condition (ii) of the previous proposition is satisfied. □

The unitary Peckness of B_n is implicit in [K, p. 317], though it probably goes back much earlier. Simple proofs may be found in [F-H, Lemma 5.1][G-L-L, p. 13]. The unitary Peckness of $L_n(q)$ is equivalent to a result of Kantor [K] (see [Sta$_1$, Thm. 2(d)]). Our proof seems simpler, since it is based on only simple structural properties of $L_n(q)$. (On the other hand, Kantor obtains a related result for *affine* subspaces which does not seem to follow directly from the methods here.)

Our final result of this section is a generalization of [Sta$_3$, Thm. 4.14]. The proof is analogous to that of [Sta$_3$, Thm. 4.14] and will be omitted. As in [Sta$_3$], given a graded poset P let $\mathcal{H}(P_{[i,j]})$ denote the Hasse graph of the rank-selected subposet

$$P_{[i,j]} = \{x \in P : i \leq \rho(x) \leq j\}.$$

Thus the vertices of $\mathcal{H}(P_{[i,j]})$ are the elements of $P_{[i,j]}$, and vertices x and y are joined by an (undirected) edge if x covers y or y covers x. Denote by $\mathrm{Ch}\,\mathcal{H}(P_{[i,j]})$ the characteristic polynomial (normalized to be monic) of the adjacency matrix of $\mathcal{H}(P_{[i,j]})$.

Regarding $\mathbf{r} = (r_0, r_1, \dots)$ as fixed, define for $1 \le a \le b-1$ and $s \ge 0$ the $(b-a+2) \times (b-a+2)$ tridiagonal matrix

$$M_{ab}^{(s)} = \begin{bmatrix} 0 & R(s,s+a-1) & 0 & 0 & & \\ 1 & 0 & R(s,s+a) & 0 & & \\ & 1 & 0 & \ddots & & \\ & & & \ddots & & \\ & & \ddots & & & \\ & & & & 0 & R(s,s+b-1) \\ & & & & 1 & 0 \end{bmatrix},$$

where

$$R(s,s+c) = r_s + r_{s+1} + \cdots + r_{s+c}.$$

Finally set

$$C_{ab}^{(s)} = \mathrm{Ch}\, M_{ab}^{(s)}.$$

2.12 THEOREM. *Let P be an $\mathbf{r}$-differential poset, and let $0 \le i \le j$. Then*

$$\mathrm{Ch}\,\mathcal{H}(P_{[i,j]}) = \prod_{s=0}^{j-i} (C_{1s}^{(j-s)})^{\Delta p_{j-s}} \cdot \prod_{s=j-i+1}^{j} (C_{s-j+i+1,s}^{(j-s)})^{\Delta p_{j-s}}. \quad \square$$

For $j - i \le 2$ Theorem 2.12 leads to the formulas

$$\mathrm{Ch}\,\mathcal{H}(P_{[j,j]}) = \lambda^{p_j}$$

$$\mathrm{Ch}\,\mathcal{H}(P_{[j-1,j]}) = \lambda^{\Delta p_j} \prod_{s=1}^{j} (\lambda^2 - (r_{j-s} + r_{j-s+1} + \cdots + r_{j-1}))^{\Delta p_{j-s}}$$

$$\mathrm{Ch}\,\mathcal{H}(P_{[j-2,j]}) = \lambda^{\Delta p_j} (\lambda^2 - r_{j-1})^{\Delta p_{j-1}} \cdot \prod_{s=2}^{j} (\lambda^3 - (2r_{j-s} + 2r_{j-s+1} + \cdots + 2r_{j-2} + r_{j-1})\lambda)^{\Delta p_{j-s}}.$$

3. Shifted partitions. In the previous section we generalized the definition of differential poset by modifying the formula $DU - UD = rI$. We could also ask for generalizations in which the definitions of U and D themselves are modified. There are now a vast number of possibilities, and if we wish to preserve interesting enumerative results then U and D cannot be too different from their original definition. Here we will consider only a single example which naturally arises in the theory of symmetric functions and tableaux. In the next section we will consider more significant alterations in the definitions of U and D, for which all enumerative results are lost but for which we can still deduce some structural properties of the poset P in the spirit of Proposition 2.10.

A *strict partition* λ of n, denoted $\lambda \models n$, is an integer sequence $\lambda = (\lambda_1, \lambda_2, \dots)$ satisfying $\lambda_1 > \lambda_2 > \cdots > \lambda_\ell > \lambda_{\ell+1} = \lambda_{\ell+2} = \cdots = 0$ and $\sum \lambda_i = n$. We also write $\lambda = (\lambda_1, \lambda_2, \dots, \lambda_\ell)$. We call the integers $\lambda_i > 0$ the *parts* of λ, and call $\ell = \ell(\lambda)$ the *length* of λ. Define the *shifted Young's lattice* $\tilde{Y}$ to be the sublattice of Young's lattice, as defined in [Sta₂, p. 168] or [Sta₃], consisting of all strict partitions (including the empty partition ϕ of of 0). $\tilde{Y}$ is a locally finite distributive lattice with $\hat{0}$ (and hence graded) with rank-generating function

$$F(\tilde{Y}, q) = \prod_{n \geq 1} (1 + q^n).$$

The rank $\rho(\lambda)$ of $\lambda \in \tilde{Y}$ is just the sum $|\lambda|$ of its parts (the same as in Y). A saturated chain $\phi = \lambda^0 \subset \lambda^1 \subset \cdots \subset \lambda^n = \lambda$ in the interval $[\phi, \lambda]$ of $\tilde{Y}$ is equivalent to a *standard shifted tableau* of shape λ. The number $e(\lambda)$ of each tableaux is often denoted g^λ. For further information concerning these concepts, see e.g. [M, Ex. 8, pp. 134–136][Sa][Ste].

Now define two continuous linear transformations $D, \tilde{U} : \hat{K}\tilde{Y} \to \hat{K}\tilde{Y}$ as follows: D is the same as before, i.e., for $\lambda \in \tilde{Y}$,

$$D\lambda = \sum_{\mu \in C^-(\lambda)} \mu,$$

summed over all μ which λ covers in $\tilde{Y}$. $\tilde{U}$ is given by

$$\tilde{U}\lambda = 2 \sum_{\substack{\mu \in C^+(\lambda) \\ \ell(\mu) = \ell(\lambda)}} \mu + \sum_{\substack{\nu \in C^+(\lambda) \\ \ell(\nu) > \ell(\lambda)}} \nu.$$

Note that if $\nu \in C^+(\lambda)$ with $\ell(\nu) > \ell(\lambda)$ and $\lambda = (\lambda_1, \dots, \lambda_\ell)$, then $\lambda_\ell \geq 2$ and $\nu = (\lambda_1, \dots, \lambda_\ell, 1)$.

3.1 PROPOSITION. *We have $D\tilde{U} - \tilde{U}D = I$.*

Proof. The proof is a straightforward verification and will be omitted. □

Unfortunately there seems to be no analogue of (2) (or Proposition 2.4) for $\tilde{U}$ and D. This means that our previous results on differential posets (or on Young's lattice) have "shifted analogues" only in certain special cases. Before discussing these results, let us first point out the connection with symmetric functions, analogous to the connection between Young's lattice and symmetric functions discussed in various places in [Sta₃]. Let $Q_\lambda(x;t)$ denote the Hall-Littlewood symmetric function indexed by λ [M, Ch. III], and write $Q_\lambda = Q_\lambda(x) = Q_\lambda(x;-1)$. The symmetric functions $Q_\lambda(x)$ were created by Schur [Sc] in connection with his investigation of projective representations of the symmetric group. For further information, see e.g. [M, Ex. 8, pp. 134-136] [Sa][Ste].

Now let $\hat{\Omega}_K$ denote the algebra of symmetric formal power series over K in the variables $x = (x_1, x_2, \dots)$ consisting of infinite linear combinations of all Q_λ where λ is a strict partition. This algebra is generated (as an algebra of formal power series) by the odd power sums $p_1, p_3, \dots$, i.e., $\hat{\Omega}_K = K[[p_1, p_3, \dots]]$. (See [M] for information on symmetric functions needed here.) Define a continuous vector space isomorphism $\sigma : \hat{K}\tilde{Y} \to \hat{\Omega}_K$ by $\sigma(\lambda) = 2^{-|\lambda|}Q_\lambda$ for $\lambda \in \tilde{Y}$. By known results concerning the Q_λ's, the following diagrams commute:

$$\begin{array}{ccc} \hat{K}\tilde{Y} & \xrightarrow{\sigma} & \hat{\Omega}_K \\ \tilde{U}\downarrow & & \downarrow p_1 \\ \hat{K}\tilde{Y} & \xrightarrow{\sigma} & \hat{\Omega}_K \end{array}$$

$$\begin{array}{ccc} \hat{K}Y & \xrightarrow{\sigma} & \hat{\Omega}_K \\ D\downarrow & & \downarrow \frac{\partial}{\partial p_1} \\ \hat{K}Y & \xrightarrow{\sigma} & \hat{\Omega}_K. \end{array}$$

Here p_1 and $\frac{\partial}{\partial p_1}$ have the same meaning as in [Sta$_3$, remark after Thm. 2.5]. Hence our results below on $\tilde{Y}$ can all be interpreted in terms of symmetric functions.

We now consider some enumerative results from [Sta$_3$](or Section 2 of this paper) which depend only on the formula $DU - UD = I$ and therefore carry over to $\tilde{Y}$ with U replaced by $\tilde{U}$. The first is [Sta$_3$, Thm. 3.7] (or Theorem 2.3 here). If w is a word in the letters $\tilde{U}$ and D, then $\langle w\phi, \lambda\rangle$ has the same value as in [Sta$_3$, Thm. 3.7] (or (13) here with each $r_i = 1$). But rather than counting Hasse walks $\phi = \lambda^0, \lambda^1, \dots, \lambda^\ell = \lambda$ with each choice $\lambda^{i-1} < \lambda^i$ or $\lambda^{i-1} > \lambda^i$ specified, now $\langle w\phi, \lambda\rangle$ counts such walks W weighted by a factor $2^{t(W)}$, where $t(W)$ is the number of steps $\lambda^{i-1} < \lambda^i$ for which $\ell(\lambda^{i-1}) = \ell(\lambda^i)$. For instance, suppose $w = D^n\tilde{U}^n$. If $\phi = \lambda^0 < \lambda^1 < \cdots < \lambda^n > \lambda^{n+1} > \cdots > \lambda^{2n} = \phi$ is a Hasse walk W, then $t(w) = |\lambda^n| - \ell(\lambda^n)$. Hence

$$\langle D^n\tilde{U}^n\phi, \phi\rangle = \sum_{\lambda \models n} 2^{n-\ell(\lambda)}(g^\lambda)^2, \tag{25}$$

where $g^\lambda = e(\lambda)$ as discussed above. Thus by Corollary 2.4 or [Sta$_3$, Cor. 3.9], we get

$$\sum_{\lambda \models n} 2^{n-\ell(\lambda)}(g^\lambda)^2 = n!,$$

a well-known formula with many combinatorial and algebraic ramifications.

Additional results from [Sta$_3$]which carry over to $\tilde{Y}$ by replacing U by $\tilde{U}$ are Theorem 3.11, Theorem 3.12, Corollary 3.14, Corollary 3.15, Corollary 3.16, as well as Theorem 2.5 from this paper. Let us state as an illustrative example the shifted analogue of [Sta$_3$, Cor. 3.14].

3.2 PROPOSITION. *Let*

$$\kappa_{2k}(n) = \sum_W 2^{t(W)},$$

summed over all closed Hasse walks $\lambda^0, \lambda^1, \ldots, \lambda^{2k} = \lambda^0$ *of length* $2k$ *in* $\tilde{Y}$ *with* $|\lambda^0| = n$, *where* $t(W)$ *is the number of steps* $\lambda^{i-1} < \lambda^i$ *for which* $\ell(\lambda^{i-1}) = \ell(\lambda^i)$. *Then*

$$\sum_{n \geq 0} \kappa_{2k}(n) q^n = \frac{(2k)!}{2^k k!} \left(\frac{1+q}{1-q}\right)^k \prod_{i \geq 1} (1+q^i). \qquad \square$$

The results in [Sta$_3$, Section 4] concerning characteristic polynomials all carry over to $\tilde{Y}$ with U replaced by $\tilde{U}$, so we will not state them explicitly here. Let us note, however, that the shifted analogue of [Sta$_3$, Cor. 4.2] is the result that D is surjective and $\tilde{U}$ is injective. However, since U is adjoint to D we also get that U is injective. Moreover, the shifted analogue of [Sta$_3$, Thm. 4.14] (or Theorem 2.12 of this paper) evaluates not the characteristic polynomial of the graph $\mathcal{H}(\tilde{Y}_{[i,j]})$, but rather the digraph $\tilde{\mathcal{H}}(\tilde{Y}_{[i,j]})$ whose vertices are the elements of $\tilde{Y}_{[i,j]}$, with one edge from λ to μ if λ covers μ or if μ covers λ and $\ell(\mu) = \ell(\lambda) + 1$, and with two edges from λ to μ if μ covers λ and $\ell(\mu) = \ell(\lambda)$.

It is natural to ask whether there is some modification of U and D for the poset $\tilde{Y}$ such that analogues of *both* (1) and (2) hold. A remarkable result of this nature was found by M. Haiman (private communication) and will now be briefly discussed. Let $\omega = (1+i)/\sqrt{2} = e^{2\pi i/8}$, $\overline{\omega} = (1-i)/\sqrt{2} = e^{-2\pi i/8}$ (where $i^2 = -1$). Define continuous linear transformations $V, E : \hat{K}\tilde{Y} \to \hat{K}\tilde{Y}$ as follows:

$$V\lambda = \sqrt{2} \sum_{\substack{\mu \in C^+(\lambda) \\ \ell(\mu) = \ell(\lambda)}} \mu + \omega \sum_{\substack{\nu \in C^+(\lambda) \\ \ell(\nu) > \ell(\lambda)}} \nu$$

$$E\lambda = \sqrt{2} \sum_{\substack{\mu \in C^-(\lambda) \\ \ell(\mu) = \ell(\lambda)}} \mu + \overline{\omega} \sum_{\substack{\nu \in C^-(\lambda) \\ \ell(\nu) > \ell(\lambda)}} \nu.$$

It is then straightforward to verify the following result.

3.3 PROPOSITION. *We have*

$$EV - VE = I$$
$$E\tilde{\mathbf{Y}} = (V + \overline{\omega})\tilde{\mathbf{Y}}. \qquad \square$$

Reasoning as in [Sta$_3$, Cor. 2.6] leads to such results as

$$e^{(V+E)t} = e^{\frac{1}{2}t^2 + Vt} e^{Et}$$
$$e^{Et} e^{Vt} = e^{t^2 + Vt} e^{Et}$$
$$e^{Et}\tilde{\mathbf{Y}} = e^{\omega t + \frac{1}{2}t^2 + Vt}\tilde{\mathbf{Y}}.$$

From this all the enumerative results in [Sta$_3$] will have shifted analogues involving V and E instead of U and D. To obtain combinatorially meaningful results one must take real and imaginary parts. As an example, consider

$$\beta(0 \to n) := \langle V^n \hat{0}, \tilde{\mathbf{Y}} \rangle \tag{26}$$
$$= \sum_{\lambda \vdash n} \omega^{\ell(\lambda)} 2^{\frac{1}{2}(n-\ell(\lambda))} g^{\lambda}.$$

We get from the techniques of [Sta$_3$] that

$$\sum_{n \geq 0} \beta(0 \to n) \frac{t^n}{n!} = e^{\omega t + \frac{1}{2}t^2}.$$

Thus

$$\sum_{n \geq 0} \mathrm{Re}\ \beta(0 \to n) \frac{t^n}{n!} = e^{\frac{t}{\sqrt{2}} + \frac{1}{2}t^2} \cos \frac{t}{\sqrt{2}} \tag{27}$$
$$\sum_{n \geq 0} \mathrm{Im}\ \beta(0 \to n) \frac{t^n}{n!} = e^{\frac{t}{\sqrt{2}} + \frac{1}{2}t^2} \sin \frac{t}{\sqrt{2}}.$$

Moreover, taking the real part of (26) yields

$$\mathrm{Re}\ \beta(0 \to n) = \sum_{\lambda \vdash n} c_\lambda g^\lambda, \tag{28}$$

where

$$c_\lambda = \begin{cases} 2^{[\frac{1}{2}(n-\ell(\lambda))]}, & \ell(\lambda) \equiv 0, 1, 7 \pmod 8 \\ -2^{[\frac{1}{2}(n-\ell(\lambda))]}, & \ell(\lambda) \equiv 3, 4, 5 \pmod 8 \\ 0, & \ell(\lambda) \equiv 2, 6 \pmod 8). \end{cases} \tag{29}$$

Combining (27), (28) and (29) yields a curious combinatorial result. Similarly we have

$$\mathrm{Im}\ \beta(0 \to n) = \sum_{\lambda \vdash n} d_\lambda g^\lambda,$$

where

$$d_\lambda = \begin{cases} 2^{[\frac{1}{2}(n-\ell(\lambda))]}, & \ell(\lambda) \equiv 1, 2, 3 \pmod 8 \\ -2^{[\frac{1}{2}(n-\ell(\lambda))]}, & \ell(\lambda) \equiv 5, 6, 7 \pmod 8 \\ 0, & \ell(\lambda) \equiv 0, 4 \pmod 8). \end{cases}$$

4. Non-enumerative variations. Our previous variations were close enough to differential posets to retain many enumerative features of them. In this section we will be concerned only with questions of injectivity and surjectivity of certain linear transformations and their applications to structural properties of P; no explicit formulas will be obtained.

Let P be a poset satisfying axioms (S1) and (S2) of Definition 2.1 (or (D1) and (D2) of [Sta$_3$, Def. 1.1]). For each $i \geq 0$, define an axiom E_i as follows:

(E_i) If $x \in P_i$, then x is covered by more elements than x covers, i.e., $\#C^+(x) > \#C^-(x)$.

4.1 THEOREM. *Let P satisfy (S1), (S2), and (E_i) for some i. Let U and D have their usual meanings. Then $DU_i : KP_i \to KP_i$ is a bijection. Hence U_i is injective and D_{i+1} is surjective, so (as in Corollary 2.7) $p_i \leq p_{i+1}$ and there is an order-matching $\mu : P_i \to P_{i+1}$.*

Proof. Given $x \in P$, let

$$d_x = \#C^+(x) - \#C^-(x).$$

The axioms (S1), (S2), and (E_i) imply that

$$(DU - UD)_i x = d_x x,$$

for all $x \in P_i$. Hence

$$DU_i = UD_i + A,$$

where A is a diagonal matrix (with respect to the basis P_i of KP_i) with positive entries d_x. Since U_{i-1} and D_i are adjoints (see [Sta$_3$, Sect. 2]), UD_i is semidefinite. Since A is positive definite, the sum $UD_i + A$ is positive definite and hence invertible. □

We now give an application of Theorem 4.1. Let p be a prime and $k, n \geq 1$. Define $L_{kn}(p)$ to be the lattice of subgroups of the abelian p-group $(\mathbb{Z}/p^k\mathbb{Z})^n$. It has been conjectured (though I cannot recall by whom) that $L_{kn}(p)$ has the Sperner property. (In general, the lattice of subgroups of a finite abelian p-group need not have the Sperner property, e.g., the group $(\mathbb{Z}/p\mathbb{Z}) \oplus (\mathbb{Z}/p^2\mathbb{Z})$.) One can also ask whether $L_{kn}(p)$ has stronger properties, such as being Peck, unitary Peck, or having a symmetric chain decomposition. (It is well-known that $L_{kn}(p)$ is rank-symmetric, and by a recent result of Butler [B] is rank-unimodal.) The case $k = 1$ is well-understood (see e.g. Corollary 2.11 and [G]); here we consider $k = 2$.

4.2 PROPOSITION. *The lattice $L = L_{2n}(p)$ has order-matchings $\mu : L_i \to L_{i+1}$ for $i < n$ and $\mu : L_{i+1} \to L_i$ for $i > n$. Hence L has the Sperner property.*

Proof. Clearly L satisfies (S1), while (S2) follows since L is modular. Now let $G = (\mathbb{Z}/p^2\mathbb{Z})^n$, and let H be a subgroup of G. It is not hard to see (see [M, (4.3), p. 93] for a much stronger result) that

$$H \cong (\mathbb{Z}/p\mathbb{Z})^j \oplus (\mathbb{Z}/p^2\mathbb{Z})^k,$$
$$G/H \cong (\mathbb{Z}/p\mathbb{Z})^j \oplus (\mathbb{Z}/p^2\mathbb{Z})^{n-j-k},$$

where $j + 2k \leq 2n$. Hence in the lattice L, H covers $(p^{j+k} - 1)/(p-1)$ elements and is covered by $(p^{n-k} - 1)/(p-1)$ elements. If $j + 2k < n$ (i.e., the rank $\rho(H)$ of H in L is $< n$), then $j + k < n - k$, so $\#C^+(H) > \#C^-(H)$. Thus (E_i) holds for $i < n$, so by the previous theorem there is an order-matching $\mu : L_i \to L_{i+1}$ for $i < n$. Since L is self-dual [M, (1.5), p. 87] (or by an argument dual to the preceding) we get an order-matching $\mu : L_{i+1} \to L_i$ for $i > n$. It is now a standard argument (viz., the matching μ partitions L into saturated chains all passing through the middle rank L_n) to deduce the Sperner property. □

Perhaps some refinement of the preceding argument can be used to show that $D^{2n-2i}U^{2n-2i} : L_i \to L_{2n-i}$ is a bijection for $0 \le i \le n$, and therefore establish that L is unitary Peck.

We now consider a variation of Theorem 4.1 where U and D are replaced by other operators. If P is a finite poset, then as in Section 2 let $J(P)$ denote the lattice of order ideals of P [Sta$_2$, Ch. 3.4]. $J(P)$ is a graded distributive lattice, with the rank $\rho(Q)$ of an order ideal Q of P given by its cardinality $\#Q$. If $Q \in J(P)$, then define

$$\begin{aligned} M(Q) &= \{x \in P : x \text{ is a maximal element of } Q\}, \\ m(Q) &= \{x \in P : x \text{ is a minimal element of } P - Q\}. \end{aligned}$$

4.3 Proposition. *Fix an integer $0 \le i < \#P$. Suppose there is a function $\phi : P \to \mathbf{R}$ satisfying the following property: for all $Q \in J(P)_i$, we have*

$$\sum_{x \in M(Q)} \phi(x) < \sum_{x \in m(Q)} \phi(x). \tag{30}$$

Then there is an order-matching $\mu : J(P)_i \to J(P)_{i+1}$.

Proof. Define linear transformations $U(\phi), D(\phi) : K{\cdot}J(P) \to K{\cdot}J(P)$ as follows: if $Q \in J(P)$ then

$$\begin{aligned} U(\phi)Q &= \sum_{Q' \in C^+(Q)} \phi(Q'-Q)Q', \\ D(\phi)Q &= \sum_{Q' \in C^-(Q)} \phi(Q-Q')Q'. \end{aligned}$$

Here if $Q' - Q = \{x\}$ then $\phi(Q'-Q) := \phi(x)$, and similarly for $\phi(Q-Q')$.

It is not difficult to check that distributivity of $J(P)$ insures that

$$(D(\phi)U(\phi) - U(\phi)D(\phi))Q = \left[\sum_{x \in m(Q)} \phi(x) - \sum_{x \in M(Q)} \phi(x)\right] Q. \tag{31}$$

Thus by (30) the expression in brackets in the right-hand side of (31) is positive for all $Q \in J(P)_i$. It now follows just as in the proof of Theorem 4.1 that $D(\phi)U(\phi)$ is positive definite. Thus $U(\phi)_i$ is injective, so by the usual arguments an order-matching $\mu : J(P)_i \to J(P)_{i+1}$ exists. □

Unfortunately we have been unable thus far to find any interesting applications of Proposition 4.3 that were not previously known.

The results and techniques of this section are closely related to work of Proctor [P$_1$] [P$_2$] [P$_4$]. Given a finite ranked poset P of rank n, call a linear transformation

$Y : KP \to KP$ a *lowering operator* if $Yx \in KP_{i-1}$ when $x \in P_i$. Also call $X : KP \to KP$ an *order-raising operator* if for all $x \in P$, $Xx = \sum c_{xy}y, c_{xy} \in K$, where y ranges over $C^+(x)$. If there exist X, Y as above satisfying for $0 \leq i \leq n$,

$$(XY - YX)_i = (2i - n)I_i, \tag{32}$$

then P is called an *sl(2)-poset*. By methods similar to those here Proctor showed that sl(2)-posets are Peck (and conversely) [P_1, Thm. 1]. In some interesting cases one can choose $X = U$ (though usually not $Y = D$) [P_1, Thm. 3]. This situation combines features of Sections 2 and 3, since (32) is analogous to (3), while U and Y replace $\tilde{U}$ and D in Section 3.

Let us also point out the similarity between our Proposition 4.3 and Proctor's concepts of *edge-labelable* posets [P_1, p. 279] and *vertex-labelable* posets [P_4, p. 105]. Indeed, a poset P is vertex-labelable if there exists a function $\phi : P \to \mathbb{Q}$ such that the equation

$$\sum_{x \in M(Q)} \phi(x) - |Q| = \sum_{y \in m(Q)} \phi(y) - |P - Q|$$

is satisfied for every order ideal $Q \in J(P)$. Hence the quantity in brackets in equation (31) is just $|P| - 2 \cdot |Q|$. From this Proctor deduces that $J(P)$ is actually an sl(2)-poset and hence Peck. Furthermore, Proctor [P_4, Thm. 1] gives an elegant classification of all vertex-labelable posets using Dynkin diagrams. If we regard the function ϕ of Proposition 4.3 as labeling the edge (I, I') of $J(P)$ by $\phi(I' - I)$,then we obtain an exact analogue of Proctor's definition of edge-labeling. One could also, as in Proctor, extend the concept to uniquely modular posets.

Finally let us mention that Proctor [P_3] contains results closely related to Proposition 2.9 for some particular sl(2)-posets.

REFERENCES

[B] L. Butler, *A unimodality result in the enumeration of subgroups of a finite abelian group*, Proc. Amer. Math. Soc., 101 (1987), pp. 771–775.

[F-H] W. Foody and A. Hedayat, *On theory and applications of BIB designs with repeated blocks*, Annals Statist., 5 (1977), pp. 932–945.

[G] J. Griggs, *Sufficient conditions for a symmetric chain order*, SIAM J. Appl. Math, 32 (1977), pp. 807–809.

[G-L-L] R.L. Graham, S.-Y. R. Li, and W.-C. W. Li, *On the structure of t-designs*, SIAM J. Alg. Disc. Meth., 1 (1980), pp. 8–14.

[K] W.M. Kantor, *On incidence matrices of finite projective and affine spaces*, Math. Z., 124 (1972), pp. 315–318.

[M] I.G. Macdonald, *Symmetric Functions and Hall Polynomials*, Oxford University Press, Oxford, 1979.

[P_1] R.A. Proctor, *Representations of $sl(2, \mathbb{C})$ on posets and the Sperner property*, SIAM J. Alg. Disc. Meth., 3 (1982), pp. 275–280.

[P_2] R.A. Proctor, *Solution of two difficult combinatorial problems with linear algebra*, Amer. Math. Monthly, 89 (1982), pp. 721–734.

[P_3] R.A. Proctor, *Product evaluations of Lefschetz determinants for Grassmannians and of determinants of multinomial coefficients*, preprint dated April, 1988.

[P_4] R.A. Proctor, *A Dynkin diagram classification theorem arising from a combinatorial problem*, Advances in Math., 62 (1986), pp. 103–117.

[Sa] B.E. Sagan, *Shifted tableaux, Schur q-functions, and a conjecture of R. Stanley*, J. Combinatorial Theory (A), 45 (1987), pp. 62–103.

[Sc] I. SCHUR, *Über die Darstellung der symmetrischen und der alternienden Gruppe durch gebrochene lineare Substitutionen*, J. reine angew. Math., 139 (1911), pp. 155–250.

[Sta_1] R. STANLEY, *Quotients of Peck posets Order*, 1, 1984, pp. 29–34.

[Sta_2] R. STANLEY, *Enumerative Combinatorics*, vol. 1, Wadsworth & Brooks/Cole, Monterey, CA, 1986.

[Sta_3] R. STANLEY, *Differential posets*, J. Amer. Math. Soc., 1 (1988), pp. 919–961.

[Ste] J.R. STEMBRIDGE, *Shifted tableaux and the projective representations of symmetric groups*, preprint.

IDEMPOTENTS
FOR
THE FREE LIE ALGEBRA AND q-ENUMERATION

F. BERGERON*, N. BERGERON† & A.M. GARSIA‡

Abstract. The nth homogeneous component of the free Lie algebra over an alphabet A may be expressed as the linear span of polynomials obtained by multiplying words of length n by a fixed idempotent of the group algebra of S_n. We construct here a number of such idempotents and prove some of their properties by means of identities obtained from the theory of P-partitions. We also construct an idempotent whose coefficients give the Taylor expansion of the reciprocal of the cyclotomic polynomial. As a by-product we obtain a new proof of the dimension formula for the free Lie algebra. The paper ends with the proof of a conjecture of R. Stanley concerning certain representations of S_n introduced by Reutenauer (Springer L.N. # 1234, pp. 267–293).

Introduction. Let $A = \{a_1, a_2, \cdots, a_N\}$ be an alphabet with N letters and let A^* and A^n, as customary, denote the collections of all A-words and respectively all A-words with n letters. The (non-commutative) algebra of all polynomials

$$f = \sum_{w \in A^*} f_w w$$

with rational coefficients will be denoted here by $Q[A^*]$. Multiplication in $Q[A^*]$ is carried out by means of the familiar concatenation product of words in A^*. The subspace of $Q[A^*]$ spanned by the words of length n will be denoted by $Q[A^n]$. The *bracket* of two polynomials f, g is defined by setting

$$[f, g] = fg - gf.$$

We shall deal here with vector subspaces of $Q[A^*]$ spanned by families of polynomials which are obtained by successive bracketings of elements of $Q[A^*]$. To be precise, given a complete binary tree T with n terminal nodes and a word $w = b_1 b_2 \cdots b_n$ of A^n, we shall let the pair (w, T) represent the configuration obtained by appending the letters of w, successively and from left to right, upon the leaves of T. Let T have T_1 and T_2 as left and right subtrees and let w_1 and w_2 be the portions of w respectively hanging from T_1 and T_2 in (w, T). This given, we recursively set

$$\text{(I.1)} \qquad b(w, T) = [b(w_1, T_1), b(w_2, T_2)],$$

with the agreement that when T consists of a single node (that is if $n = 1$ and w is a letter of A) we must set

$$\text{(I.2)} \qquad b(w, T) = w.$$

---

*Dép. Maths et Info, Université Du Québec À Montrèal, C.P. 8888, Succ. A, Montréal, H3C 3P8 Canada

†Department of Mathematics, University of California, San Diego, La Jolla, CA 92093

‡Department of Mathematics, University of California, San Diego, La Jolla, CA 92093

The subspace of $Q[A^*]$ spanned by the polynomials $b(w,T)$ will be denoted here by LIE$[A]$ and referred to as the *Free Lie Algebra* on the Alphabet A. The subspace of LIE$[A]$ spanned by the $b(w,t)$ with w of length n will be denoted by LIE$[A^n]$. We shall also denote by

$$\text{LIE}[a_1 a_2 \cdots a_n]$$

the subspace of LIE$[A]$ spanned by the polynomials $b(w,T)$ when w is restricted to vary among words which are permutations of the letters

$$a_1 a_2 \cdots a_n.$$

We shall make use here of the action of the symmetric group S_n on words of A^n. More precisely, if

$$\sigma = \begin{bmatrix} 1 & 2 & \cdots & n \\ \sigma_1 & \sigma_2 & \cdots & \sigma_n \end{bmatrix} \in S_n$$

and

$$w = b_1 b_2 \cdots b_n \in A^n$$

then we let

$$w\sigma = b_{\sigma_1} b_{\sigma_2} \cdots b_{\sigma_n}.$$

It is convenient to identify LIE$[a_1 a_2 \cdots a_n]$ with a subspace of the group algebra $A(S_n)$ by identifying the word

$$a_{\sigma_1} a_{\sigma_2} \cdots a_{\sigma_n}$$

with the permutation

$$\sigma_1 \sigma_2 \cdots \sigma_n.$$

In this vein, as we shall see, we may represent the subspace LIE$[a_1 a_2 \cdots a_n]$ as a left ideal in $A(S_n)$. More precisely LIE$[a_1 a_2 \cdots a_n]$ may be written in the form

$$\text{(I.3)} \qquad \text{LIE}[a_1 a_2 \cdots a_n] = A(S_n)\rho,$$

with ρ an idempotent element of $A(S_n)$. Here and after, such an element will be referred to as a *Lie idempotent.*

In this paper we shall present a number of Lie idempotents with remarkable combinatorial properties. It develops that the free Lie Algebra is closely related to the q-enumeration of permutations by the major index statistic. Surprisingly, a study of this statistic leads to a new family of idempotents which involve in a curious way the coefficients of the cyclotomic polynomials. This also leads to some very interesting combinatorial and representation theoretical questions. One of the by-products of our work is a rather direct calculation of the dimension of the finely homogeneous components of the free Lie algebra.

Reutenauer in [8] studied certain idempotents $\rho_n^{(k)}$, which induce representations of S_n with dimension given by the Stirling numbers $s(n,k)$. Since these representations are closely related to the Free Lie Algebra, it is natural, in the present context, to ask for a combinatorial explanation of their dimension. At the end of this paper we give a proof of a conjecture of R. Stanley suggesting a certain method of constructing them which does throw some light into this question.

There is extensive literature on the free Lie algebra, written with many different outlooks. We adopt here the combinatorial point of view. Such an approach to the free Lie algebra, including all the background material that will be needed here, may be found in [4].

1. The standard bracketing idempotents. We shall denote here by L_n an R_n the standard totally *leftist* and *rightist* trees on n leaves. In pictures

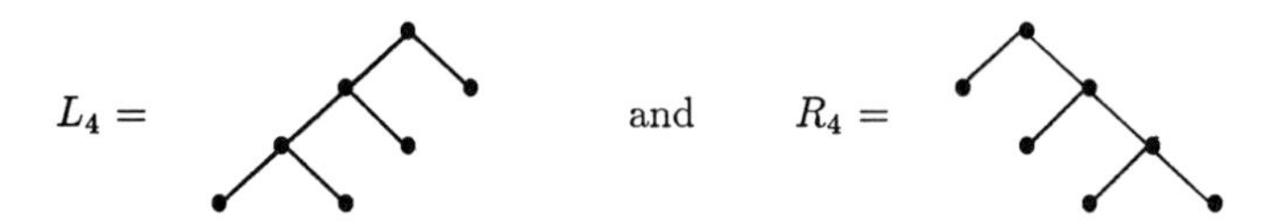

We see that if $w = b_1 b_2 b_3 b_4$ then

$$b(w, L_4) = [[[b_1, b_2], b_3], b_4] \quad \text{and} \quad b(w, R_4) = [b_1, [b_2, [b_3, b_4]]].$$

We shall refer to $b(w, L_n)$ and $b(w, R_n)$ as the standard *left* and *right* bracketings of words of length n.

By a successive application of the Jacobi identity in the form

(A B C) = (C A B) + (A B C)

we can derive (by an induction argument that every polynomial $b(w,T)$ may be expressed as a linear combination of standard left bracketings of words of A^*. A similar result holds for the standard right bracketings. Here and after we shall mostly be concerned with standard left bracketings.

It is not difficult to see that there is an element θ_n of the group algebra of S_n such that the standard left bracketing of a word $w = b_1 b_2 \cdots b_n$ is given by the expression

$$b(w, L_n) = \begin{bmatrix} 1 & 2 & \cdots & n \\ b_1 & b_2 & \cdots & b_n \end{bmatrix} \theta_n \tag{1.1}$$

Reutenauer in [8] points out that 1.1 actually holds with

$$\theta_n = (1 - \gamma_2)(1 - \gamma_3) \cdots (1 - \gamma_n)$$

where γ_i (for $i = 2, \ldots, n$) denotes the *backwards* i-cycle that is

$$\gamma_i = (i, i-1, \ldots, 2, 1).$$

To see this, let us assume by induction that the result is true for $n-1$ (it is trivially true for $n=2$). We then have

$$b(a_1a_2,\dots a_n, L_n) = b(a_1a_2\dots a_{n-1}, L_{n-1})a_n - a_n b(a_1a_2\dots a_{n-1}, L_{n-1})$$

and this is equivalent to

$$\theta_n = \sum_{\sigma\in S_{n-1}} \theta_{n-1}(\sigma)\left(\begin{bmatrix} 1 & 2 & 3 & \dots & n-1 & n \\ \sigma_1 & \sigma_2 & \sigma_3 & \dots & \sigma_{n-1} & n \end{bmatrix} - \begin{bmatrix} 1 & 2 & 3 & \dots & n \\ n & \sigma_1 & \sigma_2 & \dots & \sigma_{n-1} \end{bmatrix}\right).$$

From which we easily derive that

$$\theta_n = \theta_{n-1}(1-\gamma_n)$$

which completes the induction.

Let us recall that the *descent set* of a permutation σ is the set

$$D(\sigma) = \{i : \sigma_i \geq \sigma_{i+1}\}.$$

If S is a subset of $[1,2,\dots,n]$ let us denote by $D_{=S}$ the formal sum of all the permutations of S_n whose descents are *exactly* the elements of S. Similarly, we let $D_{\subseteq S}$ denote the sum of the permutations whose descents are *contained* in S. In other words we have

$$D_{\subseteq S} = \sum_{T\subseteq S} D_{=T} \tag{1.2}$$

and, by inclusion exclusion

$$D_{=S} = \sum_{T\subseteq S} (-1)^{|S-T|} D_{\subseteq T}. \tag{1.3}$$

Now it develops that we have the following very useful identity

THEOREM 1.1.

$$\theta_n = \sum_{k=0}^{n-1} (-1)^k D_{=[1,k]}, \tag{1.4}$$

where here $[a,b]$ is a shorthand for the interval of integers $\{a, a+1, \dots, b\}$.

Proof. We shall only outline the proof here, details may be found in [4]. We clearly have

$$\theta_n = \sum_{S\subseteq[2,n]} (-1)^{|S|} (\prod_{i\in S} \gamma_i) \tag{1.5}$$

On the other hand, it is easily seen that if $S = \{i_1 < i_2 < \cdots < i_k\}$ and $[1,n] - S = \{1 < j_2 < j_3 < \cdots < j_h\}$ then

$$(1.6)\qquad (\prod_{i\in S} \gamma_i) = \begin{bmatrix} 1 & 2 & \dots & k & k+1 & k+2 & \dots & n-1 & n \\ i_k & i_{k-1} & \dots & i_1 & 1 & j_2 & \dots & j_{h-1} & j_h \end{bmatrix}.$$

This given, grouping the terms in 1.5 with $|S| = k$, we get that

$$\sum_{|S|=k} \prod_{i\in S} \gamma_i = D_{=[1,k]},$$

which yields 1.4. □

Formula 1.3 can now be substituted in 1.4 and obtain

$$(1.7)\qquad \theta_n = \sum_{k=0}^{n-1} \sum_{T\subseteq[1,k]} (-1)^{|T|} D_{\subseteq T}.$$

This expression can be further transformed. To do this we need some notation.

If u and v are words of lengths k and $n-k$ respectively, let

$$u \sqcup\!\sqcup v$$

denote the formal sum of all the words of length n obtained by shuffling in all possible ways the letters of u with those of v. For instance

$$12 \sqcup\!\sqcup 43 = 1243 + 1423 + 1432 + 4123 + 4132 + 4312.$$

If f is an element of the group algebra of S_n it is convenient to set

$$\downarrow f = \sum_{\sigma\in S_n} f(\sigma)\sigma^{-1}.$$

Note then that for $|T| = h$ we may write

$$(1.8)\qquad \downarrow D_{\subseteq T} = E_1 \sqcup\!\sqcup E_2 \sqcup\!\sqcup \cdots \sqcup\!\sqcup E_{h+1}$$

where E_i for $i = 1, 2, \dots, h+1$ are words obtained by cutting the word

$$12356\cdots n$$

precisely at the positions corresponding to the elements of T. For example, say $n = 9$ and $T = \{2,4,7\}$. Then

$$E_1 = 12,\ E_2 = 34,\ E_3 = 567,\ E_4 = 89$$

and

$$\downarrow D_{\subseteq\{2,4,7\}} = 12 \sqcup\!\sqcup 34 \sqcup\!\sqcup 567 \sqcup\!\sqcup 89.$$

Let us now introduce a scalar product on $Q[A^*]$ by setting for each pair of words $u, v \in A^*$

$$(1.9)\qquad \langle u, v\rangle = \chi(u = v) = \begin{cases} 1 & \text{if } u = v \\ 0 & \text{otherwise.} \end{cases}$$

Let

$$H_n(u \sqcup\!\sqcup v)$$

be the subspace of $Q[A^n]$ consisting of polynomials which are linear combination of shuffles of pairs of non empty words.

This given, the following remarkable result holds true

THEOREM 1.2. *The four conditions below are equivalent for any element of $Q(A^n)$:*

(i). $f \in \mathrm{LIE}[A]$,

(ii). f *orthogonal to* $H_n(u \sqcup\!\sqcup v)$ *with respect to the scalar product 1.9,*

(iii). $fD_{=S} = (-1)^{|S|} f$ *for all subsets* $S \subseteq [1, n-1]$,

(iv). $f\theta_n = nf$.

Proof.

The equivalence of (i), (ii) is classical and quite easy to show (see [4], [7]). Since (iv) → (i) trivially, it suffices to show the implications (ii) → (iii) → (iv).

(ii) → (iii)

Formula 1.8 yields that for any $w \in A^n$ and $T \neq \varnothing$

$$w \downarrow D_{\subseteq T} \in H_n(u \sqcup\!\sqcup v).$$

Thus (ii) implies that

$$\langle f, w \downarrow D_{\subseteq T} \rangle = 0. \tag{1.10}$$

On the other hand, it is easy to see that for any $f, g \in Q(A^n)$ and any element ρ of the group algebra of S_n we have

$$\langle f\rho, g \rangle = \langle f, g \downarrow \rho \rangle.$$

Thus 1.10 may be rewritten as

$$\langle fD_{\subseteq T}, w \rangle = 0$$

Since w is arbitrary we must necessarily have that

$$fD_{\subseteq T} = 0$$

and formula 1.3 gives

$$fD_{=S} = \sum_{T \subseteq S} (-1)^{|S-T|} fD_{\subseteq T} = (-1)^{|S|} f$$

as desired.

(iii) → (iv)

From 1.4 we get that

$$f\theta_n = \sum_{k=0}^{n-1} (-1)^k fD_{=[1,k]}.$$

Thus (iii) for $T = [1, k]$ gives

$$f\theta_n = \sum_{k=0}^{n-1} (-1)^k f\ (-1)^k = nf,$$

as asserted. □

Remark 1.1. The implication (i) → (iv) of Theorem 1.2 yields in particular that for any word $w \in A^n$

$$w\theta_n\theta_n = nw\theta_n.$$

Setting $w = a_1 a_2 \cdots a_n$ yields the identity

$$\theta_n\theta_n = n\theta_n.$$

In other words

$$\frac{1}{n}(1 - \gamma_2)(1 - \gamma_3) \cdots (1 - \gamma_n)$$

is an *idempotent* element of the algebra of S_n.

This is a fact that is not obvious at all. A purely combinatorial proof of it, based on the identity 1.4. was found by M. Wachs [11].

It is not difficult to see that an independent proof of Theorem 1.2. follows from the purely combinatorial identity

$$\theta_n D_{=S} = (-1)^{|S|}\ \theta_n \tag{1.11}$$

which itself is a special case of (iii). Thus a simple direct proof of this identity would certainly be of interest here.

2. The Klyachko idempotent. In a little known pioneering paper [6], Klyachko brought himself to introduce, precisely in the Lie Algebra context, a remarkable element of the group algebra of S_n. We shall denote it by $\mathcal{K}_n$, in deference to Klyachko, it may be defined as follows:

$$\mathcal{K}_n = \frac{1}{n} \sum_{\sigma \in S_n} \omega^{\mathrm{maj}(\sigma)} \sigma$$

where $\omega = e^{2\pi i/n}$ is the primitive n^{th} root of unity, and *maj* denotes the major index statistics. That is for a permutation

$$\sigma = \sigma_1 \sigma_2 \cdots \sigma_n, \tag{2.1}$$

we set

$$\mathrm{maj}(\sigma) = \sum_{i=1}^{n-1} i\ \chi(\sigma_i > \sigma_{i+1}).$$

It is convenient to introduce here the same statistic for sets. That is for $S \subseteq \{1, 2, \ldots\}$ we let

$$\mathrm{maj}(S) = \sum_{i \in S} i.$$

This given, we see that $\mathcal{K}_n$ can be rewritten in the form

$$\mathcal{K}_n = \frac{1}{n} \sum_{S \subseteq \{1,2,\ldots,n-1\}} \omega^{\mathrm{maj}(S)} D_{=S}. \tag{2.2}$$

For an element ρ of the group algebra of S_n let

$$Q[A^n]\rho$$

denote the subspace of $Q[A^*]\rho$ spanned by the polynomials $w\rho$ as w varies among all A-words of length n.

Klyachko in [6] shows that

$$\mathrm{LIE}[A^n] = Q[A^n]\mathcal{K}_n. \tag{2.3}$$

Unfortunately, the proof of this result given in [6] is unnecessarily complicated. We shall see here that 2.3 may be easily obtained directly from the identities of section 1 combined with a simple q-enumeration result of [5].

Let us say that two elements ρ_1, ρ_2 of $A(S_n)$ are *right equivalent* if and only if

$$\begin{cases} a) & \rho_1 \rho_2 = \rho_1 \\ b) & \rho_2 \rho_1 = \rho_2 \end{cases} \tag{2.4}$$

Note that these conditions force ρ_1 and ρ_2 to be idempotents. Moreover, we see that a) and b) imply that the corresponding left ideals in $A(S_n)$ are identical. Likewise we also have

$$Q[A^n]\rho_1 = Q[A^n]\rho_2 \tag{2.5}$$

Thus to show the Klyachko result we need only verify that θ_n/n and $\mathcal{K}_n$ are right equivalent. Now, the identity

$$\theta_n \mathcal{K}_n = n\theta_n \tag{2.6}$$

is immediate. In fact, from 1.11 we get

$$\theta_n \mathcal{K}_n = \sum_{S \subseteq [1,n-1]} \omega^{\mathrm{maj}(S)} \theta_n D_{=S} = \sum_{S \subseteq [1,n-1]} \omega^{\mathrm{maj}(S)} (-1)^{|S|} \theta_n$$

which may be rewritten as

$$\theta_n \mathcal{K}_n = (1-\omega)(1-\omega^2)\cdots(1-\omega^{n-1})\theta_n = n\theta_n.$$

The latter equality being a consequence of the fact that ω is a primitive root of unity of order n.

At this point it may be good to observe that this argument actually establishes a stronger result. To state it we need to introduce some notation. Let $\phi_n(q)$ denote the nth cyclotomic polynomial, as usual $\phi(n)$ will denote the Euler ϕ-function. Let also $r_m(q)$ denote the unique polynomial of degree strictly less than $\phi(n)$ which is a solution of

$$q^m \equiv r_m(q) \qquad (\text{mod } \phi_n(q)). \tag{2.7}$$

Finally, let us set

$$\pi_n(q) = \frac{1}{n} \sum_{\sigma \in S_n} r_{\text{maj}(\sigma)}(q)\sigma$$

and

$$\mathcal{K}_n(q) = \sum_{\sigma \in S_n} q^{\text{maj}(\sigma)}\sigma.$$

Thus given, we have

THEOREM 2.1.

$$\theta_n \mathcal{K}_n(q) = (1-q)(1-q^2)\cdots(1-q^{n-1})\theta_n$$

and

$$\theta_n \pi_n(q) = \theta_n. \tag{2.8}$$

Proof. We need only observe that the polynomial

$$(1-q)(1-q^2)\cdots(1-q^{n-1})$$

is equal to n modulo $\phi_n(q)$.

To complete our proof of the Klyachko result we need to verify the identity

$$\mathcal{K}_n \theta_n = n\mathcal{K}_n. \tag{2.9}$$

In view of Theorem 1.1 this is equivalent to the statement that $\mathcal{K}_n \in \text{LIE}[A^n]$. Again from Theorem 1.1 we see that this is equivalent to showing the identity

$$\mathcal{K}_n D_{=S} = (-1)^{|S|}\mathcal{K}_n, \tag{2.10}$$

better yet, using 1.3, we could just show that

$$\mathcal{K}_n D_{\subseteq S} = 0, \qquad (\text{ for all } S \neq \varnothing). \tag{2.11}$$

It should be noted however, that even the special case

$$\mathcal{K}_n D_{=[1,k]} = (-1)^k \mathcal{K}_n, \tag{2.12}$$

of 2.10 is sufficient to yield us 2.9. Indeed, 1.4 gives

$$\mathcal{K}_n\theta_n = \sum_{k=0}^{n-1}(-1)^k \mathcal{K}_n D_{[1,k]}$$

and 2.12 gives

$$\mathcal{K}_n\theta_n = \sum_{k=0}^{n-1}(-1)^k \mathcal{K}_n\,(-1)^k = n\mathcal{K}_n.$$

In summary we can obtain 2.9 either via 2.11 or via 2.12. We shall follow both these paths because each of them leads to interesting combinatorial questions. As a matter of fact, the underlying combinatorics is better understood if we work with $\pi_n(q)$ rather than $\mathcal{K}_n$. In this light we shall obtain 2.11 as a specialization of

$$\pi_n(q)D_{\subseteq S} = 0. \qquad (\text{ for all } S \neq \varnothing) \tag{2.13}$$

Note first that the coefficient of a permutation α in the left-hand side of 2.13, modulo the cyclotomic polynomial $\phi_n(q)$ (except for a factor $1/n$), is given by

$$\mathcal{K}_n(q)D_{\subseteq S}|_\alpha = \sum_{\tau\in D_{\subseteq S}} q^{\mathrm{maj}(\alpha\tau^{-1})} = \sum_{\sigma\in\alpha\downarrow D_{\subseteq S}} q^{\mathrm{maj}(\sigma)}. \tag{2.14}$$

The meaning of the right-hand side of this identity becomes clearer through an example. Let $S = \{3,5\}$ with $n = 8$. Then

$$\downarrow D_{\{3,5\}} = 123 \sqcup\!\sqcup 45 \sqcup\!\sqcup 678,$$

and for

$$\alpha = b_1 b_2 \cdots b_8$$

we get

$$\alpha \downarrow D_{\{3,5\}} = b_1b_2b_3 \sqcup\!\sqcup b_4b_5 \sqcup\!\sqcup b_6b_7b_8.$$

Thus we see that in the general case the right-hand side of 2.14 is

$$\sum_{\sigma\in\alpha^{(1)}\sqcup\!\sqcup\,\alpha^{(2)}\sqcup\!\sqcup\cdots\sqcup\!\sqcup\,\alpha^{(k+1)}} q^{\mathrm{maj}(\sigma)}, \tag{2.15}$$

where $\alpha^{(1)}, \alpha^{(2)}, \cdots, \alpha^{(k+1)}$ are the words obtained by segmenting α according to the elements of S. We have thus been naturally lead to a problem of q-enumeration of a family of perturbations by the major index. In fact, the following formula for 2.15 has already been given in [5]:

$$\sum_{\sigma\in\alpha^{(1)}\sqcup\!\sqcup\cdots\sqcup\!\sqcup\alpha^{(k+1)}} q^{\mathrm{maj}(\sigma)} = q^{\mathrm{maj}(\alpha^{(1)})+\cdots+\mathrm{maj}(\alpha^{(k+1)})} \begin{bmatrix} & & n & \\ p_1 & p_2 & \cdots & p_{k+1} \end{bmatrix}, \tag{2.16}$$

where p_i is the length of the word $\alpha^{(i)}$ and the expression in brackets is the usual q-multinomial coefficient.

But now we immediately see that the right-hand side of 2.16 is equal to zero because all these multinomial coefficients vanish modulo the cyclotomic polynomial $\phi_n(q)$. We have thus established the following fact

THEOREM 2.2.

$$\pi_n(q) D_{=S} = (-1)^{|S|} \pi_n(q), \tag{2.17}$$

and

$$\pi_n(q) \theta_n = n \pi_n(q). \tag{2.18}$$

Proof. The argument given above establishes the equality in 2.17, modulo $\phi_n(q)$. But in fact, exact equality must hold as well, since the coefficients of permutations in both sides of 2.17 are polynomials of degree strictly less than $\phi(n)$. Formula 2.18 follows from 2.17 by Theorem 1.2. □

Remark 2.1. Computer exploration with MAPLE, carried out in connection with this proof, suggested the following beautiful evaluation of 2.16:

$$\sum_{\sigma \in \alpha^{(1)} \sqcup\!\sqcup \cdots \sqcup\!\sqcup \alpha^{(k+1)}} q^{\operatorname{maj}(\sigma)} \equiv_{\bmod q^n - 1} q^{\operatorname{maj}(\sigma)} \begin{bmatrix} & & n & \\ p_1 & p_2 & \dots & p_{k+1} \end{bmatrix}.$$

this identity was later shown to be true by M. Wachs (personal communication).

Theorems 2.1 and 2.2 have surprising consequences. For convenience let us set

$$\pi_n(q) = \delta_0 + \delta_1 q + \cdots + \delta_f q^f,$$

where $f = \phi(n) - 1$.

THEOREM 2.3. *Each of the δ_i is a Lie element. Moreover, we have the identities*

$$\delta_i \delta_j = \begin{cases} \delta_i, & \text{if } j = 0 \\ 0, & \text{otherwise} \end{cases} \tag{2.19}$$

In particular, $\pi_n(q)$ is a Lie idempotent for any value of q. More generally, the expression

$$\delta_0 + c_1 \delta_1 + \cdots + c_f \delta_f, \tag{2.20}$$

always gives a Lie idempotent.

Proof. By equating coefficients of the powers of q in 2.8 we derive that

$$\theta_n \delta_0 = \theta_n \quad \text{and} \quad \theta_n \delta_i = 0 \quad (\text{for } i = 1, 2, \dots, f). \tag{2.21}$$

Doing the same with 2.18 gives

$$\delta_i \theta_n = n \delta_i \qquad (\text{for } i = 0, 1, \dots, f). \tag{2.22}$$

Combining these identities we get

$$\delta_i \delta_0 = \left(\frac{\delta_i \theta_n}{n} \right) \delta_0 = \frac{\delta_i}{n} \theta_n \delta_0 = \frac{\delta_i}{n} \theta_n = \delta_i.$$

This gives the first part of 2.19. A similar calculation using part two of 2.21 yields the second part of 2.19. The idempotency of $\pi_n(q)$ and that of expression 2.20 is an immediate consequence of 2.19. □

For any permutation $\sigma = \sigma_1\sigma_2 \cdots \sigma_n$ let us set as in [6]

$$\mathrm{ind}(\sigma) = \sum_{i=1}^{n} \chi(\sigma_i > \sigma_{i+1}),$$

with the convention that $\sigma_{n+1} = \sigma_1$.

Clearly, this statistic only depends on the circular rearrangement class of σ. We shall refer to it as the *circular descent* index. Let us also denote by γ the n-cycle

$$\gamma = (1, 2, 3, \ldots, n)$$

Now it develops that *maj* and *ind* are related in a surprising way, namely

LEMMA 2.1. *(Klyachko)*
For any $\sigma \in S_n$:

$$(2.23) \qquad \begin{aligned} &(a) \qquad \mathrm{maj}(\sigma\gamma) = \mathrm{maj}(\sigma) - \mathrm{ind}(\sigma) \qquad (\mathrm{mod}\ n) \\ &(b) \qquad \mathrm{maj}(\gamma\sigma) = \mathrm{maj}(\sigma) - 1 \qquad (\mathrm{mod}\ n) \end{aligned}$$

Proof. The proof of these identities can be found in [6] and is quite straightforward. □

For a further study of the elements δ_i we need some properties of the polynomials $r_m(q)$ defined in 2.7.

LEMMA 2.2. *The sequence of polynomials $r_m(q)$ is periodic of period n and their generating function is given by the identity*

$$(2.24) \qquad \sum_{m \geq 0} r_m(q) x^m = \frac{1}{1 - xq} - \frac{\phi_n(q)}{1 - xq} \frac{x^{1+f}}{\phi_n(x)}.$$

Proof. The periodicity is immediate since the trivial congruence

$$q^{m+n} \equiv q^m \qquad (\mathrm{mod}\ \phi_n(q))$$

implies the congruence

$$r_{m+n} \equiv r_m \qquad (\mathrm{mod}\ \phi_n(q))$$

but the latter must reduce to an equality since both polynomials have degree less that the degree of $\phi_n(q)$.

It follows immediately from the definition of the polynomials $r_m(q)$ that there is a constant c such that

$$q r_m(q) = c\phi_n(q) + r_{m+1}(q).$$

Setting $q = 0$ and recalling that the constant term in $\phi_n(q)$ is equal to 1, we get

$$c = -r_{m+1}(0).$$

For convenience let us set

$$c_m = r_m(0).$$

We may thus write

$$r_{m+1}(q) = q r_m(q) + c_{m+1}\phi_n(q). \tag{2.25}$$

Multiplying this identity by x^{m+1} and summing gives

$$\sum_{m\geq 0} r_m(q)x^m = \frac{1}{1-xq} + \frac{\phi_n(q)}{1-xq}\sum_{m\geq 1} c_m x^m.$$

Setting $xq = t$, multiplying by $1-t$ and letting $t \to 1$ yields

$$0 = 1 + \phi_n(q)\sum_{m\geq 1} c_m q^{-m}, \tag{2.26}$$

The passage to the limit is justified for $q > 1$, and the fact that left hand side reduces to zero is an easy consequence of the periodicity of the sequence $\{r_m\}$.

Since the cyclotomic polynomial satisfies the identity.

$$x^{1+f}\phi_n(\frac{1}{x}) = \phi_n(x). \tag{2.27}$$

Formula 2.24 must hold true as asserted. □

THEOREM 2.3.

$$\gamma\pi_n(q) \equiv q\pi_n(q) \qquad (\mathrm{mod}\ \phi_n(q)) \tag{2.28}$$

Proof. Note that we can write

$$\gamma\pi_n(q) = \frac{1}{n}\sum_{\alpha\in S_n} r_{\mathrm{maj}(\gamma^{-1}\alpha)}(q)\,\alpha.$$

Using the Klyachko identity 2.23 b) and the periodicity of the r_m's we obtain

$$\gamma\pi_n(q) = \frac{1}{n}\sum_{\alpha\in S_n} r_{1+\mathrm{maj}(\alpha)}(q)\,\alpha.$$

Substituting 2.25 we finally get

$$\gamma\pi_n(q) = q\pi_n(q) + \frac{1}{n}\phi_n(q)\sum_{\alpha\in S_n} c_{1+\mathrm{maj}(\alpha)}\,\alpha \tag{2.29}$$

which yields 2.28. □

Remark 2.2. Setting $q = 0$ in 2.24 we get

$$\sum_{m\geq 0} r_m(0)x^m = \sum_{m\geq 0} c_m x^m = 1 - \frac{x^{1+f}}{\phi_n(x)}. \tag{2.30}$$

On the other hand the definition of $\pi_n(q)$ gives

$$\delta_0 = \frac{1}{n}\sum_{\sigma\in S_n} c_{\mathrm{maj}(\sigma)}\sigma. \tag{2.31}$$

Thus we see that this remarkable idempotent is intimately related to the *inverse* of the cyclotomic polynomial.

Theorem 2.3 has an interesting corollary. For convenience let us write the nth cyclotomic polynomial in the form

$$\phi_n(q) = q^{f+1} + 1 - a_1 q - a_2 q^2 - \cdots - a_f q^f. \tag{2.32}$$

This given we have

THEOREM 2.4. *For $i = 1, \ldots, f$*

$$\delta_i = \gamma^{-i}(1 - a_1\gamma - a_2\gamma^2 - \cdots - a_i\gamma^i)\delta_0. \tag{2.33}$$

Proof.

Replacing q by $1/q$ in 2.29, multiplying by q^{1+f} and setting $q = 0$ (using 2.27) we get

$$0 = \delta_f + \frac{1}{n}\sum_{\alpha\in S_n} c_{1+\mathrm{maj}(\alpha)}\alpha.$$

So we can rewrite 2.29 as

$$\begin{aligned} \gamma\pi_n(q) &= q\pi_n(q) - \phi_n(q)\delta_f \\ &= q\pi_n(q) - (q^{1+f} + 1 - a_1 q - \cdots - a_f q^f)\delta_f. \end{aligned} \tag{2.34}$$

Equating coefficients of the successive powers of q we easily get the recurrences

$$\begin{aligned} \gamma\delta_0 &= -\delta_f, \\ \gamma\delta_1 &= \delta_0 + a_1\delta_f, \\ \gamma\delta_2 &= \delta_1 + a_2\delta_f, \\ \cdots &= \cdots \\ \gamma\delta_f &= \delta_{f-1} + a_f\delta_f, \end{aligned} \tag{2.35}$$

The ith equation may be multiplied by γ^{i-1} to give

$$\gamma^i\delta_i = \gamma^{i-1}\delta_{i-1} - a_i\gamma^i\delta_0$$

from which our desired expression can be easily derived. □

Another corollary of Theorem 2.3 is the following remarkable fact

THEOREM 2.5. *For each* $i = 0, 1, \ldots, f$

$$\phi(\gamma)\delta_i = 0. \tag{2.36}$$

and moreover

$$\text{LIE}[A^n] = Q[A^n]\delta_i \tag{2.37}$$

That is, not only the idempotent δ_0, *but the nilpotent elements* δ_i *also give* LIE$[A^n]$.

Proof. Equation 2.28 gives that

$$\phi(\gamma)\pi_n(q) \equiv 0 \qquad (\text{mod } \phi_n(q))$$

However, this congruence must in fact be an equality, since the left hand side is of degree at most f in q. This gives 2.36. Now the relation 2.33 may be inverted. Indeed, the polynomial

$$1 - a_1 q - a_2 q^2 - \cdots - a_i q^i$$

is invertible modulo $\phi_n(q)$. Let $P_i(q)$ be its inverse. Multiplying 2.33 by $P_i(\gamma)$, and using 2.36 we get

$$\delta_0 = \gamma^i P_i(\gamma)\delta_i.$$

This gives the inclusion

$$\text{LIE}[A^n] \subseteq Q[A^n]\delta_i.$$

The other inclusion follows in the same manner from 2.33. □

3. q-Enumeration and the dimension of LIE$[A]$. As we shall see, the crucial step in the calculation of the dimension of the free Lie Algebra, is the proof that the Klyachko idempotent is a Lie element. As we pointed out in the last section, to prove the latter, we do not have to establish 2.10 in full, but rather only

$$\mathcal{K}_n D_{=[1,k]} = (-1)^k \mathcal{K}_n \tag{3.1}$$

A direct verification of this identity leads to some interesting q-enumeration problems. Let us work again with $\mathcal{K}_n(q)$ and write the analogue of 2.14 for $D_{=[1,k]}$. Namely

$$\mathcal{K}_n(q) D_{=[1,k]} \mid_\alpha = \sum_{\sigma \in \alpha \downarrow D_{=[1,k]}} q^{\text{maj}(\sigma)}. \tag{3.2}$$

Note that the generic element of $D_{=[1,k]}$ is of the form

$$\tau = i_1 > i_2 > \cdots > i_k > 1 < j_1 < j_2 < \cdots < j_h.$$

thus its inverse consists of $k+1$ followed by a shuffle of the words $k\,k-1 \cdots 2\,1$ and $k+2\,k+3 \cdots n$. This gives

$$\downarrow D_{=[1,k]} = k+1 \bullet (k\,k-1 \cdots 1) \sqcup\!\sqcup (k+2\,k+3 \cdots n) \tag{3.3}$$

where here the symbol $\bullet$ denotes concatenation product. Consequently

$$\alpha \downarrow D_{=[1,k]} = \alpha_{k+1} \bullet (\alpha_k \alpha_{k-1} \cdots \alpha_1) \sqcup\!\sqcup (\alpha_{k+2}\alpha_{k+3} \cdots \alpha_n)$$

Our identity 3.2 reduces then to

$$\mathcal{K}_n(q) D_{=[1,k]} \,|_\alpha = \sum_{\sigma \in \alpha_{k+1} \bullet (\alpha_k \alpha_{k-1} \cdots \alpha_1) \sqcup\!\sqcup (\alpha_{k+2}\alpha_{k+3}\cdots\alpha_n)} q^{\mathrm{maj}(\sigma)} \tag{3.4}$$

We are thus led to a special case of the general problem of q-enumeration by the major index of a set of permutations obtained by reading the labels of a tree. To be precise let T be a tree and α be a labeling of its nodes by the integers $1, 2, \ldots, n$. Let $L(T)$ denote the set of all linear extensions of the natural order of the elements T. Then the general problem is that of calculating the sum

$$\sum_{\sigma \in \alpha L(T)} q^{\mathrm{maj}(\sigma)} \tag{3.5}$$

where $\alpha L(T)$ denotes the collection of permutations described by

$$\alpha(\sigma) = \alpha(\sigma_1)\alpha(\sigma_2) \cdots \alpha(\sigma_n)$$

as σ varies in $L(T)$.

The summation in 3.4 is the special case when T consists of two total orders, of respective lengths $k+1$ and $n-k-1$, joined at their smallest elements. Unfortunately, we do not know of any simple expression, for the sum 3.5 when the tree order is *away from the root* (that is when *up* is as in the natural trees). However, if the order is *towards the root*, (that is if *up* is as in the Computer Science trees) then a *hook formula* is available. Namely, we have

THEOREM 3.1. *(Bjorner-Wachs [1])*

$$\sum_{\sigma \in \alpha L(T)} q^{\mathrm{maj}(\sigma)} = q^{\mathrm{maj}\, \alpha(T)} \frac{[n]!}{\prod_{x \in T} [h_x]}. \tag{3.6}$$

Where, $[n]!$ is the q-analogue of $n!$, h_x denotes the number of nodes in the subtree below x ($[h_x]$ is its q-analogue) and finally,

$$\mathrm{maj}\, \alpha(T) = \sum_{x \in T} h_x \chi(\alpha(x) > \alpha(p(x))) \tag{3.7}$$

here $p(x)$ denotes the parent of x.

Upturning our trees we get two total orders joined by their top elements. We are thus replacing the labelled tree on the left in the picture by the one on the right.

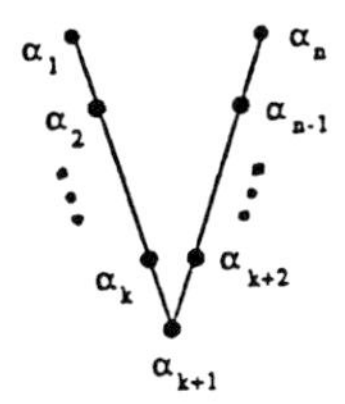

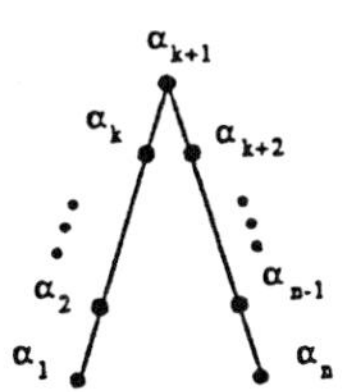

The left tree corresponds to the sum in 3.4 while the right tree corresponds to the sum

$$(3.8) \qquad \sum_{\sigma \in (\alpha_1\alpha_2\cdots\alpha_k) \sqcup\!\sqcup (\alpha_n\alpha_{n-1}\cdots\alpha_{k+2})\bullet\alpha_{k+1}} q^{\mathrm{maj}(\sigma)}.$$

The relationship between the permutation in 3.4 and those in 3.8 is very simple, the former are the reverse of the latter. More precisely, denoting by ω_0 the reversing permutation, that is

$$\omega_0 = \begin{bmatrix} 1 & 2 & \dots & n \\ n & n-1 & \dots & 1 \end{bmatrix}.$$

we can rewrite 3.2 in the form

$$(3.9) \qquad \mathcal{K}_n(q) D_{=[1,k]} \mid_\alpha = \sum_{\sigma \in (\alpha_1\alpha_2\cdots\alpha_k) \sqcup\!\sqcup (\alpha_n\alpha_{n-1}\cdots\alpha_{k+2})\cdot\alpha_{k+1}} q^{\mathrm{maj}(\sigma\omega_0)}.$$

Now, simple manipulations yield the identity

$$(3.10) \qquad \mathrm{maj}(\sigma\omega_0) = \binom{n}{2} - nd(\sigma) + \mathrm{maj}(\sigma),$$

where $d(\sigma)$ denotes the number of descents of σ.

On the other hand, the expression in 3.8, by means of the hook formula 3.6, evaluates to

(3.11)

$$\sum_{\sigma \in (\alpha_1\alpha_2\cdots\alpha_k) \sqcup\!\sqcup (\alpha_n\alpha_{n-1}\cdots\alpha_{k+2})\bullet\alpha_{k+1}} q^{\mathrm{maj}(\sigma)} = q^{\mathrm{maj}(\alpha_1\alpha_2\cdots\alpha_k) + \mathrm{maj}(\alpha_n\alpha_{n-1}\cdots\alpha_{k+1})} \frac{[n]!}{[n][k]![n-k-1]!}$$

Substituting 3.10 and 3.11 in 3.9 we obtain

$$(3.12) \quad \mathcal{K}_n(q) D_{=[1,k]} \mid_\alpha \equiv_{\mathrm{mod}\ q^n - 1} q^{\binom{n}{2} + \mathrm{maj}(\alpha_1\alpha_2\cdots\alpha_k) + \mathrm{maj}(\alpha_n\alpha_{n-1}\cdots\alpha_{k+1})} \begin{bmatrix} n-1 \\ k \end{bmatrix}.$$

Since

$$\mathrm{maj}(\alpha_1\cdots\alpha_n) = \mathrm{maj}(\alpha_1\cdots\alpha_k) + \mathrm{maj}(\alpha_{k+1}\cdots\alpha_n) + kd(\alpha_{k+1}\cdots\alpha_n)$$

and

$$\mathrm{maj}(\alpha_n\alpha_{n-1}\cdots\alpha_{k+1}) = \mathrm{maj}(\alpha_{k+1}\alpha_{k+2}\ldots\alpha_n) + \binom{n-k}{2} - (n-k)d(\alpha_{k+1}\alpha_{k+2}\cdots\alpha_n),$$

we deduce that

$$\binom{n}{2} + \mathrm{maj}(\alpha_1\alpha_2\cdots\alpha_k) + \mathrm{maj}(\alpha_n\alpha_{n-1}\cdots\alpha_{k+1}) \equiv_{\mathrm{mod}\ n} \binom{k+1}{2} + \mathrm{maj}(\alpha_1\alpha_2\cdots\alpha_n).$$

Which reduces 3.12 to our final identity

$$(3.13) \qquad \mathcal{K}_n(q)D_{=[1,k]}\,|_{\alpha} \equiv_{\text{ mod } q^n-1} q^{\binom{k+1}{2}+\text{maj}(\alpha_1\alpha_2\cdots\alpha_n)}\begin{bmatrix} n-1 \\ k \end{bmatrix}.$$

Multiplying both sides by α and summing over S_n yields

$$(3.14) \qquad \mathcal{K}_n(q)D_{=[1,k]}\,|_{\alpha} \equiv_{\text{ mod } q^n-1} q^{\binom{k+1}{2}}\begin{bmatrix} n-1 \\ k \end{bmatrix}\mathcal{K}_n(q).$$

To get 3.1 we are led to evaluate the factor of $\mathcal{K}_n(q)$, on the right side, modulo the cyclotomic polynomial. However, this is easily done since the generating function of these factors (by the q-binomial theorem) may be written as

$$\sum_{k=0}^{n-1}(-t)^k q^{\binom{k+1}{2}}\begin{bmatrix} n-1 \\ k \end{bmatrix} = (1-tq)(1-tq^2)\cdots(1-tq^{n-1}).$$

However, modulo the cyclotomic polynomial the right hand side of this identity reduces to

$$\frac{1-t^n}{1-t} = 1+t+t^2+\cdots+t^{n-1},$$

thus

$$q^{\binom{k+1}{2}}\begin{bmatrix} n-1 \\ k \end{bmatrix} \equiv_{\text{ mod } \phi_n(q)} (-1)^k.$$

this gives not only 3.1 but also

$$\pi_n(q)D_{=[1,k]} = (-1)^k\pi_n(q)$$

Remark 3.1. The identity in 3.14 gives a little bit more. Indeed, multiplying both sides by $(-1)^k$, summing and using the q-binomial theorem we get

$$\mathcal{K}_n(q)\theta_n \equiv_{\text{ mod } q^n-1} (1-q)(1-q^2)\cdots(1-q^{n-1})\mathcal{K}_n(q).$$

which is a stronger identity than 2.18.

Remark 3.2. Curiously 3.13 is actually an equality when α is the identity permutation! For in this case we can write

$$\mathcal{K}_n(q)D_{=[1,k]}\,|_{id} = \sum_{2\leq i_1<i_2<\cdots<i_k\leq n} q^{i_1-1+i_2-1+\cdots+i_k-1}.$$

Now the generating function of these polynomials is

$$\sum_{k=0}^{n-1} t^k\,\mathcal{K}_n(q)D_{=[1,k]}|_{id} = (1+tq)(1+tq^2)\cdots(1+tq^{n-1})$$

and the q-binomial theorem gives

$$\mathcal{K}_n(q)D_{=[1,k]}\,|_{id} = q^{\binom{k+1}{2}}\begin{bmatrix} n-1 \\ k \end{bmatrix}$$

This circumstance makes one wonder if for general α there is a similar exact evaluation of the left hand side of 3.13.

Remark 3.3. We should point out that if we are willing to use Stanley's theory of P-partitions, the hook formula 3.6 has a simpler derivation than that given in [1]. For sake of completeness we give here a sketch of this alternate argument.

The point of departure is Stanley's ω, P-partition formula. We are given a partially ordered set P and a labelling ω of its elements by the integers $1, 2, \ldots, n$. We let then $F_\omega(P)$ be the set of all integer valued functions f which are weakly decreasing with respect to the partial order of P and which decrease strictly when ω decreases in P. Stanley's [10] identity (see [5] for a proof) states that

$$\sum_{f \in F_\omega(P)} q^{|f|} = \frac{\sum_{\sigma \in \omega L(P)} q^{\text{maj}(\sigma)}}{(1-q)(1-q^2)\cdots(1-q^n)} \tag{3.15}$$

again here $L(P)$ denotes the set of linear extensions of P and $\omega L(P)$ denotes the set of permutations obtained by reading the labelling ω according to these linear extensions. Of course n gives the cardinality of P and

$$|f| = \sum_{x \in P} f(x)$$

Now for the case of a Computer Science tree the summation on the left of 3.15 can be very easily evaluated. The basic fact that allows this evaluation is that, for CS-trees, each $f \in F_\omega(T)$ can be uniquely decomposed as a linear combination of indicator functions of *principal* lower order ideals. Since a lower order ideal below an element x of a CS-tree T is simply the subtree x and its cardinality is h_x, we can easily derive the formula

$$\sum_{f \in F_\omega(T)} q^{|f|} = q^{\text{maj}\,\omega(T)} \prod_{x \in T} \frac{1}{1-q^{h_x}}. \tag{3.16}$$

The factor

$$q^{\text{maj}\,\omega(T)}$$

represents the contribution of the smallest f in $F_\omega(T)$. When this function is subtracted from the general element of $F_\omega(T)$ we are left with a remainder which is just weakly decreasing in T. The second factor in 3.16 accounts for the contributions of all these remainders. Combining 3.15 and 3.16 we get the hook formula 3.6.

Before proceeding we need to recall the notions of *primitive words, primitive classes* and *Lyndon words* [2]. Briefly, a word is said to be primitive if all its circular rearrangements are distinct. A circular rearrangement class is called primitive if it is generated by a primitive word. Finally, a word is said to be *Lyndon* if it is lexicographically strictly smaller than all its circular rearrangements. It is not difficult to show that Lyndon words give a set of distinct representatives of the primitive circular rearrangements classes. It will be convenient to let $L(A^n)$ denote the collection of all Lyndon words of length n in the alphabet A.

The introduction of the Klyachko idempotent gives what is perhaps the simplest justification of the dimension formula for the Free Lie Algebra. This classical theorem may be stated as follows.

THEOREM 3.2. *The dimension of* LIE$[A^n]$ *is equal to the number of primitive circular rearrangement classes of words of length* n *in the alphabet* A. *Moreover, a basis for this space is given by the set of Lie polynomials*

$$\{u\mathcal{K}_n\}_{u\in L(A^n)} \tag{3.17}$$

Proof. Since we have all the ingredients to prove the result we shall include a sketch of the argument. Clearly, it is sufficient to show the second part of statement. We shall start by showing that the polynomials in 3.17 span. To this end note that, since $\mathcal{K}_n$ is a Lie idempotent, a trivial spanning set for LIE$[A^n]$ can be obtained by taking all the polynomials

$$u\mathcal{K}_n \qquad (u \in A^n). \tag{3.18}$$

On the other hand, 2.28 evaluated at $q = \omega$ gives that for any word $u \in A^n$

$$u\gamma^s\mathcal{K}_n = \omega^s u\mathcal{K}_n$$

This implies that $\mathcal{K}_n$ kills all non-primitive words and that words in the same circular rearrangement class yield essentially the same polynomial when acted upon on the right by $\mathcal{K}_n$. Thus a spanning set if obtained by letting u in 3.18 vary in any set of distinct representatives of primitive classes.

The shortest path to independence is given by Klyachko's original line of reasoning. Simple manipulations which use 2.23 (a) yield the remarkable identity

$$\mathcal{K}_n\tau = \tau \tag{3.19}$$

where

$$\tau = \frac{1}{n}\sum_{s=0}^{n-1}\omega^{-s}\gamma^s.$$

Now, it is very easy to see that the set of polynomials

$$\{u\tau\}_{u\in L(A^n)} \tag{3.20}$$

is independent. This is because if u_1 and u_2 are in different primitive classes the two polynomials $u_1\tau$ and $u_2\tau$ do not vanish and have disjoint support. However, 3.19 shows that right multiplication by τ sends the set in 3.17 onto the set in 3.20, consequently the set in 3.17 must be independent as well. This completes the proof of the theorem. □

4. The Reutenauer representations. The group algebra element

$$\rho_n = \frac{1}{n}\sum_{d=0}^{n-1}\frac{(-1)^d}{\binom{n-1}{d}}D_{=d}$$

(here $D_{=d}$ is the sum of the permutations with d descents) was shown (indirectly) by Solomon [9] and directly by Reutenauer [8] to be a Lie idempotent. In Reutenauer's proof ρ_n is the first term of a sequence $\rho_n^{(k)}$ of idempotents that are closely connected to the Free Lie Algebra. A remarkable result proved in [8] is that the dimension of the representation induced by the action of S_n on the left ideal $A(S_n)\rho_n^{(k)}$ is given by the Stirling number of the first kind $s(n,k)$. In [8] Reutenauer asks for a combinatorial explanation of this fact. R. Stanley constructed representations with the same dimensions and conjectured (personal communication) that they might have the same character as those introduced by Reutenauer. In this section we shall give a proof of Stanley's conjecture.

The simplest way to introduce the $\rho_n^{(k)}$'s is through their generating function. More precisely, we let $\rho_n^{(k)}$ for $k = 1, 2, \ldots, n$ be the elements of the group algebra $A(S_n)$ satisfying the relation

$$\sum_{k=1}^{n} \rho_n^{(k)} x^k = \frac{1}{n!} \sum_{\sigma \in S_n} (x - d(\sigma)) \uparrow^n \sigma, \tag{4.1}$$

where $(x) \uparrow^n$ denotes the upper factorial polynomial. The proof that this sequence is the same as that studied by Reutenauer is given in [4]. We shall also need here some auxiliary facts concerning the $\rho_n^{(k)}$'s, this material including all of the necessary background may be found in [4].

Let us recall that the Frobenius image of a character χ of S_n is the symmetric polynomial

$$F\chi = \frac{1}{n!} \sum_{\sigma \in S_n} \chi(\sigma) \psi_1^{p_1(\sigma)} \psi_2^{p_2(\sigma)} \cdots \psi_n^{p_n(\sigma)}$$

where ψ_k is the kth power symmetric function and $p_k(\sigma)$ denotes the number of cycles of length k in σ. It may also be shown that the Frobenius images of the character of a module $A(S_n)\rho$ (when ρ is an idempotent) is given by the polynomial

$$P_\rho = \sum_{\sigma \in S_n} \rho(\sigma)\, \psi_1^{p_1(\sigma)}\, \psi_2^{p_2(\sigma)} \cdots \psi_n^{p_n(\sigma)}$$

We shall use these symmetric polynomials in our proof of Stanley's conjecture.

We begin with Stanley's construction. For a given set of non-negative integers $p_1, p_2, \cdots, p_n$ satisfying

$$\begin{aligned} p_1 + p_2 + \cdots + p_n &= k \\ 1p_1 + 2p_2 + \cdots + np_n &= n \end{aligned} \tag{4.2}$$

we choose and fix an arbitrary permutation σ_p with p_m cycles of length m. Let G_p denote the stabilizer of σ_p, that is

$$G_p = \{\alpha : \alpha \sigma_p \alpha^{-1} = \sigma_p\}$$

Next, we pick a one dimensional character $\chi^{(p)}$ of G_p and induce from G_p to S_n. It is seen that the dimension of the resulting representation is $n!/|G_p|$, and this is precisely the number of permutations of cycle type $1^{p_1} 2^{p_2} \cdots n^{p_n}$. Thus the direct sum of these representations over all p satisfying 4.2 (k and n remaining fixed) has dimension $s(n,k)$. Clearly the resulting character is

$$\chi^S = \sum_{\substack{p_1+p_2+\cdots+p_n=k \\ 1p_1+2p_2+\cdots+np_n=n}} \chi^{(p)} \uparrow_{G_p}^{S_n} .$$

Stanley conjectures that there is a choice of the $\chi^{(p)}$'s which makes χ^S equal to the Reutenauer character.

It appears that there is such a choice and it may be constructed as follows. We first observe that G_p is essentially the cartesian product of the wreath products $S_{p_m}[C_m]$. We shall decompose the set of integers $1,2,\cdots,n$ into successive blocks of sizes $p_m \times m$ and shall let $S_{p_m}[C_m]$ act only on the integers in its corresponding block.

The standard realization of the wreath product $S_p[C_m]$ acting on the integers $1,2,\ldots,p\times m$ has elements of the form

$$\xi = (\sigma; \gamma^{s_1}, \gamma^{s_2}, \cdots, \gamma^{s_p}) \tag{4.3}$$

where σ is an arbitrary element of S_p, γ is the cycle $(1,2,\ldots,m)$ and the exponents s_i vary between 0 and $m-1$. If we arrange the integers $1,2,\ldots,p\times m$ in rows of length m with $(i-1)m+1,(i-1)m+2,\cdots,im$ occupying the ith row. Then the element ξ in 4.3 acts on these integers by first circularly shifting each row (the ith row by s_i) and then bodily permuting the resulting rows according to the permutation σ. It is not difficult to show that the map defined by setting

$$\chi^{p,m}(\xi) = \omega^{s_1}\omega^{s_2}\cdots\omega^{s_p},$$

(where $\omega = e^{2\pi i/m}$) gives a one dimensional representation (or character) of $S_p[C_m]$. This given, our choice of $\chi^{(p)}$ will simply be the "product" of the $\chi^{p_m,m}$'s.

We shall next study the character of the Reutenauer representation. To this end we need to give a closer look at the idempotents $\rho_n^{(k)}$. It was shown in [8] that each $w \in A^n$ has the decomposition

$$w = w\rho_n^{(1)} + w\rho_n^{(2)} + \cdots w\rho_n^{(n)}. \tag{4.4}$$

It develops that the polynomial $w\rho_n^{(k)}$ gives the projection of w into a certain subspace $HS_{=k}$ of $Q[A^*]$. The nature of the components $w\rho_n^{(k)}$ was derived in [4] by combining the contents of [6] and [9]. Since this is crucial in our calculation of the Frobenius image of the Reutenauer character, we shall give a brief review of the contents of [4]. The reader is referred to [4] for proofs and further details.

Starting from any basis $\{b[u]\}$ indexed by Lyndon words (in particular the one in 3.17) we let $HS_{=k}$ be the subspace of $Q[A^*]$ spanned by the polynomials

$$s[u_1,u_2,\ldots,u_k] = \frac{1}{k!}\sum_{\sigma\in S_k} b[u_{\sigma_1}]b[u_{\sigma_2}]\cdots b[u_{\sigma_k}] \tag{4.5}$$

(It may be shown that $HS_{=k}$ is independent of the choice of the basis $\{b[u]\}$). The polynomials in 4.5 will actually give a basis for $HS_{=k}$ if we restrict the k-tuples $u_1, u_2, \cdots, u_k$ to be in some specific order, say lexicographically weakly decreasing.

We can define an action of the general linear group $GL(N)$ on $Q[A^*]$ as follows. For a matrix $T = \|t_{ij}\|$ and a letter $a_j \in A$ we set

$$Ta_j = \sum_{i=1}^{N} a_i t_{ij}.$$

This action is extended to words multiplicatively, that is for $w = b_1 b_2 \cdots b_n$ we let

$$Tw = (Tb_1)(Tb_2)\cdots(Tb_n)$$

and finally we can extend it to the whole of $Q[A^*]$ by linearity.

Clearly, this action restricted to $Q[A^n]$ is the nth tensor power of the defining representation of $GL(N)$. It is not difficult to show that each of the subspaces $HS_{=k}$ is invariant under this action. In view of the definition 4.5 we see that the action on $HS_{=k}$ is the kth symmetric power of the action on $HS_{=1}$ which is LIE$[A^*]$ itself. This implies (see for instance [4]) that the character of the action on $HS_{=k}$ is given by the plethysm of the homogeneous symmetric function h_k with the character of LIE$[A^*]$, that is the formal series

$$L(X) = \sum_{n\geq 1} \frac{1}{n} \sum_{p|n} \mu(p)\psi_p^{\frac{n}{p}}. \tag{4.6}$$

The nth homogeneous component of $HS_{=k}$ (that is the linear span of the polynomials $s[u_1, u_2, \ldots, u_k]$ in 4.5 with $u_1, u_2, \ldots, u_k$ of combined length n) is clearly also invariant under this action. Let us denote it by $HS_{=k}\mid_n$. Its character is thus given by the nth homogeneous component of the plethysm

$$h_k[L(X)]. \tag{4.7}$$

Let us denote this polynomial by

$$h_k[L(X)]\mid_n. \tag{4.8}$$

It may be shown (see for instance [4]) that if ρ is an idempotent of the group algebra $A(S_n)$ then the Frobenius image of the character of the action of S_n on the left ideal $A(S_n)\rho$ is the same symmetric polynomial as that giving the character of the action of $GL(N)$ on the subspace $Q[A^*]\rho$ of $Q[A^*]$ spanned by the polynomials $w\rho$ for $w \in A^n$. Since the Reutenauer idempotent $\rho_n^{(k)}$ projects a word $w \in A^n$ into $HS_{=k}|_n$, it follows that

$$HS_{=k}\mid_n = Q[A^n]\rho_n^{(k)}.$$

We thus obtain that the Frobenius image of the character of the Reutenauer representation must be given by the polynomial in 4.8. That is

$$P_{\rho_n^{(k)}} = h_k[L(X)]\mid_n. \tag{4.9}$$

We are left with the calculation of the Frobenius image of the Stanley character. To do this it is best to proceed combinatorially. Clearly, the product

$$\chi^{(p)} = \prod_{m=1}^{n} \chi^{p_m, m}$$

is an idempotent of S_n. Now it develops that the polynomial $P_{\chi^{(p)}}$ has an interesting combinatorial interpretation. To see this, we shall visualize words of length n with their letters written out into the cells of the sequence of $p_m \times m$ Ferrers' diagrams *(blocks).* The words obtained by reading one of the rows of one of the blocks will be referred to as the *row words* of w. This given, it can be shown [4] that the polynomial $w\chi^{(p)}$ is equal to zero if any of the row words of w is not primitive. Let us now call a word $w \in A^n$ *p-Lyndon* if each of the row words is Lyndon, and within a block the row words are in lexicographically decreasing order. It is straightforward to show that the polynomials $w\chi^{(p)}$ as w describes all p-Lyndon words give a basis for the space $Q[A^*]\chi^{(p)}$. This implies [4] that the polynomial $P_{\chi^{(p)}}$ (which by the way is also the Frobenius image of $\chi^{(p)} \uparrow_{G_p}^{S_n}$) is the Polya enumerator of the set of p-Lyndon words. Consequently, the Frobenius image of Stanley's character (with this choice of $\chi^{(p)}$) is the Polya enumerator of all words which may be factored into a product of k Lyndon words of weakly increasing lengths with factors of equal length in weakly decreasing lexicographic order.

To complete our proof of Stanley's conjecture we need only give a combinatorial interpretation for the polynomial in 4.9. However, this is almost immediate since it is well known (see for instance [3]) that the plethysm $P_1[P_2]$ of two Polya enumerators P_1 and P_2 is the Polya enumerator of the patterns enumerated by P_1 *colored* by the patterns enumerated by P_2. This implies that

$$h_k[L(X)]$$

is the enumerator of lexicographically weakly decreasing k-tuples of Lyndon words. In particular,

$$h_k[L(X)] \mid_n$$

enumerates those k-tuples that are of combined length n.

Clearly then the patterns enumerated by $F\chi^S$ are simply rearrangements of the patterns enumerated by $P_{\rho_n^{(k)}}$. Thus these two polynomials are identical and our proof is complete.

REFERENCES

[1] A. Bjorner and M. Wachs, *q-Hook Formulas for Trees and Forests*, J. Comb. Th. A. (to appear).

[2] K.T. Chen, R.H. Fox, R.C. Lyndon, *Free, differential Calculus IV. The quotient Groups of the lower central Series*, Ann. Math., V. 68 (1958), pp. 81–95.

[3] Y.M. Chen, A.M. Garsia, and J. Remmel, *Algorithms for Plethysm*, Contemporary Math., #34, *Combinatorics and Algebra*, Curtis Greene, ed. (1984).

[4] A.M. Garsia, *Combinatorics of the Free Lie Algebra and the Symmetric Group*, a Volume commemorating J. Moser's 60th birthday (to appear).

[5] A.M. GARSIA AND I. GESSEL, *Permutation statistics and partitions*, Advances in Math, 31, No. 3 (1979), pp. 288–305.

[6] A. KLYACHKO, *Lie elements in the Tensor Algebra*, Siberian Math. J. (Translation), 15, No. 6 (1974), pp. 1296–1304.

[7] M. LOTHAIRE, *Combinatorics on Words*, Encyc. Math, 17, Add. Wesley (1983).

[8] C. REUTENAUER, *Theorem of Poincaré - Birkhoff - Witt, and symmetric group representations of degrees equal to Stirling numbers*, Lecture Notes in Math., 1234, Springer-Verlag (1986), pp. 267–293.

[9] L. SOLOMON, *On the Poincaré-Birkhoff-Witt Theorem*, J. Comb. Theory, 4 (1968), pp. 363–375.

[10] R. STANLEY, *Ordered structures and partitions*, Mem. Amer. Math. Soc., 119 (1972).

[11] M. WACHS, *A combinatorial proof of idempotency for the standard bracketing operator*, U. Miami preprint (1987).

TABLEAUX IN THE REPRESENTATION THEORY OF THE CLASSICAL LIE GROUPS

SHEILA SUNDARAM*

Abstract. Column-strict tableaux were introduced by I. Schur in his study of the characters of $Gl(n)$. Following early work of Alfred Young, (who introduced partitions and tableaux in the study of the representations of the symmetric group), and Specht, Weyl used Young symmetrisers to construct the irreducible modules for the polynomial representations of the classical Lie groups. This paper surveys the connections between the combinatorics and the algebra which arise from these classical constructions.

Key words. Lie groups, representations, tableau, insertion algorithm

AMS(MOS) subject classifications. Primary 05A21; Secondary 17B10.

0. Introduction. This paper is an overview of the interplay between combinatorics and the representation theory of the classical Lie groups. We have attempted to collect together algebraic aspects of the theory which can be discussed from the viewpoint of the combinatorialist, and to show how elementary combinatorial techniques can be used to deduce module-theoretic results, without recourse to the actual modules. The combinatorial object of primary interest is the tableau. We explain the use of tableaux to index the weights of a representation (a classical construction of Schur in the case of the general linear group, having its origins in the work of Alfred Young), and the representation-theoretic interpretation of such combinatorial algorithms as Robinson-Schensted insertion. For more details on the algebraic background, the reader is encouraged to consult the survey paper by [Ha]. Throughout this paper G will stand for one of the classical Lie groups over the complex field $\mathbb{C}$:

(1) $Gl(n)$, the general linear group of all n by n non-singular matrices,

(2) $Sp(2n)$, the symplectic group of all $2n$ by $2n$ matrices which preserve any given non-degenerate skew-symmetric form on $\mathbb{C}^{2n}$,

(3) $SO(n)$, the special orthogonal group, which is the subgroup of $Gl(n)$ consisting of all complex orthogonal matrices of determinant one.

We shall be concerned only with the polynomial representations of G.

In Section 1 we review briefly the labelling and construction of the irreducible modules for the classical Lie groups. In Section 2 we discuss solutions to the problem of combinatorially describing the weights and characters of these representations. In Section 3 we introduce the tensor space, and explain how the various insertion schemes fit into the picture. We also show how the combinatorial techniques may be used to determine multiplicities in certain "branching rules". In Section 4 we state formulas for the dimensions of the irreducible representations, which are expressible

*Department of Mathematics, University of Michigan, Ann Arbor, Michigan 48109. This paper was written while the author was a post-doctoral member at the Institute for Mathematics and Its Applications.

as polynomials in the rank of the group. We describe Koike and Terada's universal characters, and some formulas for the irreducible characters due to Weyl. These are analogues of the well-known Jacobi-Trudi formulas for Schur functions. We also discuss the various Cauchy identities; combinatorics enters here in the form of generalised insertion algorithms which would establish these identities. In Section 5 we discuss the problem of decomposing the tensor product of representations into irreducibles. A classical rule due to Littlewood and Newell is presented. The computational difficulties inherent in this rule lead naturally to a discussion of King's modification rules for the characters. We conclude in Section 6 with a brief discussion of the centraliser algebras, and recent work of Hanlon and Wales, and Hans Wenzl.

1. Background from representation theory and Lie algebras.

The most accessible reference for the material in this section is [Hum] (Chapters I, II and VI).

We begin with

DEFINITION 1.1. *A polynomial representation of G is a group homomorphism $\phi : G \mapsto Gl(V)$, for some vector space V, such that the entries of $\phi(g)$ are polynomials in the entries of g, for all $g \in G$. The character of ϕ is the function $char\phi : G \mapsto C$ defined by $(char\phi)(g) = trace\ \phi(g)$.*

For more details about such topics as root systems and weights, we refer the reader to [Hum]. It is well-known that, in general, the irreducible polynomial representations of G are indexed by highest weight ω (see [Hum]) or by partitions λ with at most (rank G) parts (here rank G is the rank of the corresponding root system). Let $\omega_1, \ldots, \omega_n$ be the fundamental weights (these are formed by taking the dual vectors of a base of simple roots for the root system: see [Hum], p. 67), and $e_1, \ldots, e_n$ be standard unit coordinates.

For $Gl(n)$ and $Sp(2n)$ we have $\omega_i = \sum_{k=1}^{i} e_k$, so the partition $\lambda = \sum_{k=1}^{n} \lambda_k\, e_k$ indexes the irreducible representation corresponding to the highest weight $\omega = \sum_{k=1}^{n} a_k\, \omega_k$ where a_i is the number columns of λ of length i. Furthermore, all the irreducible representations are obtained in this manner from partitions with at most n parts.

For $SO(2n+1)$, the irreducible polynomial representations are indexed by highest weights $\omega = \sum_{k=1}^{n} a_k\, \omega_k$ where the Dynkin indices a_k must satisfy $a_n \equiv 0 \pmod 2$. Hence the irreducible representations are once again indexed by partitions with at most n parts, but this time the partition $\lambda = \sum_{k=1}^{n} \lambda_k\, e_k$ gives the irreducible representation corresponding to the highest weight $\omega = \sum_{k=1}^{n} a_k\, \omega_k$ where a_i is the number columns of λ of length i for $i < n$, and $a_n = 2\lambda_n$.

For $SO(2n)$, the irreducible polynomial representations are indexed by highest weights $\omega = \sum_{k=1}^{n} a_k\, \omega_k$ where the Dynkin indices a_k must satisfy $a_{n-1} \leq a_n$, and $a_n - a_{n-1} \equiv 0 \pmod 2$. Hence the partitions with less than n parts index the irreducibles as before, with λ corresponding to the highest weight $\omega = \sum_{k=1}^{n-1} a_k\, \omega_k$ where a_i is the number columns of λ of length i, $i < n$, and $a_n = a_{n-1}$, but if $\lambda_n \neq 0$, then the representation splits into a sum of the two irreducibles indexed by the weights ω^1, ω^2 as follows: Set a_i = number of columns of λ of length i, $i < n$, and define a_n by setting $(a_n - a_{n-1})/2 = \lambda_n$. Then $\omega^1 = \sum_{k=1}^n a_k\, e_k$ and $\omega^2 = \sum_{k=1}^{n-2} a_k\, e_k + a_n\, e_{n-1} + a_{n-1}\, e_n$.

We now describe briefly how the actual modules are constructed. Some preliminary notation and definitions are reviewed.

If $\lambda = (\lambda_1 \geq \ldots \geq \lambda_n \geq 0)$ is a partition with at most n parts, we write $\ell(\lambda)$ = (**length** of λ) = number of nonzero parts of λ, and $|\lambda| = \lambda_1 + \ldots + \lambda_n$ for the sum of the parts of λ. We also write $\lambda \vdash |\lambda|$ and say λ is a partition of $|\lambda|$.

DEFINITION 1.2. *If $\lambda \vdash k$, a* **standard Young tableau** *of shape λ is a filling of the Ferrers diagram of λ with the integers $1, \ldots, k$ such that the entries increase strictly left-to-right along rows and top-to-bottom down the columns.*

Example. A standard Young tableau of shape $(3,3,1)$ is $T = \begin{smallmatrix} 1 & 3 & 7 \\ 2 & 4 & 6 \\ 5 & & \end{smallmatrix}$.

The irreducible modules are constructed in the space $V^{\otimes k}$ where V affords the defining representation for G (i.e., the natural action of the matrices in G on a vector space of dimension equal to the order of the matrices).

We first consider the case $G = Gl(n)$. Here $V = \mathbb{C}^n$. Fix a partition λ such that $\ell(\lambda) \leq n$. The irreducible $Gl(n)$-module indexed by λ, called the Weyl module W^λ, is constructed in $V^{\otimes k}$, where $k = |\lambda|$, as follows (see [We], [Boe]):

Fix any standard Young tableau T of shape λ, and let e_T be the corresponding Young symmetriser

$$e_T = \sum_{\gamma \in C_T} (sgn \gamma)\, \gamma \sum_{\sigma \in R_T} \sigma,$$

where C_T (respectively R_T) denotes the column (respectively row) stabiliser of T, that is, the set of permutations on k letters which leave the elements in each column (respectively row) of T unchanged. Identify a basis tensor $e_{i_1} \otimes \ldots \otimes e_{i_k}$ in $V^{\otimes k}$ with an array of shape λ with entries from $1, \ldots, n$, by putting i_j in the cell occupied by j in T.

Example. For $k = 5, n = 3$ and $\lambda = (3,2)$, take $T = \begin{smallmatrix} 1 & 3 & 5 \\ 2 & 4 & \end{smallmatrix}$. Then the tensor $e_2 \otimes e_1 \otimes e_1 \otimes e_3 \otimes e_2$ is identified with the array $\begin{smallmatrix} 2 & 1 & 2 \\ 1 & 3 & \end{smallmatrix}$.

Under this identification, e_T acts on $V^{\otimes k}$, and W^λ is defined to be the image of the operator e_T. It can be shown that for a different choice of the standard tableau T, one gets no more than an isomorphic copy of W^λ. This is essentially contained in Schur's double centraliser theorem (see Theorem 2.1 and also Section 6).

Two particular irreducible modules of interest are: (for $m \geq 0$) the mth symmetric power $Sym^m(V)$ of the defining representation V of $Gl(n)$, and, for $0 \leq m \leq n = dim(V)$, the mth exterior power $\bigwedge^m(V)$. The latter is more often known as the mth fundamental representation, since it is indexed by the fundamental weight ω_m. Note that we have

$$W^{(m)} = Sym^m(V); \quad W^{(1^m)} = \bigwedge^m(V).$$

Next let $G = Sp(2n)$; fix λ, $\ell(\lambda) \leq n$. The corresponding irreducible module in $(\mathbf{C}^{2n})^{\otimes|\lambda|}$ is defined by

$$\tilde{W}^{\lambda} = W^{\lambda}/(W^{\lambda} \cap U),$$

where W^{λ} is the Weyl module for $Gl(2n)$, and, if $\{x_1, \ldots, x_n, y_1, \ldots, y_n\}$ is a basis defining the skew-symmetric form on $\mathbf{C}^{2n}$ such that $\langle x_i, y_i \rangle = -\langle y_i, x_i \rangle$, then $U = span\{\sum_{i=1}^{n}(u \otimes x_i \otimes v \otimes y_i - u \otimes y_i \otimes v \otimes x_i) : u \otimes v \in (\mathbf{C}^{2n})^{|\lambda|-2}\}$.

Note that $Sp(2n)$ fixes the vectors in U.

We remark that the mth fundamental representation ($0 \leq m \leq n$) is a quotient of two fundamental $Gl(2n)$-representations:

$$\tilde{W}^{(1^m)} = W^{(1^m)}/W^{(1^{m-2})} = \bigwedge^{m}(V)/\bigwedge^{m-2}(V), \quad dim(V) = 2n,$$

while the mth symmetric power of the defining representation of $Sp(2n)$ coincides with that of $Gl(2n)$:

$$\tilde{W}^{(m)} = W^{(m)} = Sym^m(V), \quad dim(V) = 2n.$$

Likewise for $G = SO(n)$ and λ, $\ell(\lambda) \leq \lfloor(n/2)\rfloor$, construct

$$\mathring{W}^{\lambda} = W^{\lambda}/(W^{\lambda} \cap span\{\sum_{i=1}^{\lfloor(n/2)\rfloor}(u \otimes x_i \otimes v \otimes x_i) : u \otimes v \in (\mathbf{C}^{n})^{|\lambda|-2}\},$$

where $\{x_1, \ldots, x_n\}$ is a defining basis for a symmetric form on $\mathbf{C}^n$. (Recall that if $\ell(\lambda) = n/2, n$ even, this module is not irreducible).

We remark that this time the mth symmetric power of the defining representation of $SO(n)$ is a quotient of two symmetric powers of defining representations of $Gl(n)$:

$$\mathring{W}^{(m)} = W^{(m)}/W^{(m-2)} = Sym^m(V)/Sym^{(m-2)}(V), \quad dim(V) = n,$$

while the mth exterior power of the defining representation of $SO(n)$ ($0 \leq m \leq \lfloor(n/2)\rfloor$) coincides with that of $Gl(n)$, unless n is even and $m = \frac{n}{2}$:

$$\mathring{W}^{(1^m)} = W^{(1^m)} = \bigwedge^{m}(V), \quad dim(V) = n,$$

Now let G denote an arbitrary Lie group. We record some facts about polynomial representations of G and their characters, referring the reader to [Hum] for proofs. Let ϕ be such a representation. Then:

1. $char\ \phi$ determines the representation ϕ uniquely (up to equivalence).
2. All polynomial representations of G are completely reducible; the irreducible polynomial representations are finite-dimensional.
3. If $T = \{diag(x_1, \ldots, x_n)\}$ is the subgroup of diagonal matrices in G (a maximal torus) then $\bigcup_{g \in G} gTg^{-1}$ is dense in G.

4. $(char\ \phi)(g)$ therefore depends only on the eigenvalues of any $g \in G$, and hence can be thought of as a function of the indeterminates $x_1, \ldots, x_n$.
5. If W_G is the Weyl group of (the Lie algebra of) G, then $char\ \phi$ is an element of $\mathbb{Z}[x_1, \ldots, x_n]^{W_G}$, the ring of polynomials in the variables $x_1, \ldots, x_n$, with integer coefficients, which are invariant under the action of W_G; furthermore, there is a unique multiset $M_\phi \subset \mathbb{Z}^n$ such that $(char\ \phi)(x_1, \ldots, x_n) = \sum_{\alpha \in M_\phi} x_1^{\alpha_1} \ldots x_n^{\alpha_n}$. Note that M_ϕ is simply the set of weights occurring in the representation ϕ for the Lie algebra of G. Again, see [Hum].
6. The set $\{char\ \phi : \phi$ is an irreducible representation of $G\}$ is a $\mathbb{Z}$-basis for $\mathbb{Z}[x_1, \ldots, x_n]^{W_G}$.

2. The combinatorial problem. In this section we address the following two questions: Given a Lie group G and a partition λ which gives an irreducible polynomial representation ϕ_λ of G,

- find a set $\mathcal{T}_\lambda$ of objects which index the weights of the representation ϕ_λ, so that $|\mathcal{T}_\lambda| =$ the dimension of ϕ_λ,
- find a weighting scheme for $\mathcal{T}_\lambda$ so that the weighted generating function for $\mathcal{T}_\lambda$ is the character $char\ \phi_\lambda$.

The first result in this direction is due to Schur (in his 1901 dissertation):

THEOREM 2.1. *[Sch] A basis for the irreducible $Gl(n)$-module W^λ (as defined in Section 1), $\ell(\lambda) \leq n$, is the set of all column-strict tableaux (i.e., weakly increasing left-to-right along rows, strictly increasing top to-bottom down the columns) of shape λ with entries from $\{1, 2, \ldots, n\}$.*

If T is a column-strict tableau of shape λ which is a basis element of some irreducible $Gl(n)$-module, define the weight of T to be the monomial

$$\prod_{i=1}^{n} x_i^{\text{number of } i\text{'s in} T}.$$

Example.

$T = \begin{matrix} 1 & 1 & 3 \\ 3 & 4 & 4 \\ 4 & & \end{matrix}$ is a column-strict tableau of shape $(3, 3, 1)$ in the entries $\{1, 2, 3, 4\}$, of weight $x_1^2 x_3^2 x_4^3$.

We then have

THEOREM 2.2. *The character of the irreducible representation W^λ of $Gl(n)$ is*

$$s_\lambda(x_1, \ldots, x_n) = \sum_{\substack{T \text{ column-strict tableau} \\ \text{of shape } \lambda}} weight(T).$$

The reader will recognise s_λ as the Schur function corresponding to λ (see [Mac]) Since the Weyl group of $Gl(n)$ is the symmetric group on n letters S_n, we have the well-known fact that the Schur functions s_λ, $\ell(\lambda) \leq n$, form a basis for the ring of symmetric polynomials $\mathbb{Z}[x_1, \ldots, x_n]^{S_n}$ in the variables $x_1, \ldots, x_n$.

For the remaining classical groups, sets T_λ of tableaux indexing the weights of the representation ϕ_λ were first defined by R.C.King in [Ki]. Later Proctor discovered a unified way of indexing the weights for all the classical groups, by means of Gelfand patterns [Pr1]. When translated into column-strict tableaux, for $Gl(n)$ these produce the same tableaux as those of Theorem 2.1. In [K-T], Koike and Terada also obtain sets of tableaux for the orthogonal groups, which can be easily shown to coincide with the conversion of Proctor's formulation into tableaux for $SO(2n+1)$ (see Theorem 2.6 below) and which coincide with King's tableaux for $SO(2n)$ ([Ki2]).

First we consider the symplectic group $Sp(2n)$. The irreducible representations $\tilde{\phi}_\lambda$ are indexed by partitions λ, $\ell(\lambda) \leq n$; a maximal torus (= maximal abelian subgroup) is $T = \{diag(x_1, {x_1}^{-1}, \ldots, x_n, {x_n}^{-1}) : x_i \neq 0, i = 1, \ldots, n\}$, and the Weyl group is the hyperoctahedral group B_n $(= \mathbb{Z}_2{}^n \rtimes S_n)$. Hence $char\ \phi_\lambda \in \mathbb{Z}[x_1^{\pm 1}, \ldots, x_n^{\pm 1}]^{B_n}$.

In [Ki], R.C. King describes a set of tableaux which index the weights of ϕ_λ as follows. We shall need an alphabet of $2n$ symbols $1, \bar{1}, \ldots, n, \bar{n}$. Now define a tableau T of shape λ, $\ell(\lambda) \leq n$, with entries in $1, \bar{1}, \ldots, n, \bar{n}$ to be **symplectic** if

(1) T is a column-strict tableau

(2) all entries in row i are larger than or equal to i.

Define the weight of a symplectic tableau T (for $Sp(2n)$) to be the monomial

$$\prod_{i=1}^{n} (x_i)^{\text{number of } i\text{'s in T} - \text{number of } \bar{i}\text{'s in T}}.$$

Example. $n = 3$, $\lambda = (4, 2, 1)$. Then

$$T = \begin{array}{llll} 1 & \bar{1} & \bar{1} & \bar{3} \\ \bar{2} & 3 & & \\ 3 & & & \end{array}$$

is a symplectic tableau for $Sp(6)$ of weight ${x_1}^{-1}{x_2}^{-1}x_3$.

King and El-Sharkaway can then prove

THEOREM 2.3. *[K-El] The character of the irreducible representation $\tilde{\phi}_\lambda$ is the generating function*

$$sp_\lambda(x_1^{\pm 1}, \ldots, x_n^{\pm 1}) = \sum_{\substack{T\ n-symplectic \\ tableau\ of \\ shape\ \lambda}} weight(T)$$

We shall call sp_λ the **symplectic** Schur function corresponding to λ.

Let $\tilde{W}^\lambda$ be the irreducible $Sp(2n)$-module defined in Section 1. Berele has shown explicitly that

THEOREM 2.4. *[Be2] The symplectic tableaux of shape* λ *in* $1, \bar{1}, \ldots, n, \bar{n}$ *form a basis for the irreducible module* $\tilde{W}^{\lambda}$.

Next take $G = SO(2n+1)$ and suppose λ is a partition with at most n parts. A maximal torus is now given by the subgroup $\{diag(x_1{}^{\pm 1}, \ldots, x_n{}^{\pm 1}, 1) : x_i \neq 0\}$; the Weyl group is B_n, the same as for $Sp(2n)$. This time we need an alphabet of $(2n+1)$ symbols $1 < \bar{1} < 2 < \bar{2} < \ldots < n < \bar{n} < \infty$. King then defines the tableaux of the following theorem in [Ki]:

THEOREM 2.5. *[K-El] Fill the Ferrers diagram of* λ *with entries from* $1 < \bar{1} < 2 < \bar{2} < \ldots < n < \bar{n} < \infty$ *such that*

(K1) *the resulting array is a column-strict tableau;*

(K2) *all entries in row* i *are larger than or equal to* i*;*

(K3) *if* i, $\bar{i}$ *are consecutive entries in the same row, then there is an* i *in the preceding row, immediately above the* $\bar{i}$.

If T *is a tableau satisfying (1)-(3), let* $m(T)$ *be the number of occurrences of* i *directly above* $\bar{i}$ *in the first column, with* $\bar{i}$ *in row* i*, for* $i = 1, \ldots, n$.

Then char $\overset{\circ}{\phi}_{\lambda}=$ *character of* $\overset{\circ}{W}{}^{\lambda}$

$$= \sum_{\substack{T satisfying (1)-(3) \\ shape(T)=\lambda}} 2^{m(T)}\ weight(T)$$

where as before $weight(T) = \prod_{i=1}^{n} x_i{}^{\text{number of } i\text{'s } - \text{number of } \bar{i}\text{'s}}$.

Note that *char* $\overset{\circ}{\phi}_{\lambda}$ is in $\mathbf{Z}[x_1^{\pm 1}, \ldots, x_n^{\pm 1}]^{B_n}$.

Proctor's Gelfand patterns for $SO(2n+1)$ translate into the following simple tableaux, which coincide with those described in [K-T]:

THEOREM 2.6. *Fix an alphabet of* $3n$ *symbols* $1 < \overset{\circ}{1} < \bar{1} < 2 < \overset{\circ}{2} < \bar{2} < \ldots < n < \overset{\circ}{n} < \bar{n}$. *Given a partition* λ *of length at most* n*, consider tableaux of shape* λ *obtained by filling the Ferrers diagram of* λ *with entries from this alphabet such that*

(PKT1) *the resulting array is a column-strict tableau;*

(PKT2) *all entries in row* i *are larger than or equal to* i*;*

(PKT3) *the symbol* $\overset{\circ}{i}$ *appears at most once in row* i*, and never in any other row,* $i = 1, \ldots, n$.

Then char $\overset{\circ}{\phi}_{\lambda}=$ *character of* $\overset{\circ}{W}{}^{\lambda}$

$$= \sum_{\substack{T \text{ satisfying (PKT1)-(PKT3)} \\ shape(T)=\lambda}} weight(T)$$

where as usual $weight(T) = \prod_{i=1}^{n} x_i{}^{\text{number of } i\text{'s } - \text{number of } \bar{i}\text{'s}}$.

A third set of tableaux for $SO(2n+1)$ was discovered by this author (see [Su3]):

DEFINITION 2.7. *Reverting to the alphabet of* $(2n+1)$ *symbols* $1 < \bar{1} < 2 < \bar{2} < \ldots < n < \bar{n} < \infty$, *for a partition* λ *with at most* n *parts, define an* **so-tableau** T *of shape* λ *to be a column-strict tableau of shape* λ *with entries in* $1 < \bar{1} < 2 < \bar{2} < \ldots < n < \bar{n} < \infty$ *such that*

(S1) *all entries in row* i *are larger than or equal to* i*;*

(S2) *in any row, the symbol* ∞ *appears at most once, i.e., the cells of* λ *which contain the symbol* ∞ *form a vertical strip (see [Mac]).*

It can be shown that

THEOREM 2.8. *[Su3] If* $\ell(\lambda) \leq n$*, the character of the irreducible representation* $\overset{\circ}{W}{}^{\lambda}$ *is given by*

$$so_{\lambda} = \sum_{\substack{T \text{ an so-tableau} \\ shape(T)=\lambda}} weight(T)$$

where again $weight(T) = \prod_{i=1}^{n} x_i{}^{\text{number of } i\text{'s } - \text{number of } \bar{i}\text{'s}}$.

The equivalence of these three definitions is not obvious; that is, this author knows of no simple bijections from any one of the three sets of tableaux for $SO(2n+1)$ to another.

For $G = SO(2n)$, King and El-Sharkaway show that the tableaux and generating function of Theorem 2.5 also give the character of the irreducible representation $\overset{\circ}{\phi}_{\lambda}$ if $\ell(\lambda) < n$. If $\ell(\lambda) = n$, then as we noted in Section 1, $\overset{\circ}{W}{}^{\lambda} = \overset{\circ}{W}{}^{\lambda}_{1} \oplus \overset{\circ}{W}{}^{\lambda}_{2}$ splits into two irreducible submodules, whose characters are described by

THEOREM 2.9. *[K-El]*

$$char\ \overset{\circ}{W}{}^{\lambda}_{i} = \sum_{\substack{T \\ shape(T)=\lambda}} 2^{v(T)}\ weight(T)$$

where

$$v(T) = \begin{cases} m(T) - 1, & \text{if } m(T) > 0 \\ 0, & \text{else,} \end{cases}$$

where $m(T)$ *is as in Theorem 2.5, and the sum ranges over all tableaux* T *of shape* λ *satisfying (K1)-(K3) and in addition*

(K4) if $m(T) = 0$*, i.e., if the entry in the first column of row* j *is* j *or* $\bar{j}$ *for all* $j = 1, \ldots, n$*, then the number of* j *such that the (j,1)-entry of* T *is* $\bar{j}$ *is*

$$\begin{cases} \text{even}, & \text{if } i = 1 \\ \text{odd}, & \text{if } i = 2. \end{cases}$$

As mentioned earlier, these tableaux coincide with the ones described in [K-T] (King, personal communication). The Gelfand patterns of Proctor [Pr] can also be converted to a presumably different set of $SO(2n)$-tableaux, although this author has not actually done so.

3. The tensor space and insertion schemes.

Since, for partitions of k, the irreducible G-modules were constructed inside the space carrying the kth tensor power of the defining representation of G, one can ask how the whole tensor space decomposes into irreducibles. In this section, we describe what is known about this decomposition for the different Lie groups. We consider insertion algorithms from the point of view of providing combinatorial proofs for the corresponding character identities. We also show how such an insertion algorithm can, in at least one case, be used to obtain "branching rules": i.e., to compute the multiplicities in the decomposition of a $Gl(n)$-module into irreducibles for its symplectic and orthogonal subgroups.

For $Gl(n)$, again Schur supplies the answer to the tensor space problem:

THEOREM 3.1. *[Sch] Let V be any n-dimensional vector space (over $\mathbb{C}$) and k a positive integer. Then $V^{\otimes k}$ is an $S_k \times Gl(n)$-module, (where S_k is the symmetric group on k letters) and as such, decomposes into irreducibles as follows:*

$$V^{\otimes k} = \bigoplus_{\substack{\lambda \vdash k \\ \ell(\lambda) \leq n}} (S^\lambda \otimes W^\lambda).$$

Here W^λ is the $Gl(n)$-irreducible, and S^λ is the S_k-irreducible (Specht module) corresponding to the partition λ.

COROLLARY 3.2. *As a $Gl(n)$-module, we have the decomposition*

$$V^{\otimes k} = \bigoplus_{\substack{\lambda \vdash k \\ \ell(\lambda) \leq n}} (f^\lambda \, W^\lambda),$$

where f^λ is the number of standard Young tableaux of shape λ, and also the dimension of the S_k-module S^λ.

If we now take the traces of the representations on either side of this decomposition, we get the following character identity:

$$(x_1 + \ldots + x_n)^k = \sum_{\substack{\lambda \vdash k \\ \ell(\lambda) \leq n}} f^\lambda \, s_\lambda(x_1, \ldots, x_n)$$

Enter combinatorics, via the Robinson-Schensted [Scn] insertion algorithm, also described in [S]. This famous bijection establishes the above identity by mapping a word $w_1 \ldots w_k$ in $\{1, \ldots, n\}$ to a pair (P_λ, Q_λ) where P_λ is a column-strict tableau of shape λ, with entries in $\{1, \ldots, n\}$, and Q_λ is a standard Young tableau of the same shape λ (entries in $\{1, \ldots, k\}$). Note that P_λ has weight $w_1 \ldots w_k$. Thus P_λ is a basis element of the Weyl module, while Q_λ represents a choice of Young symmetriser.

To address the analogous question for $G = Sp(2n)$, we shall need a tensoring rule believed to be known to Littlewood:

LEMMA 3.3. *Consider the defining representation of $Sp(2n)$, afforded by $V = \tilde{W}^{(1)}$ ($\dim V = 2n$). Let $\tilde{W}^{\lambda}$ be an irreducible module affording the representation of $Sp(2n)$ indexed by λ, $\ell(\lambda) \leq n$. Then the tensor product*

$$\tilde{W}^{\lambda} \otimes \tilde{W}^{(1)}$$

decomposes into irreducibles as

$$\coprod_{\substack{\mu \subset \lambda \\ \lambda/\mu=(1)}} \tilde{W}^{\mu} \coprod_{\substack{\nu \supset \lambda \\ \ell(\nu)\leq n \\ \nu/\lambda=(1)}} \tilde{W}^{\nu}.$$

Next we define:

DEFINITION 3.4. *An* **up-down** (*or* **oscillating**) *tableau of shape μ, and length k is a sequences of shapes $(\emptyset = \mu^0, \mu^1, \ldots, \mu^k = \mu)$ such that any two consecutive shapes differ by exactly one box, i.e., for all $i = 1, \ldots, k$, either the skew-shape $\mu^i/\mu^{(i-1)} = (1)$ or $\mu^{(i-1)}/\mu^i = (1)$. We shall call such a sequence n-***symplectic** if, in addition, we require that*

$$\ell(\mu^i) \leq n, \text{for all } i = 1, \ldots, k.$$

We let $\tilde{f}_k^{\mu}$ denote the total number of up-down tableaux of shape μ and length k, and $\tilde{f}_k^{\mu}(n)$ denote the number of these that are n-symplectic.

We can now state :

THEOREM 3.5. *Let V afford the defining representation for $Sp(2n)$. Then the kth tensor power of V decomposes into irreducible $Sp(2n)$-modules $\tilde{W}^{\mu}$ as follows:*

$$V^{\otimes k} = \coprod_{\substack{\mu \\ \ell(\mu)\leq n}} \tilde{f}_k^{\mu}(n)\ \tilde{W}^{\mu}$$

Proof. This follows easily by successive applications of Lemma 3.3. □

Taking the characters of the representations involved in the above decomposition, we get the symplectic Schur function identity:

$$(3.6) \qquad (x_1 + {x_1}^{-1} + \cdots + x_n + {x_n}^{-1})^k = \sum_{\substack{\mu \\ \ell(\mu)\leq n}} \tilde{f}_k^{\mu}(n)\ sp_{\mu}(x_1^{\pm 1}, \ldots, x_n^{\pm 1})$$

A bijection establishing this identity was discovered by Berele [Be1], and works by mapping a k-word $w_1 \ldots w_k$ in the alphabet $1, \bar{1}, \ldots, n, \bar{n}$ to a pair

$$(\tilde{P}_{\mu}, \tilde{Q}_{\mu}^{k}(n))$$

where $\tilde{P}_\mu$ is a symplectic tableau of shape μ (with entries in $1, \bar{1}, \ldots, n, \bar{n}$) and weight $\prod_{i=1}^k w_i$, and $\tilde{Q}_\mu^k(n)$ is an n-symplectic up-down tableau of the same shape μ, and length k.

We refer the reader to [Be1] for a detailed description of Berele's insertion scheme.

The reader may wonder of what practical use such an insertion scheme can be (besides providing a combinatorial proof of the character identity (3.6)). Before we can illustrate the algebraic applications, we shall need some more definitions.

DEFINITION 3.7. *A* **lattice permutation** *is a word $w_1 \ldots w_n$ of positive integers such that, reading the word from left to right, the number of occurrences of any integer i is greater than or equal to the number of occurrences of $i+1$. The* **weight** *of a lattice permutation w is the finite nonnegative integer vector α where $\alpha_i = |\{j : 1 \leq j \leq n, w_j = i\}|$.*

Remarks.

(1) Notice that a lattice permutation always has **partition weight**, i.e., its weight α is a partition.
(2) If the word w is a lattice permutation , so is any initial segment $w_1 \ldots w_i$, $i \leq \ell(w)$.
(3) There is a natural bijection between lattice permutations w of length n and weight α, and standard Young tableaux of shape α, defined as follows: for $i = 1, \ldots, n$, if $w_i = r$, place an i in the rth row of the Ferrers diagram of α; conversely, given a standard Young tableau of shape α, set $w_i = r$ for all i appearing in row r of α. It is clear that the lattice permutation condition, which seems rather *ad hoc* at first glance, is precisely equivalent to guaranteeing that the tableau associated to the lattice permutation in this manner is standard.

Example. The lattice permutation 12113231 has weight $(4,2,2)$ and corresponds to the standard Young tableau

```
1 3 4 8
2 6
5 7
```

.

DEFINITION 3.8. *Let T be a tableau of skew-shape λ/μ (geometrically, these are the cells in the set-theoretic difference between the diagrams of λ and μ). The* **word of T, read in lattice permutation fashion,** *is the word obtained by reading the filling T row by row, from* **top to bottom** *and* **right to left**. *A lattice permutation w* **fits** *a skew-shape λ/μ, if there is a column-strict tableau T of shape λ/μ such that the word of T, read in lattice permutation fashion, is w.*

For partitions λ, μ, ν, the **Littlewood-Richardson coefficient,** *which we denote by $c^\lambda_{\mu,\nu}$, is the number of lattice permutations of shape λ/μ and weight ν.*

Example. The lattice permutation 11211321 fits the skew-shape (5,4,3,1)/(3,1,1) as follows:

```
□ □ □ 1 1
□ 1 1 2
□ 2 3
1
```

.

Clearly $c^\lambda_{\mu,\nu} = 0$ unless $\lambda \supseteq \mu, \nu$, and $\lambda \vdash (|\mu| + |\nu|)$.

It is a well-known fact about symmetric functions (see [Mac]; also Section 5) that the Littlewood-Richardson coefficient $c^{\lambda}_{\mu,\nu}$ is the coefficient of the Schur function s_λ in the product $s_\mu \, s_\nu$.

DEFINITION 3.9. *Suppose w is a lattice permutation of weight β where β has even columns. We say w fits the skew-shape λ/μ n-**symplectically**, if in the filling of λ/μ, $(2i+1)$ appears no lower than row $(n+i)$ of λ, for all $i = 1, \ldots, \frac{1}{2}\ell(\beta)$.*

The next two results, while completely combinatorial in nature and derivation, combine to give an explicit formula for the multiplicities in the decomposition of an irreducible $Gl(2n)$-module into submodules which are irreducible with respect to the action of the subgroup $Sp(2n)$.

THEOREM 3.10. *[Su1] Let μ be a partition of length at most n, and k a positive integer. There is a bijection ψ from the set of all n-symplectic up-down tableaux $\tilde{Q}^k_\mu(n)$ of shape μ and length k to the set of all pairs $(lp^{\beta}_{\nu/\mu}(n), T_\nu)$ such that*

- *ν is a partition of k which contains μ; β is a partition of $(k-|\mu|)$ with even columns;*
- *T_ν is a standard Young tableau of shape ν, and*
- *$lp^{\beta}_{\nu/\mu}(n)$ is a lattice permutation of weight β which fits the skew-shape ν/μ n-symplectically (i.e., so that $(2i+1)$ appears no lower than row $(n+i)$).*

Thus one has the formula

$$\tilde{f}^{\mu}_{k}(n) = \sum_{|\nu|=k} f^{\nu} \sum_{\substack{\beta \vdash (k-|\mu|) \\ \beta \text{ has even columns}}} c^{\nu}_{\mu,\beta}(n)$$

where $c^{\nu}_{\mu,\beta}(n)$ is the number of lattice permutations w of weight β, such that w fits the skew-shape ν/μ n-symplectically.

We refer the reader to the Appendix for a description of this bijection.

THEOREM 3.11. *[Su1] Let $w = w_1 \ldots w_k$ be a k-word in $1, \bar{1}, \ldots, n, \bar{n}$. Suppose*

$$\text{(I)} \qquad w \longleftrightarrow (P_\lambda, Q_\lambda) \begin{pmatrix} \text{via Robinson-Schensted insertion;} \\ Q_\lambda \text{ is the standard Young tableau} \end{pmatrix}$$

$$\longleftrightarrow (\tilde{P}_\mu, \tilde{Q}^k_\mu(n)) \text{ (via Berele insertion)}$$

$$\text{(II)} \qquad \longleftrightarrow (\tilde{P}_\mu, lp^{\beta}_{\nu/\mu}(n), T_\nu) \text{ (via the bijection of the previous theorem)}$$

Then $\lambda = \nu$ and $Q_\lambda = T_\nu$.

Recall that if we interpret w as a basis element of the tensor space $\mathbb{C}^{2n}$, then

- from the decomposition of the tensor space with respect to the action of $Gl(2n)$:

$$V^{\otimes k} = \coprod_{\substack{\lambda \vdash k \\ \ell(\lambda) \leq 2n}} (f^{\lambda} \, W^{\lambda}),$$

in (I) we think of the tableau P_λ as a basis element in a weight space of the Weyl module W^λ of $Gl(2n)$, while Q_λ represents the choice of Young symmetriser for W^λ; and likewise,

- from the decomposition of the tensor space with respect to the action of $Sp(2n)$:

$$V^{\otimes k} = \coprod_{\substack{\mu \\ \ell(\mu)\leq \mathbf{n}}} \tilde{f}_k^\mu(n)\ \tilde{W}^\mu$$

$$= \coprod_{\substack{\lambda\vdash k \\ \ell(\lambda)\leq 2n}} f^\lambda \sum_{\substack{\mu\subseteq\lambda \\ \ell(\mu)\leq \mathbf{n}}} \tilde{W}^\mu \left(\sum_{\substack{\beta\vdash(|\lambda|-|\mu|) \\ \beta \text{ has even columns}}} c^\lambda_{\mu,\beta}(\mathbf{n}) \right),$$

in (II) we can think of $\tilde{P}_\mu$ as representing a basis element in a weight space of the $Sp(2n)$-irreducible $\tilde{W}^\mu$, while T_ν represents the choice of Young symmetriser for the $Gl(2n)$-module W^ν.

Hence the preceding result says that the two $Gl(2n)$-modules W^λ and W^ν are in fact the same. We can therefore extract the following rule for the multiplicities in the decomposition of W^λ into $Sp(2n)$-irreducibles:

COROLLARY 3.12. *[Su1] The $Gl(2n)$-module W^λ, $(\ell(\lambda) \leq 2n)$, when restricted to the subgroup $Sp(2n)$, decomposes into irreducibles as follows:*

$$W^\lambda \downarrow^{Gl(2n)}_{Sp(2n)} = \bigoplus_{\mu,\ell(\mu)\leq n} \tilde{W}^\mu \sum_{\substack{\beta\vdash(|\lambda|-|\mu|) \\ \beta \text{ has even columns}}} c^\lambda_{\mu,\beta}(\mathbf{n}).$$

Note that for the case $\ell(\lambda) \leq \mathbf{n}$, the multiplicity reduces to the ordinary Littlewood-Richardson coefficient, agreeing with the classical branching rule known to Weyl and Littlewood:

THEOREM 3.13. *([Li], Appendix, p. 295) If the length of λ does not exceed n, then*

$$W^\lambda \downarrow^{Gl(2n)}_{Sp(2n)} = \bigoplus_{\mu,\ell(\mu)\leq n} \tilde{W}^\mu \left(\sum_{\substack{\beta\vdash(|\lambda|-|\mu|) \\ \beta \text{ has even columns}}} c^\lambda_{\mu,\beta} \right).$$

We now discuss the tensor space decomposition and the existence of insertion schemes for the orthogonal groups. For $SO(2n)$, Littlewood and Weyl presumably knew the analogue of the tensoring rule stated for $Sp(2n)$ in Lemma 3.3, which gives the following:

THEOREM 3.14. *If $k \leq n$, the kth tensor power of the defining representation $V = \mathbb{C}^{2n}$ decomposes into $SO(2n)$-modules $\mathring{W}^\mu$ as follows:*

$$V^{\otimes k} = \coprod_{\substack{\mu \\ \ell(\mu)\leq n}} \tilde{f}_k^\mu(n)\ \mathring{W}^\mu.$$

However no combinatorial proof is known. Also, an explicit formula for the multiplicities in the case $k > n$ is not known.

Fortunately we are able to say a little more about $SO(2n+1)$:

THEOREM 3.15. *[Su3] There is an insertion algorithm S which establishes, for all k, the identity*

$$(x_1 + x_1{}^{-1} + \cdots + x_n + x_n{}^{-1} + 1)^k = \sum_{\substack{\mu \\ \ell(\mu) \leq n}} \tilde{F}_k^{\mu}(n) \ so_{\mu}(x_1^{\pm 1}, \ldots, x_n^{\pm 1}, 1)$$

where $\tilde{F}_k^{\mu}(n)$ is the number of k-sequences of shapes $(\emptyset = \mu^0, \mu^1, \ldots, \mu^k = \mu)$ such that

(1) $\ell(\mu^i) \leq n$ *for all* $i = 1, \ldots, k$,
(2) *either* $\mu^i/\mu^{(i-1)} = (1)$ *or* $\mu^{(i-1)}/\mu^i = (1)$, *or*
(3) $\ell(\mu^{(i-1)}) = n$ *and* $\mu^{(i-1)} = \mu^i$.

Note that this is an identity involving $SO(2n+1)$-characters, and hence lifts up to give the numbers $\tilde{F}_k^{\mu}(n)$ as the multiplicities in the decomposition of $V^{\otimes k}$, where $V = \mathbb{C}^{2n+1}$. Equivalently, the kth tensor power of the defining representation $V = \mathbb{C}^{2n+1}$ decomposes into $SO(2n+1)$-modules $\mathring{W}^{\mu}$ as follows:

$$V^{\otimes k} = \coprod_{\substack{\mu \\ \ell(\mu) \leq n}} \tilde{F}_k^{\mu}(n) \ \mathring{W}^{\mu}.$$

Also observe that if $k \leq n$, $\tilde{F}_k^{\mu}(n) = \tilde{f}_k^{\mu}(n) = \tilde{f}_k^{\mu}$.

There is also a classically known branching rule for the orthogonal groups $SO(m)$ (here m may be even or odd):

THEOREM 3.16. *([Li], p. 240, (II)) If $\ell(\lambda) \leq \lfloor \frac{m}{2} \rfloor$, then the multiplicity of the $SO(m)$-module $\mathring{W}^{\lambda}$ in the $Gl(m)$-module W^{λ} is given by*

$$\sum_{\substack{\gamma \vdash (|\lambda| - |\mu|) \\ \gamma \text{ has even rows}}} c_{\mu,\gamma}^{\lambda}.$$

One can ask whether the branching rule above can be extended to all partitions λ of length at most m. At present, no such extension is known.

4. Dimension formulas and some character identities.

In this section we discuss some consequences of Weyl's character formula for the various groups, and some identities of Weyl and Littlewood involving the irreducible orthogonal and symplectic characters.

The character formula of Weyl ([Hum], p.139) expresses the formal character as a quotient of two sums over the Weyl group:

THEOREM 4.1. *[We] The formal character corresponding to the highest weight vector λ is given by*

$$char\ \phi_\lambda = \frac{\sum_{\sigma\in W_G} sgn(\sigma) e^{<\sigma\,.\,(\lambda+\rho)>}}{\sum_{\sigma\in W_G} sgn(\sigma) e^{<\sigma\,.\,\rho>}}.$$

Here ρ is half the sum of the positive roots of the Lie algebra of G, and $< , >$ denotes the usual (Euclidean) inner product of n-tuples.

To get the character as we know it from Section 1, one interprets the formal exponentials e^{α_i}, where the α_i are the simple roots of the Lie algebra, as the eigenvalues of the generic group element in the maximal torus.

For $Gl(n)$ this gives the Schur function s_λ as the familiar quotient of determinants in Jacobi's original definition (see [Mac]):

THEOREM 4.2. *If λ is a partition of length at most n, then*

$$s_\lambda(x_1,\dots,x_n) = \frac{det({x_i}^{\lambda_i-i+j})}{det({x_i}^{n-j})}, 1\le i,j\le n.$$

Evaluating the principal specialisation $s_\lambda(1,q,\dots,q^{n-1})$ one obtains (see [Mac])

$$s_\lambda(1,q,q^2,\dots,q^{n-1}) = q^{n(\lambda)}\prod_{x\in\lambda}\frac{1-q^{n+c(x)}}{1-q^{h(x)}} \tag{4.3}$$

where

(1) $n(\lambda)=\sum_{i\ge1}(i-1)\lambda_i$;
(2) if $x\in\lambda$ is the cell whose matrix-style coordinates are (i,j) then $c(x)=j-i$, the content of x;
(3) for $x\in\lambda$, $h(x)=\lambda_i-i+\lambda^t_j-j+1$ is the length of the **hook** at x, (λ^t denotes the conjugate (i.e., the transposed Ferrers diagram) of the partition λ (see [Mac])).

Hence, setting $q=1$, one obtains the following formula for the dimension of an irreducible $Gl(n)$-module:

THEOREM 4.4. *The irreducible $Gl(n)$-module indexed by λ has dimension*

$$dim_{Gl}(\lambda,n) = \prod_{x\in\lambda}\frac{n+c(x)}{h(x)},$$

a polynomial in the rank n of the group.

By manipulating the Weyl character formula for the other classical groups, El-Samra and King obtained similar polynomial expressions for the dimensions:

THEOREM 4.5. *[El-K]*

(1) *For* $\ell(\lambda) \leq n$,

$$dim_{Sp}(\lambda, 2n) = \prod_{x \in \lambda} \frac{2n - r_\lambda(x)}{h(x)},$$

where

$$r_\lambda(x) = r_\lambda(i,j) = \begin{cases} \lambda_i^t + \lambda_j^t - i - j, & i \leq j \\ i + j - \lambda_i - \lambda_j - 2, & i > j. \end{cases}$$

(2) *For* $\ell(\lambda) \leq \lfloor \frac{m}{2} \rfloor$

$$dim_{SO}(\lambda, m) = \prod_{x \in \lambda} \frac{m + s_\lambda(x)}{h(x)},$$

where

$$s_\lambda(x) = s_\lambda(i,j) = \begin{cases} \lambda_i + \lambda_j - i - j, & i \geq j \\ i + j - \lambda_i^t - \lambda_j^t - 2, & i < j. \end{cases}$$

The Jacobi-Trudi identities for Schur functions (see [Mac]) can be interpreted as a polynomial expression for the characters s_λ in terms of the characters of the symmetric powers of the defining representation of $Gl(n)$ or in terms of the fundamental characters (the exterior powers of the defining representation):

THEOREM 4.6. *For* $\ell(\lambda) \leq n$,

$$\begin{aligned} s_\lambda(x_1, \dots, x_n) &= det(s_{(\lambda_i - i + j)})_{1 \leq i,j \leq \ell(\lambda)} \\ &= det(s_{(1^{\lambda_i^t - i + j})})_{1 \leq i,j \leq \ell(\lambda)} \end{aligned}$$

Likewise, there are similar formulas for the symplectic and orthogonal groups, due to Weyl. In what follows, we denote by o_λ the character of the special orthogonal group $SO(m)$ indexed by the partition λ.

THEOREM 4.7. *[We] For* $\ell(\lambda) \leq rank(G)$,

(1) *(See [We], p.219, Theorem 7.8E)*

$$(Sp_{JT}) \qquad sp_\lambda(x_1^{\pm 1}, \dots, x_n^{\pm 1}) = \frac{1}{2} det(sp_{(\lambda_i - i - j + 2)} + sp_{(\lambda_i - i + j)})_{1 \leq i,j \leq \ell(\lambda)}$$

Note that the character $sp_{(k)}$ *is the same as* $s_{(k)} = h_k(x_1^{\pm 1}, \dots, x_n^{\pm 1})$, *where* h_k *is the k-th homogeneous symmetric function (see [Mac]).*

(2) *([We], p.228 Theorem 7.9A) (See [Li], p.233, for an accessible proof.)*

$$(So_{JT}) \qquad o_\lambda = det(s_{(\lambda_i - i - j)} + s_{(\lambda_i - i + j)})_{1 \leq i,j \leq \ell(\lambda)},$$

where the (ordinary) Schur function $s_{(k)}$ *(*$=h_k$*) is to be taken in the variables* $\{x_1^{\pm 1}, \dots, x_n^{\pm 1}\}$ *or* $\{x_1^{\pm 1}, \dots, x_n^{\pm 1}, 1\}$ *according as the orthogonal group is even or odd.*

In contrast to the well-developed theory of enumerating non-intersecting lattice paths (see [GV], for example) which provides combinatorial proofs of the Jacobi-Trudi identities for the ordinary Schur functions, no bijections establishing the above formulas for the orthogonal and symplectic characters are known.

It is easily seen from the dimension formulas in Theorem 4.5 that

$$(-1)^{|\lambda|} dim_{Sp}(\lambda, 2n) = dim_{SO}(\lambda^t, -2n),$$

although of course this makes sense only if $\ell(\lambda) \leq n$ and $\ell(\lambda^t) \leq n$.

Koike and Terada [KT1] define a universal character ring which enables them to extend formally the definition of the characters to infinitely many variables (i.e., to partitions of arbitrary length), by using the determinantal expansions above. In this ring they can then define the involution ω which takes s_λ to s_{λ^t} (see [Mac]), and they obtain (for λ of arbitrary length)

$$\omega(sp_\lambda) = o_{\lambda^t},$$
$$\omega(o_\lambda) = sp_{\lambda^t}.$$

(That is, ω maps symplectic characters to orthogonal characters, and vice versa.)

Notice that this "duality" between symplectic and orthogonal characters was already reflected in the branching rules in Section 3. One can ask whether it is possible to understand this at the module level, or perhaps combinatorially (i.e., in terms of tableaux).

We shall see more manifestations of this phenomenon later.

We now discuss another set of character equations, popularly referred to as the Cauchy identities. Again, these seem to appear in [We], although heavily disguised by cumbersome notation. Before stating these identities, we remind the reader that the formal characters are functions of the eigenvalues of (a generic element in the maximal torus of) the group, which are

(1) $\{x_1, \dots, x_m\}$ for $Gl(m)$;
(2) $\{x_1{}^{\pm 1}, \dots, x_n{}^{\pm 1}\}$ for $Sp(2n), SO(2n)$;
(3) $\{x_1{}^{\pm 1}, \dots, x_n{}^{\pm 1}, 1\}$ for $SO(2n+1)$.

Now fix n, and let V be the defining representation of $Gl(n)$; let G be any of the classical Lie groups, and let W be the defining representation for G. Write $t_1, \dots, t_n$ for the eigenvalues of $Gl(n)$. Then the symmetric algebra $Sym(V \otimes W)$ of the module $V \otimes W$ is a $(Gl(n) \times G)$-module, whose character is

$$\prod_{i=1}^{n} \prod_{\substack{z \text{ runs over all} \\ \text{eigenvalues of an element in} \\ \text{the maximal torus of } G}} (1 - t_i z)^{-1}.$$

THEOREM 4.8. *Write $s_\lambda(t)$ for the Schur function s_λ in the variables $t_1, \dots, t_n$. Then*

(1) *For the general linear group $Gl(m)$:*

$$\prod_{i=1}^{n} \prod_{j=1}^{m} (1 - t_i x_j)^{-1} = \sum_{\lambda} s_\lambda(t) s_\lambda(x_1, \dots, x_m).$$

(2) *For the symplectic group $Sp(2n)$,*

$$\prod_{1\le i<j\le n}(1-t_it_j)\prod_{i=1}^{n}\prod_{\substack{z \text{ runs over all}\\ \text{eigenvalues of an element}\\ \text{in the maximal torus}}}(1-t_iz)^{-1}=\sum_{\lambda}s_\lambda(t)sp_\lambda.$$

(3) *For the orthogonal groups $SO(m)$,*

$$\prod_{1\le i\le j\le n}(1-t_it_j)\prod_{i=1}^{n}\prod_{\substack{z \text{ runs over all}\\ \text{eigenvalues of an element}\\ \text{in the maximal torus}}}(1-t_iz)^{-1}=\sum_{\lambda}s_\lambda(t)o_\lambda.$$

Proof.

(1) Various algebraic ways to prove this are given in [Mac]. A combinatorial proof is supplied by Knuth's [Kn] generalisation of Robinson-Schensted insertion, also described in [S] in this volume.

(2) We can show this very quickly by using a couple of facts about symmetric functions: Littlewood's identity ([Li]) (see also [Mac])

$$\text{(L1)}\qquad \prod_{1\le i<j\le n}(1-t_it_j)^{-1}=\sum_{\substack{\beta,\ell(\beta)\le n\\ \beta \text{ has even columns}}}s_\beta(t_1,\dots,t_n)$$

and the fact that the Littlewood-Richardson coefficient $c^\lambda_{\mu,\nu}$ is the coefficient of the Schur function s_λ in the product $s_\mu\ s_\nu$. Also the branching rule of Theorem 3.13 translates into the following character identity:

$$s_\lambda(x_1^{\pm1},\dots,x_n^{\pm1})=\sum_{\substack{\mu\\ \ell(\mu)\le n}}sp_\mu\left(\sum_{\substack{\beta\vdash(|\lambda|-|\mu|)\\ \beta \text{ has even columns}}}c^\lambda_{\mu,\beta}\right).$$

Now substitute the latter expression for $s_\lambda(x_1^{\pm1},\dots,x_n^{\pm1})$ in

$$\sum_{\substack{\lambda\\ \ell(\lambda)\le n}}s_\lambda(t_1,\dots t_n)\,s_\lambda(x_1^{\pm1},\dots,x_n^{\pm1})=\prod_{i,j=1}^{n}(1-t_ix_j{}^{\pm1})^{-1}$$

and re-arrange terms to get (2). (3) This follows exactly as in (2) above, except that we now need to use the branching rule of Theorem 3.16, and the following identity, again due to Littlewood:

$$\text{(L2)}\qquad \prod_{1\le i\le j\le n}(1-t_it_j)^{-1}=\sum_{\substack{\gamma,\ell(\gamma)\le n\\ \gamma \text{ has even rows}}}s_\gamma(t_1,\dots,t_n).\quad\square$$

Combinatorial proofs of the Cauchy identities for $Sp(2n)$, $SO(2n+1)$ were found by this author (see [Su2], [Su3]). Once again, we refer the reader to the Appendix for a brief description of the bijection establishing the Cauchy identity for $Sp(2n)$.

Each of the above identities has a dual form, which we state next. Again fix n, and let V be the defining representation of $Gl(n)$; let G be any of the classical Lie groups, and let W be the defining representation for G. Write $t_1, \ldots, t_n$ for the eigenvalues of $Gl(n)$. Then the exterior algebra $\bigwedge(V \otimes W)$ of the module $V \otimes W$ is a $(Gl(n) \times G)$-module, whose character is

$$\prod_{i=1}^{n} \prod_{\substack{z \text{ runs over all} \\ \text{eigenvalues of an element in} \\ \text{the maximal torus of } G}} (1 + t_i z).$$

THEOREM 4.9. *Write $s_\lambda(t)$ for the Schur function s_λ in the variables $t_1, \ldots, t_n$. Then*

(1) *For the general linear group $Gl(m)$:*

$$\prod_{i=1}^{n} \prod_{j=1}^{m} (1 + t_i x_j) = \sum_\lambda s_{\lambda^t}(t) s_\lambda(x_1 \ldots x_m).$$

(2) *For the symplectic group $Sp(2n)$,*

$$\prod_{1 \le i \le j \le n} (1 - t_i t_j) \prod_{i=1}^{n} \prod_{\substack{z \text{ runs over all} \\ \text{eigenvalues of an element} \\ \text{in the maximal torus}}} (1 + t_i z) - \sum_\lambda s_{\lambda^t}(t) sp_\lambda.$$

(3) *For the orthogonal groups $SO(m)$,*

$$\prod_{1 \le i < j \le n} (1 - t_i t_j) \prod_{i=1}^{n} \prod_{\substack{z \text{ runs over all} \\ \text{eigenvalues of an element} \\ \text{in the maximal torus}}} (1 + t_i z) = \sum_\lambda s_{\lambda^t}(t) o_\lambda.$$

Proof. Each of the Cauchy identities in Theorem 4.8 is a symmetric function identity in the ring $\mathbb{Z}[t_1, \ldots, t_n]^{S_n} \otimes \mathbb{Z}[z]^{W_G}$ where W_G is the appropriate Weyl group, and as usual the variables z run over all the eigenvalues of a generic element in the maximal torus of the group. By applying the automorphism ω in $\mathbb{Z}[t_1, \ldots, t_n]^{S_n}$ which takes s_λ to s_{λ^t}, one gets the corresponding dual identities. □

Combinatorial proofs of the dual identities for $Sp(2n)$, $SO(2n+1)$ were also given by this author [Su2].

The identities (L1), (L2) come with the following inverses, also due to Littlewood ([Li], p. 238). We let $\left(\begin{smallmatrix} a_1 & a_2 & \ldots & a_r \\ b_1 & b_2 & \ldots & b_r \end{smallmatrix}\right)$ denote the partition λ in Frobenius notation, so that $\lambda_i - i = a_i; \lambda_i^t - i = b_i, i = 1, \ldots, r$. If $r = 0$, interpret this to be the empty partition. Then

(L1')
$$\prod_{1\leq i<j\leq n} (1 - t_i t_j) = \sum_{\substack{r\geq 0 \\ \Gamma=\left(\begin{smallmatrix} a_1 & a_2 & \ldots & a_r \\ a_1+1 & a_2+1 & \ldots & a_r+1 \end{smallmatrix}\right)}} (-1)^{\frac{|\Gamma|}{2}} s_\Gamma(t)$$

(L2')
$$\prod_{1\leq i\leq j\leq n} (1 - t_i t_j) = \sum_{\substack{r\geq 0 \\ \Delta=\left(\begin{smallmatrix} a_1+1 & a_2+1 & \ldots & a_r+1 \\ a_1 & a_2 & \ldots & a_r \end{smallmatrix}\right)}} (-1)^{\frac{|\Delta|}{2}} s_\Delta(t).$$

Following Littlewood, we shall use these identities to invert the branching rule formulas of Section 3, thereby expressing the symplectic and orthogonal characters as integer linear combinations of the Schur functions in the variables $\{x_1^{\pm 1}, \ldots, x_n^{\pm 1}\}$.

Remark. The identities (L1'), (L2') in fact coincide with Weyl's identities for the root systems D_n, C_n. See [Mac] (pp. 46-47, Ex.9). For a unified combinatorial treatment of the four identities (L1),(L2),(L1'),(L2'), see [Bu].

We shall need a fact about Schur functions:

PROPOSITION 4.10. *[Su1], [Su3] Let $F(t)$ be a function in the ring $\bigwedge(t)$ of symmetric formal power series (over $\mathbb{Z}$) in the n variables $t_1, \ldots, t_n$, such that $F(0) = 1$ (so $F^{-1}(t)$ is also in $\bigwedge(t)$).*

Then the matrices $(< s_{\lambda/\mu}, F(t) >)_{\lambda,\mu}$ and $(< s_{\lambda/\mu}, F^{-1}(t) >)_{\lambda,\mu}$ (rows and columns indexed by $Par_n = \{\nu : \ell(\nu) \leq n\}$ with some total ordering) are inverses of each other.

Here $< .,. >$ denotes the inner product on the ring of symmetric functions with respect to which the Schur functions form an orthonormal basis (see [Mac]).

Rewriting the branching rules (Theorems 3.13, 3.16) of Section 3 in terms of characters, one gets

THEOREM 4.11. *In the ring $\mathbb{Z}[x_1^{\pm 1}, \ldots, x_n^{\pm 1}]^{B_n}$, the following identities hold:*

- *Assume λ has at most n parts. Then*

(L3)
$$s_\lambda(x_1^{\pm 1}, \ldots, x_n^{\pm 1}) = \sum_{\substack{\mu \\ \ell(\mu)\leq n}} sp_\mu(x_1^{\pm 1}, \ldots, x_n^{\pm 1}) \left(\sum_{\substack{\beta\vdash(|\lambda|-|\mu|) \\ \beta \text{ has even columns}}} c^\lambda_{\mu,\beta} \right).$$

- *Assume λ has at most n parts. Let o_μ denote the $SO(2n)$ character corresponding to μ. Then*

(L4)
$$s_\lambda(x_1^{\pm 1}, \ldots, x_n^{\pm 1}) = \sum_{\substack{\mu \\ \ell(\mu)\leq n}} o_\mu(x_1^{\pm 1}, \ldots, x_n^{\pm 1}) \left(\sum_{\substack{\gamma\vdash(|\lambda|-|\mu|) \\ \gamma \text{ has even rows}}} c^\lambda_{\mu,\gamma} \right).$$

For $SO(2n+1)$ we have $o_\mu(x_1^{\pm 1},\dots,x_n^{\pm 1}) = so_\mu(x_1^{\pm 1},\dots,x_n^{\pm 1},1)$, and hence we must also replace $s_\lambda(x_1^{\pm 1},\dots,x_n^{\pm 1})$ in the above equation by $s_\lambda(x_1^{\pm 1},\dots,x_n^{\pm 1},1)$ ($(2n+1)$ variables).

Remark. Direct combinatorial proofs of these identities are not known.

Now observe that if we let $F(t) = \prod_{1\le i<j\le n}(1-t_it_j)^{-1}$, then (L3) says that the coefficient of sp_μ in s_λ is $< s_{\lambda/\mu}, F(t) >$. We can make a similar remark for (L4). Hence by Proposition 4.10 and (L1'), (L2'), we get Littlewood's formulas:

THEOREM 4.12. *([Li], p.240, p.295) In the ring $\mathbf{Z}[x_1^{\pm 1},\dots,x_n^{\pm 1}]^{B_n}$, the following identities hold:*

- *Assume λ has at most n parts. Then*

(L3')
$$sp_\lambda(x_1^{\pm 1},\dots,x_n^{\pm 1}) = \sum_{\substack{\mu \\ \mu\subseteq\lambda}} s_\mu(x_1^{\pm 1},\dots,x_n^{\pm 1}) \left(\sum_{\substack{r\ge 0 \\ \Gamma=\begin{pmatrix} a_1 & a_2 & \dots & a_r \\ a_1+1 & a_2+1 & \dots & a_r+1\end{pmatrix}}} (-1)^{\frac{|\Gamma|}{2}} c^\lambda_{\mu,\Gamma} \right).$$

- *Assume λ has at most n parts. Let o_μ denote the $SO(2n)$ character corresponding to μ. Then*

(L4')
$$o_\lambda(x_1^{\pm 1},\dots,x_n^{\pm 1}) = \sum_{\substack{\mu \\ \mu\subseteq\lambda}} s_\mu(x_1^{\pm 1},\dots,x_n^{\pm 1}) \left(\sum_{\substack{r\ge 0 \\ \Delta=\begin{pmatrix} a_1+1 & a_2+1 & \dots & a_r+1 \\ a_1 & a_2 & \dots & a_r\end{pmatrix}}} (-1)^{\frac{|\Delta|}{2}} c^\lambda_{\mu,\Delta} \right).$$

Again for $SO(2n+1)$ we have $o_\mu(x_1^{\pm 1},\dots,x_n^{\pm 1}) = so_\mu(x_1^{\pm 1},\dots,x_n^{\pm 1},1)$, and hence we must also replace $s_\lambda(x_1^{\pm 1},\dots,x_n^{\pm 1})$ in the above equation by $s_\lambda(x_1^{\pm 1},\dots,x_n^{\pm 1},1)$ ($(2n+1)$ variables).

COROLLARY 4.13. *The Schur functions $\{s_\mu(x_1^{\pm 1},\dots,x_n^{\pm 1}) : \ell(\mu)\le n\}$ form a $\mathbf{Z}$-basis for the ring $\mathbf{Z}[x_1^{\pm 1},\dots,x_n^{\pm 1}]^{B_n}$.*

Proof. This is clear from either of the identities (L3'), (L4'), since for instance, the symplectic Schur functions $sp_\lambda, \ell(\lambda)\le n$, form a $\mathbf{Z}$-basis for the ring $\mathbf{Z}[x_1^{\pm 1},\dots,x_n^{\pm 1}]^{B_n}$. □

We close this section with some remarks about invariants. Recall that every Lie group has an adjoint action *ad* on its Lie algebra. Consider the G-modules $Sym(ad)$ and $\bigwedge(ad)$ (the symmetric and exterior algebras respectively). Then one can compute the Hilbert series of the ring of invariants of these two spaces using Littlewood's identities (L1), (L2), (L1') and (L2'). We confine ourselves to a statement of the results:

THEOREM 4.14. *([We], p. 238) (Also [Li2])*

(1) *For $Sp(2n)$ and $SO(2n+1)$ the Poincaré series (the Hilbert series of the invariants of $\bigwedge(ad)$) is*

$$F(Inv \bigwedge(ad), q) = \prod_{i=1}^{n}(1+q^{4i-1}).$$

(2) *For $SO(2n)$ the Poincaré series is*

$$F(Inv \bigwedge(ad), q) = \prod_{i=1}^{n-1}(1+q^{4i-1})\ (1+q^{2n-1}).$$

Weyl's derivation involves the computation of certain integrals; Littlewood (in [Li2]) uses the identities (L1), (L2) , (L1') and (L2') and the branching rules. In [Su1] we use the techniques of [St] in conjunction with these identities, first calculating the symmetric and exterior powers of ad as certain plethysms. (For the details of this computation for $Sp(2n)$, see [Su1]). Using these techniques we also obtain (see [Su1]) the following formula for the Hilbert series of the space of invariants of $Sym(ad)$ for $Sp(2n)$:

$$F(Inv\ Sym(ad)), q) = \prod_{k=1}^{n}(1-q^{2k})^{-1}.$$

5. Littlewood-Richardson rules and virtual characters.

Suppose $W, V,$ are irreducible G-modules. One would like to know how the tensor product $W \otimes V$ decomposes into irreducibles.

For $Gl(n)$ this question is answered by the **Littlewood-Richardson** rule:

THEOREM 5.1. *[LR] The multiplicity of the irreducible $Gl(n)$-module W^λ in the tensor product $W^\mu \otimes W^\nu$ (λ, μ, ν all of length $\leq n$) is the Littlewood-Richardson coefficient $c^\lambda_{\mu,\nu}$, i.e., the number of lattice permutations of weight ν which fit the skew-shape λ/μ. (See Section 3). In terms of characters this gives us a rule for expressing the product of two Schur functions as a linear combination of Schur functions:*

$$\text{(LR)} \qquad s_\mu(x_1,\dots,x_n)\, s_\nu(x_1,\dots,x_n) = \sum_\lambda c^\lambda_{\mu,\nu} s_\lambda(x_1,\dots,x_n).$$

A very natural and elegant combinatorial proof of (LR) was given by Thomas [Th]. The idea of the bijection is as follows: Take a pair of tableaux (P_μ, P_ν), one for each representation on the left-hand side. (So P_μ (respectively P_ν) indexes a weight of W^μ, (respectively W^ν.) Form a word out of P_ν by reading each row from left-to-right, and concatenating these, starting with the first row and moving down. Now column-insert this word into the tableau P_μ, producing a "bigger" tableau

P_λ, i.e., the shape λ contains the shape μ. Notice that precisely this procedure was carried out to obtain the bijection of Theorem 3.13 (see Appendix). The successive insertions must be suitably recorded so that the process can be reversed uniquely, and it is the labelling which keeps track of the insertions that ultimately yields the lattice permutation .

It is natural to ask whether this idea can be exploited using the insertion algorithms for the other Lie groups. Unfortunately the process seems to yield nice results only in special cases:

THEOREM 5.2. *([Su2], [Su3]) Assume* $\ell(\lambda)$, $\ell(\mu) \leq n$.

(1) *For* $Sp(2n)$ *the multiplicity of* $\tilde{W}^\mu$ *in* $\tilde{W}^\lambda \otimes \tilde{W}^k$ *is the number of partitions* $\nu, \ell(\nu) \leq n$, *such that* $\lambda \supseteq \nu \subseteq \mu$, *and*

$$\lambda/\nu, \mu/\nu \text{ are horizontal strips, } |\lambda/\nu| + |\mu/\nu| = k.$$

(2) *For* $SO(2n+1)$ *the multiplicity of* $\overset{\circ}{W}{}^\mu$ *in* $\overset{\circ}{W}{}^\lambda \otimes \overset{\circ}{W}{}^{(k)}$ *is the number of partitions* $\nu, \ell(\nu) \leq n$, *such that* $\lambda \supseteq \nu \subseteq \mu$, *and*

$\lambda/\nu, \mu/\nu$ *are horizontal strips, and*
$|\lambda/\nu| + |\mu/\nu| = k$ *or* $(k-1)$, *the latter occurring only if* $\ell(\nu) = n$.

Recall that the characters *char* ϕ_λ of a group G are defined only for partitions λ whose length does not exceed a prescribed number (the rank of the group). However, using the Jacobi-Trudi-type identities of Theorem 4.7, one can formally extend the definitions of the characters to all partitions. The new functions cannot always be thought of as characters of irreducible representations, since they can have negative (integer) coefficients. It turns out that they are **virtual** characters, and in fact it can be shown that for $\ell(\lambda)$ larger than the rank of the group, each *char* ϕ_λ is either zero or (± 1) times an ordinary irreducible character.

Having defined these virtual characters using equations (Sp_{JT}) and (SO_{JT}) of Theorem 4.7, we can state a rule of Littlewood and Newell, describing the decomposition of the tensor product of two irreducibles. We shall state their rule in terms of character identities:

THEOREM 5.3. *[BKW]*

(1) *For* $Sp(2n)$, *we have the multiplication rule*

$$sp_\mu \cdot sp_\nu = \sum_{\text{all } \lambda} sp_\lambda \left(\sum_{\xi,\eta} c^\lambda_{\xi,\eta} \sum_\tau c^\mu_{\xi,\tau} c^\nu_{\eta,\tau} \right),$$

where for $\ell(\lambda) > n, sp_\lambda$ *is defined by equation* (Sp_{JT}) *of Theorem 4.7.*

(2) *For* $SO(m)$, *we have the multiplication rule*

$$o_\mu \cdot o_\nu = \sum_{\text{all } \lambda} o_\lambda \left(\sum_{\xi,\eta} c^\lambda_{\xi,\eta} \sum_\tau c^\mu_{\xi,\tau} c^\nu_{\eta,\tau} \right),$$

where for $\ell(\lambda) > n, o_\lambda$ *is defined by equation* (SO_{JT}) *of Theorem 4.7.*

Since the above formulas involve virtual characters, it is desirable to have a formula that does not have cancellations, but identifies precisely the coefficient of an irreducible in the product.

Combinatorial rules for evaluating the virtual characters of $Sp(2n)$, $SO(m)$, have been worked out by King [Ki2]. We now present these "modification" rules for the symplectic and orthogonal groups:

THEOREM 5.4. *[Ki2] Let $\ell(\lambda) > n$. Let h be any positive integer. Starting from the lowest row and first column of λ, mark off the longest possible connected region R_λ of corner-cells of λ having no two-by-two squares (such a region of λ is called a* **border-strip** *(see [Mac]). If this maximal border-strip R_λ has length strictly less than h, set $R_\lambda(h) = \varnothing$.*

Otherwise, let $R_\lambda(h)$ be the sub-border-strip of R_λ which starts at the cell in the first column and the lowest row, and has length h; let c be the number of columns occupied by $R_\lambda(h)$, and lastly, let μ be the shape obtained by removing $R_\lambda(h)$ from λ, i.e.,

$$\mu \text{ is such that } \lambda/\mu = R_\lambda(h).$$

Then

Let $h = 2\ell(\lambda) - 2(n+1)$.

$$\text{(Sp(2n))} \qquad \begin{aligned} & sp_\lambda(x_1^{\pm 1}, \dots, x_n^{\pm 1}) \\ &= \begin{cases} 0, & \text{if } R_\lambda(h) = \varnothing \\ (-1)^c \, sp_\mu, & \text{otherwise.} \end{cases} \end{aligned}$$

Let $h = 2\ell(\lambda) - 2n$.

$$\text{(SO(2n))} \qquad \begin{aligned} & o_\lambda(x_1^{\pm 1}, \dots, x_n^{\pm 1}) \\ &= \begin{cases} 0, & \text{if } R_\lambda(h) = \varnothing \\ (-1)^{c+1} \, o_\mu, & \text{otherwise.} \end{cases} \end{aligned}$$

Let $h = 2\ell(\lambda) - (2n+1)$.

$$\text{(SO(2n+1))} \qquad \begin{aligned} & so_\lambda(x_1^{\pm 1}, \dots, x_n^{\pm 1}, 1) \\ &= \begin{cases} 0, & \text{if } R_\lambda(h) = \varnothing \\ (-1)^{c+1} \, o_\mu, & \text{otherwise.} \end{cases} \end{aligned}$$

Example. We illustrate the rule for $Sp(6)$, taking $\lambda = (3, 2, 1, 1, 1)$. Here $n = 3$, and thus $h = 2\ell(\lambda) - 2(4) = 2$. The unique border-strip of length 2 which starts at the lowest square in the first column consists of the crossed boxes in the Ferrers diagram

□□□
□□
□
⊠
⊠

of λ, and occupies one column, so $c = 1$. Clearly removing the cells in the border-strip leaves the shape $\mu = (3, 2, 1)$ and hence $sp_{(3,2,1,1,1)} = (-1)\, sp_{(3,2,1)}$.

For $Sp(4)$ (i.e., if we took $n = 2$), $h = 4$ and there is no border-strip of size 4 which starts at the lowest cell of the first column, as is clear from the four crossed boxes in the diagram □□□ / ⊠□ / ⊠ / ⊠ / ⊠. Hence as an $Sp(4)$-character, $sp_{(3,2,1,1,1)}$ vanishes.

Another set of modification rules is described in [KT1]; the equivalence between these two rules is not clear.

Observe that we could also use the identities (L3'), (L4') to extend the definitions of sp_λ, o_λ to partitions of arbitrary length. It would be interesting to prove combinatorially the equivalence of these two definitions. As a specific example, we consider the symplectic characters. Then from algebraic considerations, one knows that for any partition λ, the expressions $S^1_\lambda(x_1^{\pm 1}, \ldots, x_n^{\pm 1})$

$$= \frac{1}{2} det(sp_{(\lambda_i - i - j + 2)} + sp_{(\lambda_i - i + j)})_{1 \le i,j \le \ell(\lambda)}$$

and $S^2_\lambda(x_1^{\pm 1}, \ldots, x_n^{\pm 1})$

$$= \sum_{\substack{\mu \\ \mu \subseteq \lambda}} s_\mu(x_1^{\pm 1}, \ldots, x_n^{\pm 1}) \left(\sum_{\substack{r \ge 0 \\ \Gamma = \begin{pmatrix} a_1 & a_2 & \ldots & a_r \\ a_1+1 & a_2+1 & \ldots & a_r+1 \end{pmatrix}}} (-1)^{\frac{|\lambda|}{2}} c^\lambda_{\mu,\Gamma} \right)$$

define the same function in the ring $\mathbb{Z}[x_1^{\pm 1}, \ldots, x_n^{\pm 1}]^{B_n}$, namely the sp_λ which satisfies King's modification rules in Theorem (5.4). That is, one can show that

- For $\ell(\lambda) \le n$, $S^1_\lambda = S^2_\lambda = sp_\lambda =$ an irreducible $Sp(2n)$-character.
- For $\ell(\lambda) > n, S^1_\lambda$ obeys the modification laws in (1) of Theorem 5.4. (See [Ki2]).
- For all λ, the functions S^2_λ satisfy the branching rule equation

$$s_\lambda(x_1^{\pm 1}, \ldots, x_n^{\pm 1})$$

$$= \sum_{\substack{\mu \\ \ell(\mu) \le n}} S^2_\mu(x_1^{\pm 1}, \ldots, x_n^{\pm 1}) \left(\sum_{\substack{\beta \vdash (|\lambda| - |\mu|) \\ \beta \text{ has even columns}}} c^\lambda_{\mu,\beta} \right).$$

This follows from (L1), (L1') and Proposition 4.10.

There are possibly very interesting combinatorial reasons explaining why $S^1_\lambda = S^2_\lambda$, or why S^2_λ should also obey the modification rules of Theorem 5.4.

6. The centraliser algebras.

The theorem of Schur with which we began Section 3 is also known as the double centraliser theorem and may be stated in the following alternative form. Let $\mathbb{C}S_k$ denote the group algebra of the symmetric group S_k, let V be the defining representation of $Gl(n)$ and let $End_{Gl(n)}(V^{\otimes k})$ be the centraliser algebra (also known as the commuting algebra) of $Gl(n)$, i.e., the space of all endomorphisms of $V^{\otimes k}$ which commute with the action of $Gl(n)$.

Then

THEOREM 6.1. *There is a surjective algebra homomorphism*

$$\Psi : \mathbb{C}S_k \mapsto End_{Gl(n)}(V^{\otimes k})$$

with kernel $\bigoplus_{\substack{\lambda \vdash k \\ \ell(\lambda) > n}} I_\lambda$, *where* I_λ *is the minimal two-sided ideal of the group algebra* $\mathbb{C}S_k$ *corresponding to the partition* λ *of* k.

We refer the reader to [Boe] for facts on semisimple algebras. Using the well-known fact from the representation theory of S_k that the group algebra of S_k is the direct sum of its minimal two-sided ideals (which are indexed by all the partitions of k) this gives the decomposition

$$End_{Gl(n)}(V^{\otimes k}) \approx \bigoplus_{\substack{\lambda \vdash k \\ \ell(\lambda) \leq n}} I_\lambda; \tag{6.2}$$

that is, the semisimple quotient of the map Ψ is the algebra $\bigoplus_{\substack{\lambda \vdash k \\ \ell(\lambda) \leq n}} I_\lambda$. Hence for $k \leq n$,

$$End_{Gl(n)}(V^{\otimes k}) \approx \mathbb{C}S_k.$$

Richard Brauer started the investigation of the analogous situation for the other classical groups in [Br]. (The ultimate aim was to "find" the irreducible modules for these groups in the tensor space, mimicking Schur's decomposition result of Theorem 3.1.) He constructed the Brauer centraliser algebras $\mathcal{A}_k^{(x)}$, $\mathcal{B}_k^{(x)}$. These algebras, which depend on an indeterminate x, have the property that they map homomorphically onto the centraliser algebras $End_{O(x)}(V^{\otimes k})$, $End_{Sp(x)}(V^{\otimes k})$ respectively, when x is a positive integer:

$$\begin{aligned} \Psi_{O(x)} &: \mathcal{A}_k^{(x)} \mapsto End_{O(x)}(V^{\otimes k}) \mapsto 0; \\ \Psi_{Sp(x)} &: \mathcal{B}_k^{(x)} \mapsto End_{Sp(x)}(V^{\otimes k}) \mapsto 0 \quad (dim V = x). \end{aligned}$$

He also showed that these maps were isomorphisms if $k \leq \lfloor \frac{x}{2} \rfloor$, x a positive integer, and that, for x an indeterminate,

$$\mathcal{A}_k^{(x)} \approx \mathcal{B}_k^{(-x)};$$

this latter statement is another manifestation of the duality between the orthogonal and symplectic groups. (In Section 4 we noticed that $(-1)^{|\lambda|} dim_{Sp}(\lambda, 2n) = dim_{SO}(\lambda^t, -2n)$.)

Hanlon and Wales, in a series of papers [HW], have attempted to answer questions about the algebra $\mathcal{A}_k^{(x)}$ by studying its representations. One problem is to identify the semisimple quotient $\mathcal{A}_k^{(x)}/ker(\Psi_{O(x)})$ of the Brauer algebra. This is semisimple because it is isomorphic to the centraliser algebra of $O(x)$. In general, $\mathcal{A}_k^{(x)}$ need not be semisimple; however, Hanlon and Wales conjectured, and some time thereafter Hans Wenzl proved, that

THEOREM 6.3. *[Wz] If x is not an integer, $\mathcal{A}_k^{(x)}$ is semisimple.*

Wenzl also describes the kernel of the mapping $\Psi_{O(x)}$, in terms of the annihilator of a certain trace form.

Theorem 6.1 and Equation 6.2 account for the existence of the Frobenius characteristic map ch, which is an isometry from the character ring of S_k to the character ring $\mathbf{Z}[x_1, \ldots, x_n]^{S_k}$, for $n \geq k$, defined by :

If χ^λ is an irreducible S_k-character (so $\lambda \vdash k$), then $ch(\chi^\lambda) = s_\lambda(x_1, \ldots, x_n)$. (See [Mac]).

In what follows we give Schur's argument to explain this connection. For clarity we re-state Theorem 3.1. (As usual, V affords the defining representation for $Gl(n)$, so has dimension n.) Under the action of $S_k \times Gl(n)$ described below, $V^{\otimes k}$ decomposes into irreducibles as follows:

(DC)
$$V^{\otimes k} = \bigoplus_{\substack{\lambda \vdash k \\ \ell(\lambda) \leq n}} (S^\lambda \otimes W^\lambda).$$

Here W^λ is the $Gl(n)$-irreducible, and S^λ is the S_k-irreducible (Specht module) corresponding to the partition λ.

The action in question is simply the natural action of $Gl(n)$, followed by the action of S_k which permutes the k tensor positions. Specifically, let $\sigma \in S_k$ and $A \in Gl(n)$, and let $\{e_1, \ldots, e_n\}$ be the usual coordinate basis for V. Then for any multiset of indices $1 \leq j_1, \ldots, j_k \leq n$, $e_{j_1} \otimes \ldots \otimes e_{j_k}$ is a basis element of the tensor space and the $Gl(n)$-action is simply the diagonal action defined by

$$A\,(e_{j_1} \otimes \ldots \otimes e_{j_k}) = \sum_{1 \leq i_1, \ldots, i_k \leq n} e_{i_1} \otimes \ldots \otimes e_{i_k} \prod_{r=1}^{k} a_{i_r j_r},$$

while the S_k action is clearly given by

$$\sigma\,(e_{j_1} \otimes \ldots \otimes e_{j_k}) = e_{j_{\sigma^{-1}(1)}} \otimes \ldots \otimes e_{j_{\sigma^{-1}(k)}}.$$

One verifies easily that

$$\sigma\,(A\,(e_{j_1} \otimes \ldots \otimes e_{j_k})) = \sum_{1 \leq i_1, \ldots, i_k \leq n} e_{i_{\sigma^{-1}(1)}} \otimes \ldots \otimes e_{i_{\sigma^{-1}(k)}} \prod_{r=1}^{k} a_{i_r j_r},$$

and likewise

$$\begin{aligned} A\,(\sigma\,(e_{j_1} \otimes \ldots \otimes e_{j_k})) &= A\; e_{j_{\sigma^{-1}(1)}} \otimes \ldots \otimes e_{j_{\sigma^{-1}(k)}} \\ &= \sum_{1 \leq i_1, \ldots, i_k \leq n} e_{i_1} \otimes \ldots \otimes e_{i_k} \prod_{r=1}^{k} a_{i_r j_{\sigma^{-1}(r)}} \\ &= \sum_{1 \leq i_1, \ldots, i_k \leq n} e_{i_1} \otimes \ldots \otimes e_{i_k} \prod_{r=1}^{k} a_{i_{\sigma(r)} j_r} \end{aligned}$$

Hence $\sigma \circ A = A \circ \sigma$, i.e., the two actions commute, so that $(\sigma, A) \mapsto \sigma \circ A$ defines an action of $S_k \times Gl(n)$ on $V^{\otimes k}$.

Now take A to be in the maximal torus, so that A is diagonal and the ith diagonal element is x_i, say. Suppose σ in S_k has cycle-type (type σ) $= \mu$. We compute the trace of the element (σ, A) of $S_k \times Gl(n)$ acting on $V^{\otimes k}$ in two different ways:

(1) From (DC), we have

$$\begin{aligned} tr(\sigma, A) &= \sum_{\substack{\lambda \vdash k \\ \ell(\lambda) \leq n}} tr(\sigma \circ A) \downarrow_{S^\lambda \otimes W^\lambda} \\ &= \sum_{\substack{\lambda \vdash k \\ \ell(\lambda) \leq n}} tr(\sigma \downarrow_{S^\lambda}) \,.\, tr(A \downarrow_{W^\lambda}) \\ &= \sum_{\substack{\lambda \vdash k \\ \ell(\lambda) \leq n}} \chi^\lambda(\sigma) \,.\, s_\lambda(x_1, \ldots, x_n) \\ &= \sum_{\substack{\lambda \vdash k \\ \ell(\lambda) \leq n}} \chi^\lambda(\mu) \,.\, s_\lambda(x_1, \ldots, x_n) \end{aligned}$$

(where χ^λ is the S_k-character for S^λ)

(2) On the other hand, by direct computation, one has (for any k, n)

$$\sigma\, A\, (e_{j_1} \otimes \ldots \otimes e_{j_k}) = e_{j_{\sigma^{-1}(1)}} \otimes \ldots \otimes e_{j_{\sigma^{-1}(k)}} \prod_{r=1}^{k} x_{j_r},$$

giving

$$\begin{aligned} tr(\sigma\, A) &= \sum_{\substack{1 \leq j_1, \ldots, j_k \leq n \\ j_r = j_{\sigma^{-1}(r)}, 1 \leq r \leq k}} \prod_{r=1}^{k} x_{j_r} \\ &= \sum_{1 \leq c_1, \ldots, c_{\ell(\text{type } \sigma)} \leq n} \prod_{r=1}^{\ell(\mu)} x_{c_r}^{\mu_r} \end{aligned}$$

since the condition $j_r = j_{\sigma^{-1}(r)}, 1 \leq r \leq k$ means that the subscripts j_r are the same for r's in a cycle of σ; also recall that (type σ) $= \mu$.

$= p_\mu(x_1, \ldots, x_n)$

where p_μ is the power-sum symmetric function (see [Mac]) defined by $p_\mu = \prod_i (x_1^{\mu_i} + \ldots + x_n^{\mu_i})$.

Hence we get

$$p_\mu(x_1, \ldots, x_n) = \sum_{\substack{\lambda \vdash k \\ \ell(\lambda) \leq n}} \chi^\lambda(\mu) \,.\, s_\lambda(x_1, \ldots, x_n)$$

so that for $k \leq n$, one has

$$p_\mu(x_1, \ldots, x_n) = \sum_{\lambda \vdash k} \chi^\lambda(\mu) \cdot s_\lambda(x_1, \ldots, x_n)$$

Finally, using well-known facts about the character table $(\chi^\lambda(\mu))$ of S_k, one can invert the latter matrix to get

$$s_\lambda(x_1, \ldots, x_n) = \sum_{\sigma \in S_k} \chi^\lambda(\sigma) \cdot p_{\text{type } \sigma}(x_1, \ldots, x_n).$$

This shows that the Frobenius characteristic map ch, defined in [Mac] by

$$ch(f) = \sum_{\sigma \in S_k} f(\sigma) \cdot p_{\text{type } \sigma}(x_1, \ldots, x_n)$$

for all class functions f on S_k, sends the S_k-irreducible character χ^λ to the $Gl(n)$-character s_λ (if $k \geq n$).

An analogue of the Frobenius characteristic map for any other Lie group G would map the character ring of the semi-simple quotient of the Brauer algebra into the character ring

$$\mathbb{Z}[z : z \text{ is an eigenvalue of an element in the maximal torus}]^{W_G},$$

where W_G is the Weyl group of G. It is clear from the above discussion that the problem of finding such a map is intimately related to the problem of understanding the representations of the semi-simple quotient of the Brauer algebra; one would use the decomposition analogous to (DC) :

$$V^{\otimes k} = \bigoplus_{\substack{\lambda \vdash k \\ \ell(\lambda) \leq n}} (\mathfrak{A}^\lambda_{k,n} \otimes \bar{W}^\lambda),$$

where $\bar{W}^\lambda$ is the irreducible G-module indexed by λ, and $\mathfrak{A}^\lambda_{k,n}$ is the irreducible representation (of the semi-simple quotient of the Brauer algebra) indexed by λ. (n=dim V.) Not much is known about such an analogue.

Appendix: Some bijections.

(A1) The bijection of Theorem 3.10.

Recall that this correspondence gives the following enumerative formula:

$$\tilde{f}^\mu_k(n) = \sum_{|\nu|=k} f^\nu \sum_{\substack{\beta \vdash (k-|\mu|) \\ \beta \text{ has even columns}}} c^\nu_{\mu,\beta}(n)$$

We illustrate the bijection of this theorem by example. The reader should simultaneously consult Figure (A1.1). Given an up-down sequence of shapes $\tilde{Q}^k_\mu$, the first step consists of associating to it a sequence of **standard** tableaux T_i, such that T_i has shape $\mu^i, i = 1, \ldots, k$. Here "standard" simply means that each tableau is both row- and column- strict; the labels used come from the integers $\{1, \dots, k\}$. Observe that any standard Young tableau T of shape λ can be thought of as a sequence of $|\lambda|$ shapes, and conversely: the ith shape is the shape of the sub-tableau of T formed by the entries which are $\leq i$. This map works for our up-down sequence $\tilde{Q}^k_\mu$ until the first step where a square has been removed from the preceding shape.

Figure (A1.1).

$$Q^{10}_{(2,1,1)}$$

$$= \left(\begin{array}{cccccccccc} \square & \begin{smallmatrix}\square\\\square\end{smallmatrix} & \begin{smallmatrix}\square\square\\\square\end{smallmatrix} & \begin{smallmatrix}\square\square\\\square\\\square\end{smallmatrix} & \begin{smallmatrix}\square\\\square\\\square\end{smallmatrix} & \begin{smallmatrix}\square\\\square\end{smallmatrix} & \begin{smallmatrix}\square\square\\\square\end{smallmatrix} & \begin{smallmatrix}\square\square\\\square\square\end{smallmatrix} & \begin{smallmatrix}\square\square\\\square\square\\\square\end{smallmatrix} & \begin{smallmatrix}\square\square\\\square\\\square\end{smallmatrix} \\ 1 & \begin{smallmatrix}1\\2\end{smallmatrix} & \begin{smallmatrix}1\,3\\2\end{smallmatrix} & \begin{smallmatrix}1\,3\\2\\4\end{smallmatrix} & \begin{smallmatrix}1\\3\\4\end{smallmatrix} & \begin{smallmatrix}1\\3\end{smallmatrix} & \begin{smallmatrix}1\,7\\3\end{smallmatrix} & \begin{smallmatrix}1\,7\\3\,8\end{smallmatrix} & \begin{smallmatrix}1\,7\\3\,8\\9\end{smallmatrix} & \begin{smallmatrix}1\,7\\8\\9\end{smallmatrix} \\ & & & & \left\{\begin{smallmatrix}5\\2\end{smallmatrix}\right\} & \left\{\begin{smallmatrix}6\\4\end{smallmatrix}\right\} & & & & \left\{\begin{smallmatrix}10\\3\end{smallmatrix}\right\} \end{array} \right)$$

$$\mapsto \left(\left\{ \begin{smallmatrix} 5 & 6 & 10 \\ 2 & 4 & 3 \end{smallmatrix} \right\}, \begin{smallmatrix} 1\,7 \\ 8 \\ 9 \end{smallmatrix} \right)$$

In Figure (A1.1), this happens at step 5, where we go from $\mu^4 = (2,1,1)$ to the shape $\mu^5 = (1,1,1)$. At this point, we do the following. The square that was removed is the square labelled '3' in the standard tableau $T_4 = \begin{smallmatrix} 1\,3 \\ 2 \\ 4 \end{smallmatrix}$. We column-remove the '3', bumping out entry '2', and leaving the tableau $T_5 = \begin{smallmatrix} 1 \\ 3 \\ 4 \end{smallmatrix}$ which now has shape $\mu^5 = (1,1,1)$. (Thus column-inserting the '2' into T_5 gives us back T_4). We record the loss of the entry **2** at step **5** by forming the (vertical) pair $\left\{\begin{smallmatrix}5\\2\end{smallmatrix}\right\}$. Continuing in this manner, we end up with a final tableau $T_{10} = \begin{smallmatrix} 1\,7 \\ 8 \\ 9 \end{smallmatrix}$ of shape $\mu^{10} = \mu = (2,1,1)$, the shape of the up-down tableau, and a set of pairs which we interpret as the transpositions of a fixed-point-free involution on the removed letters: the resulting permutation (on the letters 2,3,4,5,6,10) is 5 10 6 2 4 3.

Now (see Figure (A1.2)) map the fixed-point-free involution into its image under Robinson-Schensted insertion; a well-known result (see [Kn]) says that the resulting tableau will have all its columns even. In Figure (A1.2) below, the up-down tableau goes to the pair of tableaux (Q_β, T_{10}), where $Q_\beta = \begin{smallmatrix} 2\,3 \\ 4\,6 \\ 5 \\ 10 \end{smallmatrix}$, and thus $\beta = (2,2,1,1)$.

Figure (A1.2).

$$\mathcal{L} = \left\{ \begin{smallmatrix} 5 & 6 & 10 \\ 2 & 4 & 3 \end{smallmatrix} \right\} \longleftrightarrow (2\ 5)(4\ 6)(3\ 10) \longleftrightarrow \begin{pmatrix} 2 & 3 & 4 & 5 & 6 & 10 \\ 5 & 10 & 6 & 2 & 4 & 3 \end{pmatrix} \longleftrightarrow \begin{smallmatrix} 2\,3 \\ 4\,6 \\ 5 \\ 10 \end{smallmatrix}$$

Thus

$$Q^{10}_{(2,1^2)} \longleftrightarrow \left(\begin{smallmatrix} 2\,3 \\ 4\,6 \\ 5 \\ 10 \end{smallmatrix} \quad \begin{smallmatrix} 1\,7 \\ 8 \\ 9 \end{smallmatrix} \right)$$

The final step (recall Thomas' proof of the Littlewood-Richardson rule, (Section 5)) consists of column-inserting each row of Q_β, (largest element first) starting from the top row and moving down, into the tableau T_{10}. In the example of Figure (A1.2) the sequence of insertions is

$$(10 \to (5 \to (4\ 6 \to (2\ 3 \to T_{10})))),$$

where we denote the single column-insertion of w into a tableau P by $w \to P$. This clearly produces a tableau T_ν of a shape ν which contains the original shape μ. (In this case, we have $\nu = (3,3,2,1,1)$). To get the lattice permutation $lp^{\beta}_{\nu/\mu}$ of the theorem, we only need to label the squares in the skew-shape ν/μ in a manner which will keep track of the sequence of insertions. This is done as follows: A single square x is added to the preceding shape as the result of each insertion of a letter w from a row of Q_β; to each square; if w comes from row r of Q_β, we put an r in the newly-added square x. The resulting skew-shape is then filled with a lattice permutation , in this case 121234, of weight $\beta = (2,2,1,1)$. In detail, the successive insertions produce the following sequence of shapes, starting at $\mu = (2,1,1)$ and ending at $\nu = (3,3,1,1)$:

$$\left(\begin{array}{lllllll}
\begin{array}{ll}\square&\square\\ \square& \\ \square& \end{array} &
\begin{array}{ll}\square&\square\\ \square&1\\ \square& \end{array} &
\begin{array}{lll}\square&\square&1\\ \square&1& \\ \square&& \end{array} &
\begin{array}{lll}\square&\square&1\\ \square&1& \\ \square&2& \end{array} &
\begin{array}{lll}\square&\square&1\\ \square&1&2\\ \square&2& \end{array} &
\begin{array}{lll}\square&\square&1\\ \square&1&2\\ \square&2& \\ 3&& \end{array} &
\begin{array}{lll}\square&\square&1\\ \square&1&2\\ \square&2& \\ 3&& \\ 4&& \end{array}
\end{array}\right)$$

In our example, the given up-down tableau maps to the pair

$$\left(lp^{\beta}_{\nu/\mu} = \begin{array}{lll}\square&\square&1\\ \square&1&2\\ \square&2& \\ 3&& \\ 4&& \end{array}, T_\nu = \begin{array}{ll}1&7\\ 8& \\ 9& \end{array} \right).$$

Note that the lattice permutation is n-symplectic iff $n \geq 3$, which agrees with the fact that the length of the longest shape in the up-down sequence is 3.

(A2) The bijection for the Cauchy identity for $Sp(2n)$.

¿From the enumerative viewpoint it clearly simplifies matters to re-write the identity as

$$\prod_{i,j=1}^{n} (1 - t_i x_j)^{-1}(1 - t_i {x_j}^{-1})^{-1}$$

$$= \sum_{\substack{\mu \\ \ell(\mu)\leq n}} sp_\mu(x_1^{\pm 1}, \ldots, x_n^{\pm 1}) s_\mu(t) \left(\sum_{\beta \text{ (even columns)}} s_\beta(t) \right)$$

where we have used the identity [L1] of Section 4.

We may enumerate the left-hand side by means of Knuth (see [Kn]) two-line arrays

$$\mathcal{T} = \begin{pmatrix} t_{i_1} & \ldots & t_{i_k} \\ y_{i_1} & \ldots & y_{i_k} \end{pmatrix}$$

where the y_{i_j}'s are in the set $\{x_1^{\pm 1}, \ldots, x_n^{\pm 1}\}$; such a two-line array would correspond to the term $(t_{i_1} \ldots t_{i_k} y_{i_1} \ldots y_{i_k})$ in the expansion of the left side as a formal power series. We impose the (usual) lexicographic ordering on the arrays, viz., $t_{i_1} \leq \ldots \leq t_{i_k}$, and $t_{i_j} = t_{i_{j+1}}$ implies $y_{i_j} \leq y_{i_{j+1}}$.

The right-hand side clearly counts the set of all triples

$$(\tilde{P}_\mu(x), P_\mu(t), P_\beta(t)), \text{where}$$

$P_\mu(x)$ is a symplectic tableau of shape μ with entries in $1, \bar{1}, \ldots, n, \bar{n}$, corresponding to the variables $\{x_1^{\pm 1}, \ldots, x_n^{\pm 1}\}$, $P_\mu(t)$ is an ordinary tableau with entries in $[n]$ of the same shape μ as P_μ, and $P_\beta(t)$ is an ordinary tableau of shape β with entries in $[n]$, where β has even columns.

We proceed by applying the Berele algorithm to the word $y_{i_1} \ldots y_{i_k}$ in the bottom row of $\mathcal{T}$, so that if at the jth step we have built up a sequence of pairs of tableaux $(\tilde{P}_j, P_j)$, at step $j+1$ we set

$$\tilde{P}_{j+1} \text{ to be the result of inserting } y_{i_{j+1}} \text{ into } \tilde{P}_j \text{ via Berele.}$$

If the effect of Berele insertion is to add augment the tableau, set $P_{j+1} = P_j$ with $t_{i_{j+1}}$ added in the unique position so as to force shape$(P_{j+1}) =$ shape$(\tilde{P}_{j+1})$; otherwise, shape$(\tilde{P}_{j+1}) = \mu^{j+1}$, say, has one box less than shape$(\tilde{P}_j) = \mu^j$, and we get P_{j+1} from P_j as follows:

- bump out the extra entry of P_j (the one in the unique square of μ^j which is not a square of μ^{j+1}) by columns (i.e. inverse Schensted column insertion) to get a tableau P_{j+1} of shape μ^{j+1}, and a letter x. This means that by column-inserting x into P_{j+1} we would retrieve the previous larger tableau P_j of shape μ^j.
- We record the fact that a removal has occurred at step j by putting the pair (t_{i_j}, x) into a two-line array $\mathcal{L}$, with t_{i_j} on top.

We continue this process to the end of the word $y_{i_1} \ldots y_{i_k}$, arranging the two-line array $\mathcal{L}$ so that the top row is weakly increasing.

Example. The word $t_1 x_1{}^{-1} t_1 x_2 t_2 x_1{}^{-1} t_2 x_2 (t_4 x_1)^2 t_4 x_1{}^{-1} t_5 x_1$ corresponds to the two-line array

$$\mathcal{T} = \begin{pmatrix} 1 & 1 & 2 & 2 & 4 & 4 & 4 & 5 \\ \bar{1} & 2 & \bar{1} & 2 & 1 & 1 & \bar{1} & 1 \end{pmatrix}$$

Start Berele insertion of the word in the bottom row of $\mathcal{T}$, from left to right as usual. In the schematic that follows (Figure (A2)), the computation is arranged so that:

(1) the first row contains the successive symplectic tableaux, ending in the final tableau $\tilde{P}_\mu$,

(2) the second row encodes the up-down tableau resulting from the Berele insertion, ending in the final tableau P_μ,

(3) the third row encodes the removals in the form of pairs which, at the end of the process, may be put together into a two-line array $\mathcal{L}$, so that ultimately the pair $(P_\mu, \mathcal{L})$ contains all the information to specify completely and uniquely the up-down tableau. See Figure (A2).

Figure (A2).

$$\mathcal{T} \to \begin{pmatrix} \bar{1} & \bar{1}\,2 & \begin{smallmatrix} \bar{1}\,\bar{1} \\ 2 \end{smallmatrix} & \begin{smallmatrix} \bar{1}\,\bar{1}\,2 \\ 2 \end{smallmatrix} & \begin{smallmatrix} \bar{1}\,2 \\ 2 \end{smallmatrix} & 2\,2 & \begin{smallmatrix} \bar{1}\,2 \\ 2 \end{smallmatrix} & 2\,2 & \tilde{P}_\mu \\ 1 & 1\,1 & \begin{smallmatrix} 1\,1 \\ 2 \end{smallmatrix} & \begin{smallmatrix} 1\,1\,2 \\ 2 \end{smallmatrix} & \begin{smallmatrix} 1\,2 \\ 2 \end{smallmatrix} & 1\,2 & \begin{smallmatrix} 1\,2 \\ 4 \end{smallmatrix} & 1\,2 & P_\mu \\ & & & & \left\{\begin{smallmatrix} 4 \\ 1 \end{smallmatrix}\right\} & \left\{\begin{smallmatrix} 4 \\ 2 \end{smallmatrix}\right\} & & \left\{\begin{smallmatrix} 5 \\ 4 \end{smallmatrix}\right\} & \mathcal{L} \end{pmatrix}$$

$$\to \left(\ 2\,2,\quad 1\,2,\quad \left\{\begin{smallmatrix} 4\,4\,5 \\ 1\,2\,4 \end{smallmatrix}\right\}\right)$$

$$\to \left(\ 2\,2,\quad 1\,2,\quad \begin{smallmatrix} 1\,2\,4 \\ 4\,4\,5 \end{smallmatrix}\right) = (\tilde{P}_{(2)}, P_{(2)}, P_{(3,3)}).$$

This scheme maps a two-line array $\mathcal{T}$ to a triple $(\tilde{P}_\mu, P_\mu, \mathcal{L})$ where $\tilde{P}_\mu$ is a symplectic tableau of shape μ, P_μ is a column-strict ($Gl(n)$-)tableau of the same shape, and $\mathcal{L}$ is a two-line array

$$\begin{pmatrix} j_1 & \dots & j_r \\ i_1 & \dots & i_r \end{pmatrix}$$

where $j_1 \leq \ldots \leq j_r$ and $j_k \geq i_k$, all $k = 1, \ldots, r$, with entries from the top row of labels t; it turns out that $j_k > i_k$.

The final step is to exhibit a bijection between two-line arrays $\mathcal{L}$ and ordinary tableaux with even-columned shapes; such a correspondence (which we mentioned earlier (in Section 4) in connection with the identity (L1) of Littlewood) was given by Burge in [Bu].

(A3) The insertion algorithm for $SO(2n+1)$.

Recall that this would prove the identity

$$(x_1 + x_1{}^{-1} + \cdots + x_n + x_n{}^{-1} + 1)^k = \sum_{\substack{\mu \\ \ell(\mu) \leq n}} \tilde{F}_k^\mu(n)\ so_\mu(x_1^{\pm 1}, \ldots, x_n^{\pm 1}, 1)$$

where the up-down sequences counted by $\tilde{F}_k^\mu(n)$ have the additional property that consecutive shapes may be the same, provided they are of length exactly n.

The insertion is obtained by making the obvious modification to Berele's algorithm suggested by the very definition of so-tableaux (Definition 2.7). We use all the aspects of Berele insertion restricted to the letters $1, \bar{1}, \ldots, n, \bar{n}$, and we deal with the extra symbol ∞ (again, see Definition 2.7) as follows:

(*) ∞ always bumps ∞.

This new bumping rule makes it possible to obtain a shape of length $n+1$ in the bumping process, and it is easily seen that this happens only when an ∞ was bumped down into the $(n+1)$st row from the nth row (by another ∞). To take care of this, when this happens we simply erase the ∞ in the $(n+1)$st row and declare the result of the insertion to be the remaining tableau. Clearly a repetition occurs at this point in the sequence of shapes. In Figure (A3) we have worked out the result of inserting the word $w = \infty\bar{1}\infty\infty 2$ for $n = 3$. (For details and proof, see [Su3]).

Figure (A3).

$$\mathcal{S}(w) = \left(\infty \quad \begin{smallmatrix} \bar{1} \\ \infty \end{smallmatrix} \quad \begin{smallmatrix} \bar{1}\ \infty \\ \infty \end{smallmatrix} \quad \begin{smallmatrix} \bar{1}\ \infty \\ \infty \\ (\infty) \end{smallmatrix} \quad \begin{smallmatrix} \bar{1}\ 2 \\ \infty \\ (\infty) \end{smallmatrix} \right)$$

$$= \left(\left(\square \quad \begin{smallmatrix} \square \\ \square \end{smallmatrix} \quad \begin{smallmatrix} \square\square \\ \square \end{smallmatrix} \quad \begin{smallmatrix} \square\square \\ \square \end{smallmatrix} \quad \begin{smallmatrix} \square\square \\ \square \end{smallmatrix} \right) \quad \begin{smallmatrix} \bar{1}\ 2 \\ \infty \end{smallmatrix} \right)$$

Acknowledgment. The author would like to thank Richard Stanley and Phil Hanlon, from whom she learnt much of the subject, and Dennis Stanton for many helpful suggestions during the preparation of this paper.

REFERENCES

[Be1] A. BERELE, *A Schensted-type correspondence for the Symplectic Group*, J. of Combinatorial Theory (A), 43 (1986), pp. 320–328.

[Be2] ———, *Construction of Sp-modules by Tableaux*, Linear and Multilinear Algebra, 19 (1986), pp. 299–307.

[BKW] G. R. E. BLACK, R. C. KING AND B. G. WHYBOURNE, *Kronecker products for compact semisimple Lie groups*, J. Phys. A: Math Gen., 16 (1983), pp. 1555–1589.

[Boe] H. BOERNER, *Representations of Groups*, North-Holland Publishing Company, Amsterdam, 1963.

[Br] R. BRAUER, *On Algebras which are connected with the semi-simple continuous groups*, Annals of Math., 38 (1937), pp. 854–872.

[Bu] W. H. BURGE, *Four Correspondences between Graphs and Generalized Young Tableaux*, J. of Combinatorial Theory (A), 17 (1974), pp. 12-30.

[EK] N. EL-SAMRA AND R. C. KING, *Dimensions of irreducible representations of the classical Lie groups*, No. 12, J. Phys. A: Math Gen., 12 (1979), pp. 2317–2328.

[Ha] P. HANLON, *An Introduction to the Complex Representations of the Symmetric Group and General Linear Lie Algebra*, in *Contemporary Mathematics 34*, 1984.

[HW] P. HANLON AND D. WALES, *On the decomposition of Brauer's centralizer algebras*, J. of Algebra (to appear).

[Hum] J. E. HUMPHREYS, *Introduction to Lie Algebras and Representation Theory*, Springer-Verlag, New York, 1972.

[Ki1] R. C. KING, *Weight Multiplicities for the Classical Lie Groups*, in *Lecture Notes in Physics 50*, Springer, New York, 1975, pp. 490–499.

[Ki2] ———, *Modification Rules and Products of Irreducible Representations of the Unitary, Orthogonal and Symplectic Groups*, J. Math. Phys., 12 (1971), pp. 1588–1598.

[Ki3] ———, *personal communication, March 1988.*

[KEl] R. C. KING AND N. G. I. EL-SHARKAWAY, *Standard Young Tableaux and Weight Multiplicities of the Classical Lie Groups*, J. Phys. A: Math Gen., 16 (1983), pp. 3153–3177.

[Kn] D. E. KNUTH, *Permutations, Matrices and Generalized Young Tableaux*, Pacific J. of Math, 34, No. 3 (1970), pp..

[KT1] K. KOIKE AND I. TERADA, *Young-Diagrammatic Methods for the Representation Theory of the Classical Groups of type B_n, C_n, D_n*, J. of Algebra, 107, No. 2 (1987), pp. 466–511.

[KT2] ———, *Young diagrammatic methods for the restriction of representations of complex classical Lie groups to reductive subgroups of maximal rank*, preprint, 1987.

[Li] D. E. LITTLEWOOD, *The Theory of Group Characters*, 2nd. ed., Oxford University Press, 1950.

[Li2] D. E. LITTLEWOOD, *On the Poincaré polynomials of the classical groups*, J. London Math. Soc., 28 (1953), pp. 494–500.

[LR] D. E. LITTLEWOOD AND A. R. RICHARDSON, *Group Characters and Algebra*, Royal Society of London - Phil. Trans., Series A, 233 (1934), pp. 99–141.

[Mac] I. G. MACDONALD, *Symmetric Functions and Hall Polynomials*, Oxford University Press, 1979.

[Pr] R. A. PROCTOR, *Classical Gelfand-Young patterns*, research announcement, May 1985.

[S] B. E. SAGAN, *The Ubiquitous Young Tableau*, this volume.

[Scn] C. SCHENSTED, *Longest Increasing and Decreasing Sequences*, Can. J. Math., XIII (1961), pp. 179–191.

[Sch] I. SCHUR, *Über eine Klasse von Matrizen die sich gegebenen Matrix zuordnen classen*, Dissertation, Berlin 1901, reprinted in Gesammelte Abhandlungen, 1, pp. 1-72.

[St] R. P. STANLEY, *$Gl(n, \mathbb{C})$ for Combinatorialists*, in Surveys in Combinatorics, ed. by E. K. Lloyd, in *London Math. Society Lecture Notes*, Cambridge University Press, 1983, pp. 187–199.

[Su1] S. SUNDARAM, *On the Combinatorics of Representations of $Sp(2n, \mathbb{C})$*, Thesis, Massachusetts Institute of Technology, April 1986.

[Su2] ———————, *The Cauchy identity for $Sp(2n)$*, submitted to J. of Combinatorial Theory, (A).

[Su3] ———————, *Orthogonal tableaux and an insertion scheme for $SO(2n+1)$*, submitted to J. of Combinatorial Theory, (A).

[Th] G. THOMAS, *On Schensted's Construction and the Multiplication of Schur Functions*, Advances in Math., 30 (1978), pp. 8-32.

[We] H. WEYL, *The Classical Groups*, 2nd. ed., Princeton University Press, 1946.

[Wz] H. WENZL, *On the Structure of Brauer's Centralizer Algebras*, submitted to Annals of Math., 1987.

S–FUNCTIONS AND CHARACTERS OF LIE ALGEBRAS AND SUPERALGEBRAS

RONALD C. KING†

Abstract. Schur functions and Young diagrammatic methods are exploited to define and manipulate characters of irreducible representations of the Lie algebras $gl(N), o(N)$ and $sp(N)$. As far as possible the treatment is made N–independent by using universal characters and modification rules. A key role is played by certain infinite series of Schur functions. They are used not only in the derivation of the Newell–Littlewood product rule for $o(N)$ and $sp(N)$ but also in a new approach to the Macdonald identities which leads to expansions in terms of universal characters. Finally some of these S-function techniques are extended to a Lie superalgebra context. Some of the difficulties encountered in trying to arrive at a character formula encompassing both typical and atypical irreducible representations of $gl(M/N)$ and $osp(M/N)$ are described.

0. Introduction. The aim in this article is to describe the use of Schur function (S–function) techniques in dealing with the characters of irreducible representations (irreps) of the Lie algebras, $gl(N), o(N)$ and $sp(N)$, and the Lie superalgebras, $gl(M/N)$ and $osp(M/N)$. An attempt is made to express the characters and the associated tensor product and branching rules in a manner which is essentially independent of M and N. This involves the notion of universal characters [KT,CK3] and the subsequent use of modification rules [Mu,N,Ki2,KT] appropriate to particular values of M and N. Apart from leaning heavily on the usual algebra of S-functions [Li1,Ma2], stress is laid on the use of certain infinite series of S-functions [Li1,Ki3].

As far as the Lie algebras are concerned the underlying formula on which all else is based is Weyl's character formula [W1]. In the case of $gl(N)$ this leads directly to the fact that the characters of the covariant tensor irreps are what are now known as S-functions. These S-functions can be given a combinatorial interpretation [Sta] in terms of column strict Young tableaux. The famous Littlewood-Richardson rule [LR] may then be used to decompose tensor products of irreps and to evaluate the branching of an irrep on restriction from $gl(M+N)$ to $gl(M)+gl(N)$. All this is dealt with in Section 1.

In Section 2 the analysis is extended to the case of irreps of $gl(N)$ associated with certain composite Young diagrams [AK,Ki1]. Their introduction allows universal characters to be defined and the corresponding modification rule to be described in a straightforward way [Ki2]. The tensor product rule and branching rule for these irreps can then be stated entirely in terms of S-function products and quotients, all governed by the ubiquitous Littlewood-Richardson rule. The proof of the tensor product rule [AK,Ki2] is deferred until Section 4.

In order to extend this approach to the case of $o(N)$ and $sp(N)$ Section 3 starts with the statement of the S-function expansion of a set of q-dependent generating functions [Li1]. These generating functions occur, in their $q = 1$ form, in the formulae used here to define characters of irreps of $o(N)$ and $sp(N)$ [Mu,W2,Li1]. Littlewood's expansions in terms of infinite series of S-functions are then exploited

†Mathematics Department, University of Southampton, Southampton, SO9 5NH, England

to relate these characters to those of $gl(N)$, and conversely to define the branching rules for the restriction of $gl(N)$ irreps to $o(N)$ and to $sp(N)$ [Li1]. Once again it is advantageous to use universal characters and modification rules expressed in terms of Young diagrams [Ki2]. This shows itself most clearly in Section 4 in which is derived the remarkably simple Newell-Littlewood rule [N,Li2] for decomposing tensor products of irreps of both $o(N)$ and $sp(N)$. The Section closes with the promised derivation of the Abramsky-King rule [AK,Ki2] for decomposing the tensor products of irreps of $gl(N)$ associated with composite Young diagrams.

An unexpected bonus arising from consideration of the algebra of S-function series and the Lemmas necessary to prove the validity of both the Newell–Littlewood and the Abramsky-King rules is that they may be further extended to cope first with the division of one such infinite series by another and then with the evaluation of infinite products of such infinite series. In this way a set of identities is established in Section 5 which represents a new statement of the Macdonald identities [Ma1].

Finally Section 6 is concerned with characters of finite–dimensional irreps of the Lie superalgebras $gl(M/N)$ and $osp(M/N)$. A major problem is provided by the existence of reducible but indecomposable representations. The stumbling block is the existence of atypical irreps which do not split in every representation in which they occur. Kac [Ka2] derived the criterion for distinguishing between typical and atypical irreps and gave a character formula for the former. No such generalisation of Weyl's character formula exists as yet for the atypical irreps. The status of recent attempts [BL,Le,VdJ,HK,PS] to find such a generalisation is reported upon in Section 6, along with a discussion of the use of Young diagrams [DJ,BB1,BB2] and S-function techniques [BR,Ki4] in this context.

1. Characters and S-functions. Let G be a reductive Lie algebra over C with Cartan subalgebra H. Each finite dimensional irreducible representation (irrep) of G may be specified by means of its highest weight vector $\Lambda \in H^*$. The character of this irrep is given by Weyl's formula [W1]

$$ch(\Lambda) = \sum_{w \in W} \epsilon(w)\ e^{w(\Lambda+\rho)} \Big/ \sum_{w \in W} \epsilon(w)\ e^{w\rho}, \tag{1.1}$$

where W is the Weyl group of G, $\epsilon(w)$ is the parity of the element $w \in W$, and ρ is half the sum of the positive roots of G. The positive roots, Δ^+ , of the classical Lie algebras are given in the standard basis by

(1.2a)	$sl(n)$	$\Delta^+ = \{\epsilon_i - \epsilon_j\}$	with $\epsilon_1 + \epsilon_2 + \cdots + \epsilon_n = 0$
(1.2b)	$so(2n+1)$	$\Delta^+ = \{\epsilon_i \pm \epsilon_j,\ \epsilon_i\}$	
(1.2c)	$sp(2n)$	$\Delta^+ = \{\epsilon_i \pm \epsilon_j,\ 2\epsilon_i\}$	
(1.2d)	$so(2n)$	$\Delta^+ = \{\epsilon_i \pm \epsilon_j\}$	
(1.2e)	$gl(n)$	$\Delta^+ = \{\epsilon_i \pm \epsilon_j\}$	

where $1 \leq i < j \leq n$ and $1 \leq i \leq n$. The inner product on H^* is such that $(\epsilon_i|\epsilon_j) = \delta_{ij}$ and it is convenient to denote each formal exponential e^{ϵ_i} by an indeterminate x_i.

With this notation, an arbitrary weight vector $\mathbf{M} \in H^*$ with $\mathbf{M} = m_1\epsilon_1 + \cdots + m_n\epsilon_n$ takes the form $e^{\mathbf{M}} = x_1^{m_1} x_2^{m_2} \ldots x_n^{m_n}$. Finally by way of shorthand we signify the indeterminates by $\mathbf{x} = (x_1, x_2, \ldots, x_n)$ and also write $\mathbf{M} = (m_1, m_2, \ldots, m_n)$. For each of the above classical Lie algebras the Weyl group, W, contains as a subgroup the symmetric group S_n. Hence with the ordering of the standard basis implied by (1.2) each highest weight vector $\Lambda = (\Lambda_1, \Lambda_2, \ldots, \Lambda_n)$ is such that $\Lambda_1 \geq \Lambda_2 \geq \cdots \geq \Lambda_n$.

An important class of irreps of $gl(N)$ are those for which the highest weight vector $\Lambda = (\lambda)_N = (\lambda_1, \lambda_2, \ldots, \lambda_p, 0, 0, \ldots, 0)$ where $\lambda = (\lambda_1, \lambda_2, \ldots, \lambda_p)$ is a partition and $p \leq N$. The notation is such that $\lambda_i \in \mathbb{N}$ for $i = 1, 2, \ldots, p$ and $\lambda_i \geq \lambda_{i+1}$ for $i = 1, 2, \ldots, p-1$. The partition λ is said to have weight $|\lambda| = \lambda_1 + \lambda_2 + \cdots + \lambda_p$ and length $l(\lambda) = p$. With this notation the character of the irrep of $gl(N)$ with highest weight vector λ is given by

$$ch(\lambda)_N = \sum_{\pi \in S_N} \epsilon(\pi) e^{\pi(\lambda+\rho)} \Big/ \sum_{\pi \in S_N} \epsilon(\pi) e^{\pi\rho}, \tag{1.3}$$

with $2\rho = (N-1, N-3, \ldots, -N+3, -N+1)$. Multiplying the numerator and denominator by $(x_1 x_2 \ldots x_N)^{(N-1)/2}$ then gives

$$ch(\lambda)_N = \sum_{\pi \in S_N} \epsilon(\pi) x_{\pi_1}^{\lambda_1+N-1} x_{\pi_2}^{\lambda_2+N-2} \ldots x_{\pi_N}^{\lambda_N} \Big/ \sum_{\pi \in S_N} \epsilon(\pi) x_{\pi_1}^{N-1} x_{\pi_2}^{N-2} \ldots x_{\pi_N}^{0}. \tag{1.4}$$

Hence

$$ch(\lambda)_N = \left| x_i^{\lambda_j+N-j} \right| \Big/ \left| x_i^{N-j} \right| = s_\lambda(x_1, x_2, \ldots, x_N) \tag{1.5}$$

where the resulting ratio of bialternants is nothing other than the well known Schur function [Li1,Ma2].

To each partition λ there corresponds a Young diagram or frame F^λ consisting of $l(\lambda)$ left–adjusted rows of boxes of lengths λ for $i = 1, 2, \ldots, l(\lambda)$. The Schur function (1.5) can be given a combinatorial definition by invoking the notion of a standard Young tableau, T^λ , which is a numbering of the boxes of the Young diagram F^λ in which

(i) the entries are taken from the set $\{\mathsf{E} = 1, 2, \ldots, N\}$ ordered so that $1 < 2 < \cdots < N$,

(ii) the entries are weakly increasing from left to right across each row,

(iii) the entries are strictly increasing from top to bottom down each column.

For example, the Young diagram F^λ and a standard Young tableau T^λ are illustrated in the case $\lambda = (531^2)$ by

$$F^\lambda = \begin{array}{l} \square\square\square\square\square \\ \square\square\square \\ \square \\ \square \end{array} \qquad T^\lambda = \begin{array}{l} 1\,1\,2\,4\,4 \\ 2\,3\,3 \\ 4 \\ 5 \end{array}$$

To each such standard Young tableau, T^λ, there corresponds a weight vector $\mathbf{M} = (m_1, m_2, \ldots, m_N)$, where m_j is the number of entries j in the tableau. With this terminology the Schur function (1.5) can be expressed in the form [Sta]

$$s_\lambda(x_1, x_2, \ldots, x_N) = \sum_{T^\lambda} x_1^{m_1} x_2^{m_2} \ldots x_N^{m_N}. \tag{1.6}$$

It follows that the irrep of $gl(N)$ in question has character

$$ch(\lambda)_N = \{\lambda\}(\mathbf{x})_N = s_\lambda(\mathbf{x})_N = s_\lambda(x_1, x_2, \ldots, x_N), \tag{1.7}$$

where a variety of notations has been used to specify the character. It follows from (1.6) that

$$\{\lambda\}(\mathbf{x})_N = s_\lambda(\mathbf{x})_N = 0 \qquad \text{if} \quad l(\lambda) > N\ , \tag{1.8}$$

as can be seen from condition (iii) for the standardness of the Young tableaux. This is our first example, albeit a trivial one, of a modification rule. Furthermore on restriction from $gl(N)$ to $sl(N)$ each irrep remains irreducible and its character is unchanged save for the fact that the condition $x_1 x_2 \ldots x_N = 1$ now applies by virtue of the constraint in (1.2a) on the standard basis vectors. Since every column of length N in a standard Young tableaux contains the entries $1, 2, \ldots, N$ this leads to the additional $sl(N)$ modification rule

$$\{\lambda\}(\mathbf{x})_N = \{\mu\}(\mathbf{x})_N \quad \text{where } \mu_i = \lambda_i - \lambda_N \text{ for } i = 1, 2, \ldots, N \tag{1.9}$$

corresponding to the removal of all columns of length N from the Young diagram F^λ.

The characters (1.7) may be generalised to give universal characters defined by taking the inverse limit [Ma2]

$$\{\lambda\}(\mathbf{x}) = s_\lambda(\mathbf{x}) = \lim_{\leftarrow N} \ \{\lambda\}(\mathbf{x})_N \tag{1.10}$$

The characters of $sl(N)$ and $gl(N)$ for any finite N are then recovered from the universal characters by means of the specialisations

(1.11a)

$$sl(N) \quad \{\lambda\}(\mathbf{x})_N = \{\lambda\}(x_1, x_2, \ldots, x_N, 0, 0, \ldots) \quad \text{with } x_1 x_2 \ldots x_N = 1$$

(1.11b)

$$gl(N) \quad \{\lambda\}(\mathbf{x})_N = \{\lambda\}(x_1, x_2, \ldots, x_N, 0, 0, \ldots).$$

The algebra of universal characters then coincides with the algebra of Schur functions $s_\lambda(\mathbf{x})$ of arbitrarily many indeterminates $x_1, x_2, \ldots$. This algebra is such that products are given by

$$s_{\mu.\nu}(\mathbf{x}) = s_\mu(\mathbf{x}) s_\nu(\mathbf{x}) = \sum_\lambda c^\lambda_{\mu\nu} s_\lambda(\mathbf{x}) \tag{1.12}$$

where the coefficients $c^\lambda_{\mu\nu}$ are the famous Littlewood–Richardson coefficients [LR,Li1,Ma2]. Thus the tensor product formula appropriate to gl(N) takes the form

$$\{\mu\}_N \cdot \{\nu\}_N = \sum_\lambda c^\lambda_{\mu\nu} \{\lambda\}_N \ , \tag{1.13}$$

where for typographical convenience the dependence on $(\mathbf{x})$ has been suppressed in the notation. The application of the modification rule (1.8) gives the final result

$$ch(\mu)_N \ ch(\nu)_N = \sum_{\lambda, l(\lambda) \le N} c^\lambda_{\mu\nu} ch(\lambda)_N \ . \tag{1.14}$$

It is almost as straightforward to deal with the branching rule appropriate to the restriction from $gl(M+N)$ to $gl(M) + gl(N)$. The relevant S–function identity [Li1,Sta,Ma2] takes the form

$$s_\lambda(\mathbf{x}, \mathbf{y})_{M+N} = \sum_\mu s_\mu(\mathbf{x})_M s_{\lambda/\mu}(\mathbf{y})_N = \sum_{\mu\nu} c^\lambda_{\mu\nu} \ s_\mu(\mathbf{x})_M \ s_\nu(\mathbf{y})_N, \tag{1.15}$$

where use has been made of the notation and combinatorics of skew Young diagrams $F^{\lambda/\mu}$ to which there correspond standard skew Young tableaux $T^{\lambda/\mu}$ [Sta,Ma2]. Remarkably everything is governed yet again by the Littlewood-Richardson coefficients since

$$s_{\lambda/\mu}(\mathbf{y})_N = \sum_\nu c^\lambda_{\mu\nu} \ s_\nu(\mathbf{y})_N \ . \tag{1.16}$$

One can then write the branching rule in the universal form [Ki2]

$$\{\lambda\}_{M+N} = \sum_\mu \{\mu\}_M \times \{\lambda/\mu\}_N = \sum_{\mu,\nu} c^\lambda_{\mu\nu} \{\mu\}_M \times \{\nu\}_N. \tag{1.17}$$

Finally the application of the modification rule (1.8) gives

$$ch(\lambda)_{M+N} = \sum_{\mu, l(\mu) \le M} \ \sum_{\nu, l(\nu) \le N} c^\lambda_{\mu\nu} ch(\mu)_M \ ch(\nu)_N. \tag{1.18}$$

In what follows it will be convenient to generalise the notation somewhat, exploiting the bilinear nature of S-function products and quotients. To simplify expressions we write, in a natural extension of (1.12) and (1.16),

$$\sum_\lambda c^\lambda_{\mu\nu} s_{\ldots\lambda\ldots}(\mathbf{x}) = s_{\ldots\mu.\nu\ldots}(\mathbf{x}) \ , \tag{1.19a}$$

$$\sum_\nu c^\lambda_{\mu\nu} s_{\ldots\nu\ldots}(\mathbf{x}) = s_{\ldots\lambda/\mu\ldots}(\mathbf{x}) \ , \tag{1.19b}$$

and even more generally for an arbitrary linear sum of S–functions

$$X(\mathbf{x}) = \sum_\lambda a_X^\lambda s_\lambda(\mathbf{x}) \quad \text{we write} \quad \sum_\lambda a_X^\lambda s_{\ldots\lambda\ldots}(\mathbf{x}) = s_{\ldots X\ldots}(\mathbf{x}). \tag{1.19c}$$

2. Composite Young diagrams and modification rules. Quite apart from the irreps of $gl(N)$ whose highest weight vectors are partitions, as in (1.4), there exist other inequivalent irreps with highest weight vectors of the form

$$\Lambda = (\overline{\nu};\mu)_N = (\mu_1, \mu_2, \ldots, \mu_p, 0, 0, \ldots, 0, -\nu_q, \ldots, -\nu_2, -\nu_1) \tag{2.1}$$

where $\mu = (\mu_1, \mu_2, \ldots, \mu_p)$ and $\nu = (\nu_1, \nu_2, \ldots, \nu_q)$ are partitions of lengths p and q, respectively, with $p + q \leq N$. Thanks to Weyl's character formula (1.1) it follows that the character of such an irrep is given by

$$ch(\overline{\nu};\mu)_N = \sum_{\pi\in S_N} \epsilon(\pi) x_{\pi_1}^{\mu_1+N-1} x_{\pi_2}^{\mu_2+N-2} \cdots x_{\pi_N}^{-\nu_1} \Big/ \sum_{\pi\in S_N} \epsilon(\pi) x_{\pi_1}^{N-1} x_{\pi_2}^{N-2} \cdots x_{\pi_N}^{0}. \tag{2.2}$$

Just as each partition λ with $l(\lambda) \leq N$ labels both an irrep of $gl(N)$ with character $ch(\lambda)_N$ and a Young diagram F^λ , so each pair of partitions μ and ν with $l(\mu) + l(\nu) \leq N$ labels both an irrep of $gl(N)$ with character $ch(\overline{\nu};\mu)_N$ and a generalised or composite Young diagram $F^{\overline{\nu};\mu}$ formed from F^μ and F^ν [AK,Ki1]. It is convenient in doing this to reflect the portion F^ν first in its leftmost line and then in its topmost line to give $F^{\overline{\nu}}$ and then to join it to F^μ corner–to–corner. This is illustrated in the case $\Lambda = (2^2 1^2 0 \ldots \quad \ldots 0 - 2 - 3)$ by the diagram

$F^{\overline{\nu};\mu} =$

```
 □□
□□□
   □□
   □□
   □
   □
```

Comparison of (2.2) with (1.4) shows that

$$ch(\overline{\nu};\mu)_N = (x_1 x_2 \ldots x_N)^{-\nu_1} ch(\lambda)_N \tag{2.3}$$

where $\lambda = (\mu_1 + \nu_1, \mu_2 + \nu_1, \ldots, \nu_1 - \nu_2, 0)$. This correspondence between $(\overline{\nu};\mu)$ and λ is such that F^λ is obtained from $F^{\overline{\nu};\mu}$ by taking the complement in a column of length N of each of the ν_1 columns which constitute $F^{\overline{\nu}}$ and adjoining them to the remaining μ_1 columns which constitute F^μ . This is illustrated in the case $N = 7$ and $\Lambda = (2^2 1^2 0 - 2 - 3)$ by

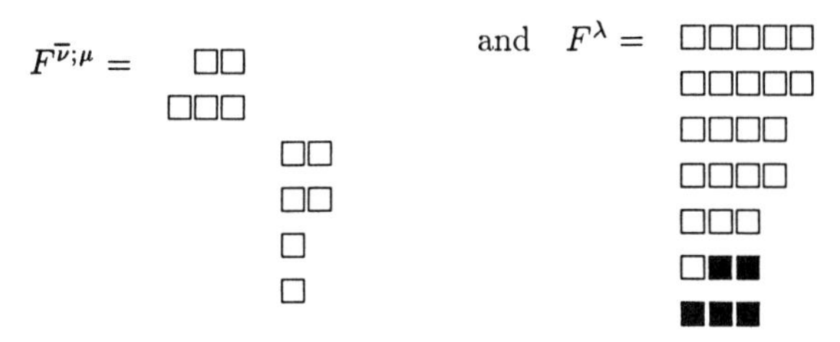

where $\mu = (2^2 1^2)$ and $\nu = (32)$, leading to $\lambda = (5^2 4^2 31)$.

Corresponding to each composite Young diagram $F^{\overline{\nu};\mu}$ there exist standard composite Young tableaux $T^{\overline{\nu};\mu}$ [Ki4,KElS,Ste2]. Each of these tableaux is a numbering of $F^{\overline{\nu};\mu}$ in which

(i) the entries in the portion F^{μ} are taken from the set $\mathsf{E} = \{1, 2, \ldots, N\}$ ordered so that $1 < 2 < \cdots < N$,

(ii) the entries in the portion $F^{\overline{\nu}}$ are taken from the set $\mathsf{E} = \{\overline{1}, \overline{2}, \ldots, \overline{N}\}$ ordered so that $\overline{N} < \cdots < \overline{2} < \overline{1}$,

(iii) the entries are strictly increasing from top to bottom down each column,

(iv) the entries are weakly increasing from left to right across each row

(v) for each $j = 1, 2, \ldots, N$ if entries j and $\overline{j}$ appear in row $r(j)$ of F^{μ} and row $\overline{r}(j)$ of $F^{\overline{\nu}}$, respectively, then $r(j) + \overline{r}(j) \leq j$.

These rules are not arbitrary. They are a consequence of the correspondence between the diagrams $F^{\overline{\nu};\mu}$ and F^{λ} dictated by the identity (2.3). This correspondence can be extended to tableaux in the obvious way by taking a complement in $12\ldots N$ of the entries in each of the first ν_1 columns of $T^{\overline{\nu};\mu}$. Typically the correspondence is illustrated by

$$T^{\overline{\nu};\mu} = \begin{array}{ccccc} & \overline{5} & \overline{5} & & \\ \overline{4} & \overline{3} & \overline{2} & & \\ & & & 1 & 2 \\ & & & 3 & 3 \\ & & & 4 & \\ & & & 7 & \end{array} \qquad T^{\lambda} = \begin{array}{ccccc} 1 & 1 & 1 & 1 & 2 \\ 2 & 2 & 3 & 3 & 3 \\ 3 & 4 & 4 & 4 & \\ 5 & 6 & 6 & 7 & \\ 6 & 7 & 7 & & \\ 7 & . & . & & \\ . & . & . & & \end{array}$$

To each such standard tableau $T^{\overline{\nu};\mu}$ there now corresponds a weight vector $\mathbf{M} = (m_1, m_2, \ldots, m_N)$ with $m_j = n_j - \overline{n}_j$ where n_j and n_j are the number of entries j and $\overline{j}$, respectively, in $T^{\overline{\nu};\mu}$. This allows us to define

$$s_{\overline{\nu};\mu}(x_1, x_2, \ldots, x_N) = \sum_{T^{\overline{\nu};\mu}} x_1^{m_1} x_2^{m_2} \ldots x_n^{m_N} \tag{2.4}$$

for $l(\mu) + l(\nu) \leq N$. It follows that the character of the $gl(N)$ irrep having highest weight (2.1) is given by

$$ch(\overline{\nu};\mu)_N = \{\overline{\nu};\mu\}(\mathbf{x})_N = s_{\overline{\nu};\mu}(\mathbf{x})_N = s_{\overline{\nu};\mu}(x_1, x_2, \ldots, x_N) \tag{2.5}$$

where, as in (1.7), a variety of notations has been used to specify the character. It should be noted that

$$s_{\overline{0};\mu}(\mathbf{x})_N = s_{\mu}(\mathbf{x})_N \quad \text{and} \quad s_{\overline{\nu};0}(\mathbf{x})_N = s_{\nu}(\overline{\mathbf{x}})_N, \tag{2.6}$$

where $\overline{\mathbf{x}} = (\overline{x}_1, \overline{x}_2, \ldots, \overline{x}_N) = (x_1^{-1}, x_2^{-1}, \ldots, x_N^{-1})$

All the above has been carried through on the assumption that $l(\mu) + l(\nu) \leq N$. However it is worth extending these notions in such a way that universal characters $ch(\nu;\mu)$ may be defined, from which the characters $ch(\nu;\mu)$ of $gl(N)$ may be recovered for each finite value of N. This can be accomplished by making use of determinantal identities associated with the column structure of F^{λ} and $F^{\overline{\lambda};\mu}$.

The partition conjugate to λ is denoted by λ', and is defined to be the partition corresponding to the Young diagram $F^{\lambda'}$ obtained from F^{λ} by interchanging rows and columns. With this notation there exists the following determinantal identity [Li1,Ma2]

$$
s_{\lambda}(\mathbf{x}) = \{\lambda\}(\mathbf{x}) = \left|\{1^{\lambda'_j+i-j}\}(\mathbf{x})\right| \quad , \tag{2.7}
$$

where $1 \leq i,j \leq \lambda_1$. The generalisation which serves to define the required universal characters takes the form [BB2]

$$
s_{\overline{\nu};\mu}(\mathbf{x}) = \begin{vmatrix} \{\overline{1}^{\nu'_l+k-l}\}(\mathbf{x}) & \vdots & \{1^{\mu'_j-k-j+1}\}(\mathbf{x}) \\ \cdots\cdots & \cdots & \cdots\cdots \\ \{\overline{1}^{\nu'_l-i-l+1}\}(\mathbf{x}) & \vdots & \{1^{\mu'_j+i-j}\}(\mathbf{x}) \end{vmatrix} \tag{2.8}
$$

where $1 \leq i,j \leq \mu_1$ and $1 \leq k,l \leq \nu_1$ and the indices i,j,k and l range from top to bottom, left to right, bottom to top and right to left, respectively.

The passage from $\{\overline{\nu};\mu\}(\mathbf{x})$ to $\{\overline{\nu};\mu\}(\mathbf{x})_N$ is effected by applying the specialisations

$$
\{\sigma\}(\mathbf{x})_N = \{\sigma\}(x_1, x_2, \ldots, x_N, 0, 0, \ldots) \tag{2.9a}
$$

$$
\{\tau\}(\overline{\mathbf{x}})_N = \{\tau\}(\overline{x}_1, \overline{x}_2, \ldots, \overline{x}_N, 0, 0, \ldots). \tag{2.9b}
$$

to all the elements $\{1^s\}(\mathbf{x})$ and $\{\overline{1}^t\}(\mathbf{x}) = \{1^t\}(\overline{\mathbf{x}})$ of (2.7). In doing this the indeterminates $\mathbf{x}$ and $\overline{\mathbf{x}}$ are to be treated as if they were independent.

If $l(\mu) + l(\nu) \leq N$ this yields the character $\{\overline{\nu};\mu\}(\mathbf{x})_N$ as originally defined by (2.3) and (2.4). This may be seen by noting that the application of the identity

$$
\{\overline{1}^t\}(\mathbf{x})_N = (x_1 x_2 \ldots x_N)^{-1}\{1^{N-t}\}(\mathbf{x})_N \tag{2.10}
$$

to the specialised form of (2.8) gives precisely the specialised form of (2.7) multiplied by the factor $(x_1 x_2 \ldots x_N)^{-\nu_1}$ with λ related to μ and ν as in (2.3).

Unfortunately if $l(\mu) + l(\nu) > N$ a difficulty arises. This can be seen by taking $N = 3$ rather than $N = 7$ in the previous example. This gives

$F^{\overline{\nu};\mu} =$
 □□
□□□
  □□
  □□
  □
  □

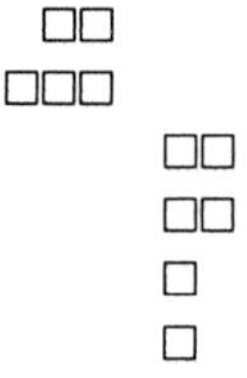

and $F^{\lambda} =$
□□□□□
□ □□
   □
   □

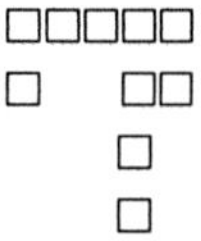

with F^{λ} non-standard. However a simple rearrangement of the columns of the determinant (2.7) can be carried out and then (2.3) used again to recover a character of the type $\{\overline{\tau};\sigma\}(\mathbf{x})_N$. Diagrammatically

```
□□□□□                         □□□□□                                □□□
□  □□          gives +        □□□□□        and hence    F^{τ̄;σ} =     □□
   □                                                                  □□
   □
```

Hence

$$\{\overline{32};2^21^2\}(\mathbf{x})_3 = +\{\overline{3};2^2\}(\mathbf{x})_3 \ . \tag{2.11}$$

The above procedure can be systematised. To this end it is convenient to define an operation [Ki2] on a Young diagram F^{μ} corresponding to the removal of a continuous boundary strip of boxes of length h starting at the foot of the first column and extending over c columns. The resulting diagram is denoted by $F^{\mu-h}$. If this diagram is itself a Young diagram specified by some partition σ then the symbol $\mu - h$ is identified with σ, otherwise $\mu - h$ is said to be null. For example in the case $\mu = (2^21^2)$ the following diagrams correspond to $F^{\mu-h}$ for $h = 0, 1, \ldots, 5$:

```
□□   □□   □□   □□   □□   □■   ...
□□   □□   □□   ■□   ■■   ■■
□    □    ■    ■    ■    ■
□    ■    ■    ■    ■    ■
```

where □ represents a box and ■ a box that has been removed, so that in these cases:

$(\mu - h)$	$=$	(2^21^2)	(2^21)	(2^2)	ϕ	(2)	(1)
h	$=$	0	1	2	3	4	5
c	$=$	0	1	1	1	2	2

For $h > 5$ it is not possible to remove the boundary strip and the result is therefore null.

Just as $\{\lambda\}(\mathbf{x})_N$ is standard if and only if $l(\lambda) \leq N$ but requires modification in accordance with the rule (1.8) if $l(\lambda) > N$, so $\{\overline{\nu};\mu\}(\mathbf{x})_N$ is standard if and only if $l(\mu) + l(\nu) \leq N$ but requires modification if $l(\mu) + l(\nu) > N$ in accordance with the rule [Ki2]

$$\{\overline{\nu};\mu\}(\mathbf{x})_N = (-1)^{c+\overline{c}+1}\{\overline{\nu - h};\mu - h\}(\mathbf{x})_N \qquad \text{where} \quad h = l(\mu) + l(\nu) - N - 1 \ . \tag{2.12}$$

This modification rule is considerably more complicated than (1.8), involving as it does two applications of the boundary strip removal procedure covering c and $\overline{c}$

columns of F^{μ} and $F^{\overline{\nu}}$, respectively. However in practice it is easy to apply. For example, in the case $\mu = (2^2 1^2), \nu = (32)$ and $N = 3$ referred to above, one has $l(\mu) = 4$, and $l(\nu) = 2$ so that $h = 2$, leading to the diagram modification

$$F^{\overline{\nu};\mu} = \begin{array}{ll} \square\square & \\ \square\square\square & \\ & \square\square \\ & \square\square \\ & \square \\ & \square \end{array} \quad \longrightarrow \quad F^{\overline{\nu - h};\mu - h} = \begin{array}{ll} \blacksquare\blacksquare & \\ \square\square\square & \\ & \square\square \\ & \square\square \\ & \blacksquare \\ & \blacksquare \end{array}$$

and hence to the result (2.11) given earlier. In general this modification procedure may have to be repeated more than once in order to reduce the original composite diagram either to one which occupies a total of no more than N rows and is thus standard, or to a null shape.

As a computational tool the determinantal definition (2.8) of universal characters does not seem very promising. However the Laplace expansion of this determinant yields the formula [Ki1]

$$s_{\overline{\nu};\mu}(\mathbf{x}) = \sum_{\zeta} (-1)^{|\zeta|} s_{\nu/\zeta}(\overline{\mathbf{x}}) s_{\mu/\zeta'}(\mathbf{x}) \,. \tag{2.13}$$

Conversely [AK,Ki2]

$$s_{\nu}(\overline{\mathbf{x}}) s_{\mu}(\mathbf{x}) = \sum_{\zeta} s_{\overline{\nu/\zeta};\mu/\zeta}(\mathbf{x}) \,. \tag{2.14}$$

Formula (2.13) serves to give what is probably the best definition of the universal characters, namely

$$gl(N) \quad \{\overline{\nu};\mu\}(\mathbf{x})_N = \sum_{\zeta} (-1)^{|\zeta|} \{\nu/\zeta\}(\overline{\mathbf{x}})_N \{\mu/\zeta'\}(\mathbf{x})_N, \tag{2.15}$$

where the terms in $\mathbf{x}$ and $\overline{\mathbf{x}}$ are defined by the independent specialisations (2.9a) and (2.9b), respectively.

These formulae may be exploited to derive results generalising the tensor product rule (1.13) and the branching rule (1.17). The tensor product rule takes the universal form [AK,Ki2]

$$\{\overline{\nu};\mu\}_N \times \{\overline{\lambda};\kappa\}_N = \sum_{\sigma,\tau} \{\overline{(\nu/\sigma).(\lambda/\tau)};(\mu/\tau).(\kappa/\sigma)\}_N, \tag{2.15}$$

subject to the modification rule (2.12).

By way of illustration (2.15) yields

$$\begin{aligned} \{\overline{1};1\}_N \times \{\overline{1};1\}_N = \{\overline{2};2\}_N + \{\overline{2};1^2\}_N + \{\overline{1}^2;2\}_N + \{\overline{1}^2;1^2\}_N \\ + 2\{\overline{1};1\}_N + \{0\}_N \,. \end{aligned}$$

In the case of $gl(3)$ the modification rule (2.12) implies that $\{\overline{2};1^2\}_3 = 0$ so that

$$\{\overline{1};1\}_3 \times \{\overline{1};1\}_3 = \{\overline{2};2\}_3 + \{\overline{2};1^2\}_3 + \{\overline{1}^2;2\}_3 + 2\{\overline{1};1\}_3 + \{0\}_3 \; .$$

Similarly for $gl(2)$ we have $\{\overline{2};1^2\}_2 = \{\overline{1}^2;2\}_2 = 0$ and $\{\overline{1}^2;1^2\}_2 = -\{\overline{1};1\}_2$ so that

$$\{\overline{1};1\}_2 \times \{\overline{1};1\}_2 = \{\overline{2};2\}_2 + \{\overline{1};1\}_2 + \{0\}_2 \; .$$

In the same way the branching rule appropriate to the restriction from $gl(M+N)$ to $gl(M)+gl(N)$ takes the universal form [AK,Ki1,Ki3]

$$\{\overline{\nu};\mu\}_{M+N} = \sum_{\rho,\sigma,\tau} \{\overline{\nu/\sigma};\mu/\tau\}_M \times \{\overline{\sigma/\rho};\tau/\rho\}_N \; , \tag{2.16}$$

subject of course to the modification rule (2.12).

3. Infinite S-function series and characters of o(N) and sp(N). Littlewood [Li1] gave a number of generating functions for infinite series of S-functions, and the list can be extended [Ki3] to give

(3.1a)
$$A_q(\mathbf{x}) = \prod_{i<j}(1 - qx_ix_j) = \sum_{\alpha\in A}(-1)^{|\alpha|/2}\, q^{|\alpha|/2}\, s_\alpha(\mathbf{x})$$

(3.1b)
$$B_q(\mathbf{x}) = \prod_{i<j}(1 - qx_ix_j)^{-1} = \sum_{\beta\in B} q^{|\beta|/2}\, s_\beta(\mathbf{x})$$

(3.1c)
$$C_q(\mathbf{x}) = \prod_{i\leq j}(1 - qx_ix_j) = \sum_{\gamma\in C}(-1)^{|\gamma|/2}\, q^{|\gamma|/2}\, s_\gamma(\mathbf{x})$$

(3.1d)
$$D_q(\mathbf{x}) = \prod_{i\leq j}(1 - qx_ix_j)^{-1} = \sum_{\delta\in D} q^{|\delta|/2}\, s_\delta(\mathbf{x})$$

(3.1e)
$$E_q(\mathbf{x}) = \prod_{i}(1 - qx_i)\prod_{i<j}(1 - q^2x_ix_j) = \sum_{\epsilon\in E}(-1)^{(|\epsilon|+r)/2}q^{|\epsilon|}s_\epsilon(\mathbf{x})$$

(3.1f)
$$F_q(\mathbf{x}) = \prod_{i}(1 - qx_i)^{-1}\prod_{i<j}(1 - q^2x_ix_j)^{-1} = \sum_{\zeta\in F} q^{|\zeta|}s_\zeta(\mathbf{x})$$

(3.1g)
$$G_q(\mathbf{x}) = \prod_{i}(1 + qx_i)\prod_{i<j}(1 - q^2x_ix_j) = \sum_{\epsilon\in E}(-1)^{(|\epsilon|-r)/2}q^{|\epsilon|}s_\epsilon(\mathbf{x})$$

(3.1h)
$$H_q(\mathbf{x}) = \prod_{i}(1 + qx_i)^{-1}\prod_{i<j}(1 - q^2x_ix_j)^{-1} = \sum_{\zeta\in F}(-1)^{|\zeta|}\, q^{|\zeta|}s_\zeta(\mathbf{x})$$

(3.1i)
$$I_q(\mathbf{x};\mathbf{y}) = \prod_{i,a}(1 - qx_iy_a) = \sum_{\zeta\in F}(-1)^{|\zeta|}\, q^{|\zeta|}\, s_\zeta(\mathbf{x})\, s_{\zeta'}(\mathbf{y})$$

(3.1j)

$$J_q(\mathbf{x};\mathbf{y}) = \prod_{i,a}(1 - qx_iy_a)^{-1} = \sum_{\zeta\in F} q^{|\zeta|}\ s_\zeta(\mathbf{x})\ s_\zeta(\mathbf{y})$$

(3.1l)

$$L_q(\mathbf{x}) = \prod_i(1 - qx_i) = \sum_m(-1)^m\ q^m\ s_{1^m}(\mathbf{x})$$

(3.1m)

$$M_q(\mathbf{x}) = \prod_i(1 - qx_i)^{-1} = \sum_m\ q^m\ s_m(\mathbf{x})$$

(3.1p)

$$P_q(\mathbf{x}) = \prod_i(1 + qx_i)^{-1} = \sum_m(-1)^m\ q^m\ s_m(\mathbf{x})$$

(3.1q)

$$Q_q(\mathbf{x}) = \prod_i(1 + qx_i) = \sum_m\ q^m\ s_{1^m}(\mathbf{x})$$

where, in Frobenius notation, A, C and E are the sets of partitions of the form $\begin{pmatrix} a_1 & a_2 & \dots \\ a_1+1 & a_2+1 & \dots \end{pmatrix}$, $\begin{pmatrix} a_1+1 & a_2+1 & \dots \\ a_1 & a_2 & \dots \end{pmatrix}$ and $\begin{pmatrix} a_1 & a_2 \dots \\ a_1 & a_2 \dots \end{pmatrix}$ respectively, D is the set of partitions all of whose parts are even, B is the set of partitions all of whose distinct parts are repeated an even number of times and F is the set of all partitions. The trivial partition 0 is taken to be a member of each of the sets $A, B, \dots, F$, and by definition $s_0(\mathbf{x}) = s_0(\mathbf{y}) = 1$, so that each of the above series has as its first term 1. The Frobenius rank of the self-conjugate partition ϵ has been denoted by r. It is convenient in what follows to denote the $q = 1$ series $A_1(\mathbf{x})$ by $A(\mathbf{x}), B_1(\mathbf{x})$ by $B(\mathbf{x})$ and so on, and from time to time to suppress the explicit dependence upon $(\mathbf{x})$.

These expansions are universally valid in the sense that if the products on the left are taken over all positive integer values of i and j , then on the right the S-functions $s_\lambda(\mathbf{x})$ are the universal S-functions, (1.10), involving an infinite number of variables $(x_1, x_2, \dots)$. Restricting the domain of the indices i and j on the left to $(1, 2, \dots, N)$ leads on the right to S-functions $s_\lambda(x_1, x_2, \dots, x_N)$ which are of course subject to the modification rule (1.8). Similarly products over all positive integer values of a lead to universal S-functions $s_\zeta(\mathbf{y})$, which are subject to the same modification if the values of a are restricted.

Just as $gl(N)$ possesses finite dimensional irreps whose highest weights are specified by partitions, so do both $o(N)$ and $sp(N)$. The notation used for the corresponding group characters [Mu,Li1,Ki3] is :

(3.2a)	$gl(N)$	$ch(\lambda)_N = \{\lambda\}(\mathbf{x})_N$
(3.2b)	$o(N)$	$ch(\lambda)_N = [\lambda](\mathbf{x})_N$
(3.2c)	$sp(N)$	$ch(\lambda)_N = \langle\lambda\rangle(\mathbf{x})_N.$

Making use of Weyl's character formula (1.1) it is possible to derive generating formulae [Mu,W2,Li1] whose universal extensions take the form:

$$\prod_{i,a}(1 - x_i y_a)^{-1} = \sum_{\lambda} \{\lambda\}(\mathbf{x})\ \{\lambda\}(\mathbf{y}) \tag{3.3}$$

$$\prod_{i,a}(1 - x_i y_a)^{-1} \prod_{a \leq b}(1 - y_a y_b) = \sum_{\lambda} [\lambda](\mathbf{x})\ \{\lambda\}(\mathbf{y}) \tag{3.4}$$

$$\prod_{i,a}(1 - x_i y_a)^{-1} \prod_{a<b}(1 - y_a y_b) = \sum_{\lambda} \langle\lambda\rangle(\mathbf{x})\ \{\lambda\}(\mathbf{y}) \tag{3.5}$$

where the summations are over all partitions λ.

The specialisations appropriate to each of the classical Lie algebras are such that

(3.6a) $sl(n)$ $\{\lambda\}(\mathbf{x})_n = \{\lambda\}(x_1, x_2, \ldots, x_n, 0, \ldots, 0)$ with $x_1 x_2 \ldots x_n = 1$

(3.6b) $so(2n+1)$ $[\lambda](\mathbf{x})_{2n+1} = [\lambda](x_1, x_2, \ldots, x_n, \overline{x}_1, \overline{x}_2, \ldots, \overline{x}_n, 1, 0, \ldots, 0)$

(3.6c) $sp(2n)$ $\langle\lambda\rangle(\mathbf{x})_{2n} = \langle\lambda\rangle(x_1, x_2, \ldots, x_n, \overline{x}_1, \overline{x}_2, \ldots, \overline{x}_n, 0, \ldots, 0)$

(3.6d) $so(2n)$ $[\lambda](\mathbf{x})_{2n} = [\lambda](x_1, x_2, \ldots, x_n, \overline{x}_1, \overline{x}_2, \ldots, \overline{x}_n, 0, \ldots, 0)$

(3.6e) $gl\ (n)$ $\{\lambda\}(\mathbf{x})_n = \{\lambda\}(x_1, x_2, \ldots, x_n, 0, \ldots, 0)$.

where the indeterminates, as explained in Section 1, should be thought of as formal exponentials $x_i = \overline{x}_i^{-1} = e^{\epsilon_i}$.

Returning for the moment to the universal characters defined by the generating functions (3.3)-(3.5), it is a remarkable fact that they are related to one another by the following [Li1]

THEOREM 3.1. *Let $\{\lambda/X\}(\mathbf{x}) = s_{\lambda/X}(\mathbf{x})$ for any series X as in (1.19), then*

$$[\lambda](\mathbf{x}) = \{\lambda/C\}(\mathbf{x}) \tag{3.7a}$$

$$\langle\lambda\rangle(\mathbf{x}) = \{\lambda/A\}(\mathbf{x}) \tag{3.7b}$$

Proof. Substituting (3.3) into (3.4) and using the series expansion (3.1c) for $C(y)$ gives

$$\begin{aligned}\prod_{i,a}(1 - x_i y_a)^{-1} \prod_{a \leq b}(1 - y_a y_b) &= \sum_{\mu} s_\mu(\mathbf{x}) s_\mu(\mathbf{y}) \sum_{\gamma} (-1)^{|\gamma|/2} s_\gamma(\mathbf{y}) \\ &= \sum_{\mu,\gamma,\lambda} (-1)^{|\gamma|/2} c^{\lambda}_{\mu\gamma}\ s_\mu(\mathbf{x})\ s_\lambda(\mathbf{y}).\end{aligned}$$

Comparing coefficients of $\{\lambda\}(\mathbf{y}) = s_\lambda(\mathbf{y})$ in this expression and (3.4) then gives

$$[\lambda](\mathbf{x}) = \sum_{\mu,\gamma} (-1)^{|\gamma|/2}\ c^{\lambda}_{\mu\gamma} s_\gamma(\mathbf{x}) = \sum_{\gamma} (-1)^{|\gamma|/2}\ s_{\lambda/\gamma}(\mathbf{x})\ .$$

This is precisely what is meant by $\{\lambda/C\}(\mathbf{x})$. This proves (3.7a), and (3.7b) can be proved in the exactly the same way. □

It is then a straightforward matter to use these characters to derive the branching rules for the restriction from $gl(N)$ to $o(N)$ and from $gl(N)$ to $sp(N)$. The result is [Li1]

THEOREM 3.2. *Under the specialisations (3.6) appropriate to the subalgebras in question*

(3.8a) $\qquad gl(N) \longrightarrow o(N) \qquad \{\lambda\}(\mathbf{x})_N = [\lambda/D](\mathbf{x})_N$

(3.8b) $\qquad gl(N) \longrightarrow sp(N) \qquad \{\lambda\}(\mathbf{x})_N = \langle\lambda/B\rangle(\mathbf{x})_N$

Proof. By virtue of their definitions, (3.1c) and (3.1d), $D.C = 1$. Hence $\{\lambda\}(\mathbf{x}) = \{\lambda/DC\}(\mathbf{x}) = [\lambda/D](\mathbf{x})$ where the last step is a consequence of (3.7a). The use of $B.A = 1$ gives a similar result. The application of the specialisations (3.6) then completes the proof. □

The significance of the specialisations inherent in these universal branching rules (3.8) is that for particular values of N it is necessary to invoke associated modification rules analogous to those encountered already in dealing with characters of $gl(N)$. These modification rules now take the form [Ki2]

(3.9a)
$$o(N) \quad [\lambda](\mathbf{x})_N = (-1)^{c-1}[\lambda - h](\mathbf{x})_N \quad \text{with} \quad h = 2l(\lambda) - N \geq 1$$
(3.9b)
$$sp(N) \quad \langle\lambda\rangle(\mathbf{x})_N = (-1)^{c}\langle\lambda - h\rangle(\mathbf{x})_N \quad \text{with} \quad h = 2l(\lambda) - N - 2 \geq 0$$

where once again a strip removal procedure is involved. They may be derived by using certain determinantal expansions of the characters [Ki2].

To illustrate the role of modification rules consider the following example of the application of the branching rule for the restriction from $gl(N)$ to $sp(N)$ given by (3.8b) of Theorem 3.2

$$\begin{aligned}\{2^2 1^2\}_N &= \langle 2^2 1^2/B\rangle = \langle 2^2 1^2/(0 + 1^2 + 2^2 + 1^4 \ldots)\rangle_N \\ &= \langle 2^2 1^2\rangle_N + \langle 2^2\rangle_N + \langle 21^2\rangle_N + \langle 1^4\rangle_N + 2\langle 1^2\rangle_N + \langle 1\rangle_N ,\end{aligned}$$

where the dependence on $(\mathbf{x})$ has been suppressed. This result is universal. However in the case of $N = 4$, for example, the modification rule (3.9b) gives $\langle 2^2 1^2\rangle_4 = -\langle 2^2\rangle_4$, $\langle 21^2\rangle_4 = -\langle 21^2\rangle_4 = 0$ and $\langle 1^4\rangle_4 = -\langle 1^2\rangle_4$, so that we have on restriction from gl(4) to sp(4) the branching

$$\{2^2 1^2\}_4 = \langle 1^2\rangle_4 + \langle 1\rangle_4 .$$

The modification rules may be avoided if the original branching rules of Theorem 3.2 are altered to take into account each specific value of N. Two ways have recently been proposed for doing this in the case of (3.8b) by Sundaram [Su1] and Tokuyama [T]. Both methods involve the introduction of new tableaux whose enumeration amounts to a refinement of the Littlewood-Richardson rule for evaluating the quotients appearing in $\langle\lambda/B\rangle$. Sundaram's method involves the most attractive change to the original rule and is described elsewhere in these proceedings [Su2].

4. Universal tensor product rules. As stressed earlier tensor products of irreps $gl(N)$ may be decomposed by making use of universal characters. The relevant universal product rule, (1.13), involves just the Littlewood-Richardson coefficients. In the case of any particular value of N it is then only necessary to apply to the output of the universal product rule the modification rule (1.8). An exactly analogous procedure can be used to deal with tensor products of irreps of both $o(N)$ and $sp(N)$. The universal product rules are given by the following:

THEOREM 4.1 (NEWELL-LITTLEWOOD) [NE,LI2].

$$(4.1a) \qquad o(N) \qquad [\lambda]_N \times [\mu]_N = \sum_{\zeta} [(\lambda/\zeta).(\mu/\zeta)]_N$$

$$(4.1b) \qquad sp(N) \qquad \langle\lambda\rangle_N \times \langle\mu\rangle_N = \sum_{\zeta} \langle(\lambda/\zeta).(\mu/\zeta)\rangle_N$$

where $\times$ *indicates a tensor product, or equivalently a product of characters, . indicates an S-function product (1.12) and / indicates an S-function quotient (1.16). The brackets* $[\ \ ]_N$ *and* $\langle\ \ \rangle_N$ *signify characters of* $o(N)$ *and* $sp(N)$*, respectively, as in (3.2). The dependence on* $(\mathbf{x})_N$ *has been suppressed but it is to be understood that* $(\mathbf{x})$ *is specialised in accordance with (3.6) and that the characters are subject to the modification rules (3.9).*

In order to prove this Theorem it is necessary to prove a succession of small Lemmas.

LEMMA 4.2 (FOULKES) [F].

$$(4.2) \qquad s_{(\mu.\nu)/\rho}(\mathbf{x}) = \sum_{\sigma} s_{(\mu/\sigma).(\nu/(\rho/\sigma))}(\mathbf{x}) \ .$$

Proof. It suffices to expand a product of S-functions of $(\mathbf{z}) = (\mathbf{x},\mathbf{y})$ in two different ways. Firstly

$$\begin{aligned} s_\mu(\mathbf{x},\mathbf{y})s_\nu(\mathbf{x},\mathbf{y}) &= \sum_{\lambda} c^{\lambda}_{\mu\nu} s_\lambda(\mathbf{x},\mathbf{y}) && \text{from (1.12)} \\ &= \sum_{\lambda,\rho} c^{\lambda}_{\mu\nu} s_\rho(\mathbf{x}) s_{\lambda/\rho}(\mathbf{y}) && \text{from (1.15)} \\ &= \sum_{\rho} s_\rho(\mathbf{x}) s_{(\mu.\nu)/\rho}(\mathbf{y}) && \text{from (1.19a).} \end{aligned}$$

Secondly

$$\begin{aligned} s_\mu(\mathbf{x},\mathbf{y})s_\nu(\mathbf{x},\mathbf{y}) &= \sum_{\sigma,\tau} s_\sigma(\mathbf{x}) s_{\mu/\sigma}(\mathbf{y}) s_\tau(\mathbf{x}) s_{\nu/\tau}(\mathbf{y}) && \text{from (1.15)} \\ &= \sum_{\sigma,\tau,\rho} c^{\rho}_{\sigma\tau} s_\rho(\mathbf{x}) s_{\mu/\sigma}(\mathbf{y}) s_{\nu/\tau}(\mathbf{y}) && \text{from (1.12)} \\ &= \sum_{\rho,\sigma} s_\rho(\mathbf{x}) s_{\mu/\sigma}(\mathbf{y}) s_{(\nu/(\rho/\sigma))}(\mathbf{y}) && \text{from (1.16)} \end{aligned}$$

The required result (4.2) then follows by comparing the coefficients of $s_\rho(\mathbf{x})$ in these two expansions. □

LEMMA 4.3.

$$s_{D/\mu}(\mathbf{x}) = s_{(\mu/D).D}(\mathbf{x}) \tag{4.3}$$

Proof. This time one expands $D(\mathbf{x},\mathbf{y})$ in two ways.

$$\begin{aligned}
D(\mathbf{x},\mathbf{y}) &= \prod_{i\le j}(1-x_ix_j)^{-1}\prod_{i,a}(1-x_iy_a)^{-1}\prod_{a\le b}(1-y_ay_b)^{-1} \\
&= D(\mathbf{x})\ D(\mathbf{y})\sum_\lambda s_\lambda(\mathbf{x})s_\lambda(\mathbf{y}) && \text{from (3.3)} \\
&= \sum_\lambda D(\mathbf{x})s_\lambda(\mathbf{x})\sum_{\delta\in D} s_\delta(\mathbf{y})s_\lambda(\mathbf{y}) && \text{from (3.1d)} \\
&= \sum_\lambda D(\mathbf{x})s_\lambda(\mathbf{x})\sum_{\delta\in D,\mu} c^\mu_{\delta\lambda}s_\mu(\mathbf{y}) && \text{from (1.12)} \\
&= \sum_\lambda D(\mathbf{x})s_{\mu/D}(\mathbf{x})s_\mu(\mathbf{y}) && \text{from (1.19c)}
\end{aligned}$$

and

$$\begin{aligned}
D(\mathbf{x},\mathbf{y}) = \sum_{\delta\in D} s_\delta(\mathbf{x},\mathbf{y}) &= \sum_{\delta\in D,\mu} s_{\delta/\mu}(\mathbf{x})s_\mu(\mathbf{y}) && \text{from (1.15)} \\
&= \sum_\mu s_{D/\mu}(\mathbf{x})s_\mu(\mathbf{y}). && \text{from (1.19c)}
\end{aligned}$$

This time a comparison of the coefficients of $s_\mu(\mathbf{y})$ proves the required result. □

LEMMA 4.4.

$$s_{(\mu.\nu)/D}(\mathbf{x}) = \sum_\zeta s_{(\mu/(\zeta.D))}(\mathbf{x})s_{(\nu/(\zeta.D))}(\mathbf{x}) \tag{4.4}$$

Proof. In this case one can work entirely in terms of S-functions of $(\mathbf{x})$, which for convenience may be dropped from the notation on the understanding that all subsequent S-functions depend on these variables. Then

$$\begin{aligned}
s_{(\mu.\nu)/D} &= \sum_{\delta\in D} s_{(\mu.\nu)/\delta} = \sum_{\delta\in D,\sigma} s_{(\mu/\sigma)}s_{(\nu/(\delta/\sigma))} && \text{from Lemma 4.2} \\
&= \sum_\sigma s_{(\mu/\sigma)}s_{(\nu/(D/\sigma))} = \sum_\sigma s_{(\mu/\sigma)}s_{(\nu/((\sigma/D).D))} && \text{from Lemma 4.3} \\
&= \sum_{\delta\in D,\sigma} s_{(\mu/\sigma)}s_{(\nu/((\sigma/\delta).D))} \ \text{(from (1.19c))} = \sum_{\delta\in D,\sigma,\zeta} c^\sigma_{\zeta\delta}s_{(\mu/\sigma)}s_{(\nu/(\zeta.D))} && \text{from (1.19b)} \\
&= \sum_{\delta\in D,\zeta} s_{(\mu/(\zeta.\delta))}s_{(\nu/(\zeta.D))} \ \text{(from (1.19b))} = \sum_\zeta s_{(\mu/(\zeta.D))}s_{(\nu/(\zeta.D))}. && \text{from (1.19c)}
\end{aligned}$$

This proves the Lemma. □

Now we are in a position to prove the Newell-Littlewood Theorem as follows [C,CK3]:

$$\begin{aligned}
[\lambda]\times[\mu] &= s_{\lambda/C}.s_{\mu/C} = s_{(\lambda/C).(\mu/C)} && \text{from (3.7a)}\\
&= s_{((\lambda/C).(\mu/C))/DC} && \text{since } DC=1\\
&= \sum_{\varsigma} s_{((\lambda/(C\varsigma D)).(\mu/(C\varsigma D)))/C} && \text{from Lemma 4.4}\\
&= \sum_{\varsigma} s_{((\lambda/\varsigma).(\mu/\varsigma))/C} && \text{since } DC=1\\
&= \sum_{\varsigma} [(\lambda/\varsigma).(\mu/\varsigma)] && \text{from (3.7a).}
\end{aligned}$$

Applying the usual specialisation (3.6) then gives (4.1a). Similarly, replacing C and D by A and B throughout gives (4.1b), and the proof of Theorem 4.1 is complete. □

In making use of Theorem 4.1 for $o(N)$ and $sp(N)$ for some fixed N it is of course also necessary to use in addition the modification rules (3.11).

Analogous methods may be used to derive the corresponding result for tensor products of $gl(N)$:

THEOREM 4.5 (ABRAMSKY-KING) [AK,KI2].

(4.5) $$gl(N) \quad \{\overline{\nu};\mu\}_N \times \{\overline{\lambda};\kappa\}_N = \sum_{\sigma,\tau} \{\overline{(\nu/\sigma).(\lambda/\tau)};(\mu/\tau).(\kappa/\sigma)\}_N.$$

Once again it is necessary to proceed via a sequence of Lemmas. Lemma 4.6 (Littlewood) [Li1]

(4.6a) $$\sum_{\alpha,\beta}(-1)^{|\beta|}s_{\mu/(\alpha.\beta)}(\mathbf{x})s_{\nu/(\alpha.\beta')}(\mathbf{x}) = s_\mu(\mathbf{x})\ s_\nu(\mathbf{x}).$$

Proof. From (3.1i) and (3.1j), in the case $q=1$,

$$\begin{aligned}
J(\mathbf{x};\mathbf{y})\ I(\mathbf{x};\mathbf{y}) &= \sum_{\alpha,\beta}(-1)^{|\beta|}s_\alpha(\mathbf{x})s_\alpha(\mathbf{y})\ s_\beta(\mathbf{x})\ s_{\beta'}(\mathbf{y})\\
&= \sum_{\alpha\beta\sigma\tau}(-1)^{|\beta|}c^\sigma_{\alpha\beta}c^\tau_{\alpha\beta'}\ s_\sigma(\mathbf{x})\ s_\tau(\mathbf{y})
\end{aligned}$$

But $J(\mathbf{x};\mathbf{y})\ I(\mathbf{x};\mathbf{y}) = 1$ so that

(4.6b) $$\sum_{\alpha,\beta}(-1)^{|\beta|}c^\sigma_{\alpha\beta}\ c^\tau_{\alpha\beta'} = \delta_{\sigma 0}\ \delta_{\tau 0}.$$

Hence

$$\sum_{\alpha,\beta}(-1)^{|\beta|}s_{\mu/(\alpha.\beta)}\mathbf{x})\ s_{\nu/(\alpha.\beta')}(\mathbf{y})$$

$$= \sum_{\alpha\beta\sigma\tau}(-1)^{|\beta|}c^\sigma_{\alpha\beta}c^\tau_{\alpha\beta'}s_{\mu/\sigma}(\mathbf{x})\ s_{\nu/\tau}(\mathbf{y}) = s_\mu(\mathbf{x})\ s_\nu(\mathbf{y})\ □$$

It is worth pointing out that the strange identity (4.6b) is just what is required to justify the claim made earlier that (2.14) is the converse of (2.13).

LEMMA 4.7.

$$\sum_{\zeta} s_{(\nu.\lambda)/\zeta}(\mathbf{x})\ s_{(\mu.\kappa)/\zeta}(\mathbf{y})$$

(4.7a)
$$= \sum_{\gamma\delta\sigma\tau} s_{\nu/(\gamma.\sigma)}(\mathbf{x})\ s_{\lambda/(\delta.\tau)}(\mathbf{x})\ s_{\mu/(\gamma.\tau)}(\mathbf{y})\ s_{\kappa/(\delta.\sigma)}(\mathbf{y}).$$

Proof. Just as in the proof of Lemma 4.3 one expands one of the series (3.1) in two ways, this time $J(\mathbf{x},\mathbf{u};\mathbf{y},\mathbf{v})$:

$$\begin{aligned} J(\mathbf{x},\mathbf{u};\mathbf{y},\mathbf{v}) &= \prod_{i,a}(1-x_iy_a)^{-1}\prod_{i,b}(1-x_iv_b)^{-1}\prod_{j,a}(1-u_jy_a)^{-1}\prod_{j,b}(1-u_jv_b)^{-1} \\ &= \sum_{\gamma\delta\sigma\tau} s_\gamma(\mathbf{x})s_\gamma(\mathbf{y})\ s_\sigma(\mathbf{x})\ s_\sigma(\mathbf{v})s_\tau(\mathbf{u})\ s_\tau(\mathbf{y})\ s_\delta(\mathbf{u})\ s_\delta(\mathbf{v}) \\ &= \sum_{\gamma\delta\sigma\tau}\sum_{\vartheta\phi\psi\omega} c^{\vartheta}_{\gamma\sigma}\ c^{\phi}_{\gamma\tau}\ c^{\psi}_{\delta\tau}c^{\omega}_{\delta\sigma}\ s_\vartheta(\mathbf{x})\ s_\phi(\mathbf{y})\ s_\psi(\mathbf{u})s_\omega(\mathbf{v}) \\ &\qquad \text{from (1.12)} \end{aligned}$$

and

$$\begin{aligned} J(\mathbf{x},\mathbf{u};\mathbf{y},\mathbf{v}) &= \sum_{\zeta} s_\zeta(\mathbf{x},\mathbf{u})s_\zeta(\mathbf{y},\mathbf{v}) && \text{from (3.1j)} \\ &= \sum_{\zeta}\sum_{\vartheta\phi\psi\omega} c^{\zeta}_{\vartheta\psi}\ c^{\zeta}_{\phi\omega}\ s_\vartheta(\mathbf{x})\ s_\phi(\mathbf{y})\ s_\psi(\mathbf{u})s_\omega(\mathbf{v})\ . \\ & && \text{from (1.15)} \end{aligned}$$

Comparing coefficients of $s_\vartheta(\mathbf{x})s_\phi(\mathbf{y})s_\psi(\mathbf{u})s_\omega(\mathbf{v})$ gives

(4.7b)
$$\sum_{\zeta} c^{\zeta}_{\vartheta\psi}\ c^{\zeta}_{\phi\omega} = \sum_{\gamma\delta\sigma\tau} c^{\vartheta}_{\gamma\sigma}\ c^{\phi}_{\gamma\sigma}\ c^{\psi}_{\delta\tau}\ c^{\omega}_{\delta\sigma}$$

Hence

$$\begin{aligned} &\sum_{\zeta} s_{(\nu.\lambda)/\zeta}(\mathbf{x})\ s_{(\mu.\kappa)/\sigma}(\mathbf{y}) \\ &= \sum_{\zeta}\sum_{\vartheta\phi\psi\omega} c^{\zeta}_{\vartheta\psi}\ c^{\zeta}_{\phi\omega}\ s_{(\nu/\vartheta).(\lambda/\psi)}(\mathbf{x})\ s_{(\mu/\phi).(\kappa/\omega)}(\mathbf{y}) \\ &\qquad \text{from Lemma 4.2 and (1.16)} \\ &= \sum_{\gamma\delta\sigma\tau}\sum_{\vartheta\phi\psi\omega} c^{\vartheta}_{\gamma\sigma}\ c^{\phi}_{\gamma\tau}\ c^{\psi}_{\delta\tau}\ c^{\omega}_{\delta\sigma}\ s_{(\nu/\vartheta)}(\mathbf{x})\ s_{(\lambda/\psi)}(\mathbf{x})\ s_{(\mu/\phi)}(\mathbf{y})\ s_{(\kappa/\omega)}(\mathbf{y}) \\ &\qquad \text{from 4.7b} \\ &= \sum_{\gamma\delta\sigma\tau} s_{\nu/(\gamma.\sigma)}(\mathbf{x})\ s_{\lambda/(\delta.\tau)}(bx)\ s_{\mu/(\gamma.\tau)}(\mathbf{y})\ s_{\kappa/(\delta.\sigma)}(\mathbf{y}) \\ &\qquad \text{from 1.12}\ . \qquad \square \end{aligned}$$

These Lemmas enable us to prove Theorem 4.2 as follows:

$$s_{\overline{\nu};\mu}(\mathbf{x})\ s_{\overline{\lambda};\kappa}(\mathbf{x}) = \sum_{\xi,\eta}(-1)^{|\xi|+|\eta|} s_{\nu/\xi}(\overline{\mathbf{x}})\ s_{\lambda/\eta}(\overline{\mathbf{x}})\ s_{\mu/\xi'}(\mathbf{x})\ s_{\kappa/\eta'}(x)$$

from (2.13)

$$= \sum_{\xi\eta\zeta}(-1)^{|\xi|+|\eta|} s_{\overline{((\nu/\xi).(\lambda/\eta))/\zeta};((\mu/\xi').(\kappa/\eta'))/\zeta}(\mathbf{x})$$

from (2.14)

$$= \sum_{\xi\eta\zeta}\sum_{\gamma\delta\sigma\tau}(-1)^{|\xi|+|\eta|} s_{\overline{((\nu/(\xi\gamma\sigma)).(\lambda/(\eta\delta\tau))};(\mu/(\xi'\gamma\tau)).(\kappa/(\eta'\delta\sigma))}(\mathbf{x})$$

from Lemma 4.7

$$= \sum_{\sigma\tau} s_{\overline{((\nu/\sigma).(\lambda/\tau))};((\mu/\tau).(\kappa/\sigma))}(\mathbf{x}) \quad \text{from Lemma 4.6}\ .$$

Applying the specialisation described at the end of Section 2 then gives (4.5) as required. □

5. The Macdonald identities. The infinite series of S-functions encountered in the last section have a role to play in studying the Macdonald identities [Ma1]. Indeed a number of these identities can be recast in the form of expansions of certain infinite products of infinite series of S-functions in terms of infinite sums of universal characters. Modification rules are then required to recover the Macdonald identities appropriate to a finite number of variables. To make the connection we are seeking it is convenient to note that the Macdonald identity associated with a simple Lie algebra, L, can be written in the form

$$\prod_{k=1}^{\infty}\left\{(1-q^k)^n \prod_{\alpha\in R}(1-q^k e^{\alpha})\right\} = \sum_{m\in M} q^{c(m)}\ ch(m) \tag{5.1}$$

where n is the rank of L, R is the set of roots of L, and M is a lattice generated by the roots of L, suitably scaled. Whilst $c(m) = \{(m+\rho, m+\rho)-(\rho,\rho)\}/\{(\phi+\rho,\phi+\rho)-(\rho,\rho)\}$ is the eigenvalue of a second order Casimir operator, where ϕ is the highest root and ρ is half the sum of the positive roots. Finally $ch(m)$ is defined by Weyl's character formula (1.1). It should be borne in mind that the vector m in the lattice M will not in general be a highest weight vector. Weyl reflections are required in order to determine the highest weight of the corresponding irrep.

An example of such an identity (5.1) is provided by the case $L = so(3)$ for which $n = 1,\ R = \{\alpha, -\alpha\},\ \phi = \alpha$ and $M = \{m\alpha : m \in 2\mathbb{Z}\}$. One obtains

$$\prod_{k=1}^{\infty}(1-q^k)\ (1-q^k e^{\alpha})\ (1-q^k e^{-\alpha}) = \sum_{m\in 2\mathbb{Z}} q^{m(m+1)/2}\ ch(m), \tag{5.2}$$

where $ch(m) = (e^{m\alpha} - e^{-(m+1)\alpha})/(1-e^{-\alpha}) = -ch(-m-1)$ so that

$$\prod_{k=1}^{\infty}(1-q^k)\ (1-q^k e^{\alpha})\ (1-q^k e^{-\alpha}) = \sum_{m=0}^{\infty}(-1)^m\ q^{m(m+1)/2} ch(m)\ . \tag{5.3}$$

The specialisation of this for which $e^\alpha = e^{-\alpha} = 1$ yields Jacobi's identity

(5.4)
$$\prod_{k=1}^{\infty}(1-q^k)^3 = \sum_{m=0}^{\infty}(-1)^m\ (2m+1)\ q^{m(m+1)/2}\ .$$

The form of the $so(3)$ identity which will be generalised here is that given in (5.3). In the case of $so(2n+1)$ the root system (1.2b) is such that

$$\prod_{k=1}^{\infty}\left\{(1-q^k)^n \prod_{\alpha\in R}(1-q^k e^\alpha)\right\} = \prod_{k=1}^{\infty}\left\{(1-q^k)^n \prod_{1\leq i\leq n}(1-q^k x_i)(1-q^k x_i^{-1})\ .\right.$$

$$\left.\prod_{1\leq i<j\leq n}(1-q^k x_i x_j)(1-q^k x_i x_j^{-1})(1-q^k x_i^{-1} x_j)(1-q^k x_i^{-1} x_j^{-1})\right\}$$

(5.5)
$$= \prod_{k=1}^{\infty}\left\{\prod_{1\leq i<j\leq N}(1-q^k x_i x_j)\right\} = \prod_{k=1}^{\infty} A_{q^k}(\mathbf{x})_N\ ,$$

where $N = 2n+1$ and $x_{n+i} = x_i^{-1}$ for $i = 1, 2, \ldots, n$ and $x_{2n+1} = 1$ as in (3.6b). A subscript N has been added to the notation of (3.1a) to indicate that the indeterminates are finite in number. However it is important to remember that they have also been specialised in accordance with (3.6b). This means that an expansion of the form

(5.6)
$$\prod_{k=1}^{\infty} A_{q^k}(\mathbf{x})_N = \sum_{\lambda} g_\lambda(q)\ [\lambda](\mathbf{x})_N$$

must be valid. The problem is to determine the coefficients $g_\lambda(q)$. This can be done by using (3.1a), the S-function product rule (1.12), Theorem 3.1 and the S-function quotient rule (1.16). Calculations along these lines strongly suggested the validity of the very remarkable formula

(5.7)
$$\prod_{k=1}^{\infty} A_{q^k}(\mathbf{x})_N = \sum_{\alpha\in A}(-1)^{|\alpha|/2} q^{|\alpha|/2}[\alpha](\mathbf{x})_N.$$

Thus one starts from a product over k of a power series in q^k whose coefficients are S-functions and obtains the same series but this time in q with coefficients which are $so(N)$ characters. On the basis of these and similar calculations one is led to

THEOREM 5.1. *With the notation of (3.1) and (3.7),*

(5.8a)
$$\prod_{k=1}^{\infty} A_{q^k}(\mathbf{x})_N = \sum_{\alpha\in A}(-1)^{|\alpha|/2} q^{|\alpha|/2}[\alpha](\mathbf{x})_N$$

(5.8b)
$$\prod_{k=1}^{\infty} C_{q^k}(\mathbf{x})_N = \sum_{\gamma\in C}(-1)^{|\gamma|/2} q^{|\gamma|/2}\langle\gamma\rangle(\mathbf{x})_N$$

$$\prod_{k=1}^{\infty} A_{q^k}(\mathbf{x})_N L_{q^k}(\mathbf{x})_N = \sum_{\alpha\in A} (-1)^{|\alpha|/2} q^{|\alpha|/2} [\alpha](\mathbf{x})_{N+1} \tag{5.8c}$$

$$\prod_{k=1}^{\infty} C_{q^k}(\mathbf{x})_N P_{q^k}(\mathbf{x})_N = \sum_{\gamma\in C} (-1)^{|\gamma|/2} q^{|\gamma|/2} [\gamma](\mathbf{x})_{N+1} \tag{5.8d}$$

$$\prod_{k=1}^{\infty} A_{q^k}(\mathbf{x})_N \; Q_{q^k}(\mathbf{x})_N = \sum_{\alpha\in A} (-1)^{|\alpha|/2} \; q^{|\alpha|/2} \langle\alpha\rangle(\mathbf{x})_{N-1} \tag{5.8e}$$

$$\prod_{k=1}^{\infty} C_{q^k}(\mathbf{x})_N M_{q^k}(\mathbf{x})_N = \sum_{\gamma\in C} (-1)^{|\gamma|/2} \; q^{|\gamma|/2} \langle\gamma\rangle(\mathbf{x})_{N-1} \tag{5.8f}$$

$$\prod_{k=1}^{\infty} E_{q^k}(\mathbf{x})_N M_{q^{2k}}(\mathbf{x})_N = \sum_{\epsilon\in E} (-1)^{(|\epsilon|+r)/2} \; q^{|\epsilon|} \; [\epsilon](\mathbf{x})_N \tag{5.8g}$$

$$\prod_{k=1}^{\infty} E_{q^k}(\mathbf{x})_N \; Q_{q^{2k}}(\mathbf{x})_N = \sum_{\epsilon\in E} (-1)^{(|\epsilon|+r)/2} \; q^{|\epsilon|} \langle\epsilon\rangle(\mathbf{x})_N \tag{5.8h}$$

$$\prod_{k=1}^{\infty} G_{q^k}(\mathbf{x})_N \; P_{q^{2k}}(\mathbf{x})_N = \sum_{\epsilon\in G} (-1)^{(|\epsilon|-r)/2} \; q^{|\epsilon|} \; [\epsilon](\mathbf{x})_N \tag{5.8i}$$

$$\prod_{k=1}^{\infty} G_{q^k}(\mathbf{x})_N \; L_{q^{2k}}(\mathbf{x})_N = \sum_{\epsilon\in E} (-1)^{(|\epsilon|-r)/2} \; q^{|\epsilon|} \langle\epsilon\rangle(\mathbf{x})_N . \tag{5.8j}$$

Introducing yet another infinite series, $K_q(\mathbf{x})$, similar calculations suggest the validity of

THEOREM 5.2. *Let the list of series (3.1) be extended to include*

$$K_q(\mathbf{x}) = \prod_{i,j} (1 - q \; x_i x_j^{-1}) \; (1-q)^{-1} \; . \tag{5.9a}$$

Then

$$\prod_{k=1}^{\infty} K_{q^k}(\mathbf{x})_N = \sum_{\zeta\in F} (-1)^{|\zeta|} \; q^{|\zeta|} \; \{\overline{\zeta}'; \zeta\}(\mathbf{x})_N \; . \tag{5.9b}$$

In all the identities of Theorems 5.1 and 5.2 the indeterminates must be restricted as is appropriate for the characters indicated on the right-hand side of each identity. However all the results may be viewed as involving universal characters with the final restrictions being applied through the use of the relevant modification rules (3.11) or (2.12). To illustrate this, consider the first identity (5.8a) applied in the case $N = 5$ so that the group characters on the right-hand side are those of $so(5)$. Thus

$$\prod_{k=1}^{\infty} A_{q^k}(\mathbf{x})_5 = 1 - q[1^2](\mathbf{x})_5 + q^2[21^2](\mathbf{x})_5 - \ldots\ldots + q^8[4^2 3^2 2](\mathbf{x})_5 - \ldots \tag{5.10a}$$

All the terms except the first two are non-standard characters of $so(5)$ and must be modified through the use of (3.9a) with $N = 5$. Remembering that all the terms in the resulting series are labelled by partitions $\alpha \in A$, the Frobenius symbol of which has $b_1 = a_1 + 1$, it follows that the continuous boundary strip to be removed has length $h = 2(a_1 + 2) - N = h_{11} - (N - 2)$ where h_{11} is the hook length of the (1,1)-box of F^α. Removing the strip then leaves along the right-hand boundary a continuous boundary strip of length $(N - 2) = 3$, in the case of our example. The resulting character may still be non-standard in which case the process must be repeated. For example dealing with the non-standard terms exhibit in (5.10) gives the diagrams

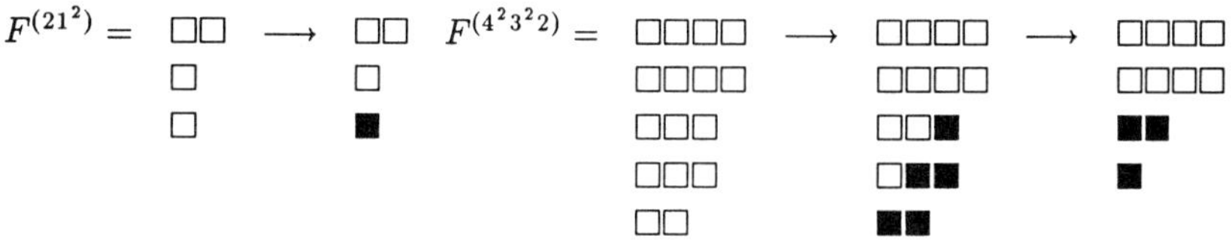

$F^{(21^2)} =$

```
□□  ⟶  □□
□      □
□      ■
```

$F^{(4^2 3^2 2)} =$

```
□□□□  ⟶  □□□□  ⟶  □□□□
□□□□      □□□□      □□□□
□□□       □□■       ■■
□□□       □■■       ■
□□        ■■
```

so that, taking into account sign factors,

$$\prod_{k=1}^{\infty} A_{q^k}(\mathbf{x})_5 = 1 - q[1^2](\mathbf{x})_5 + q^2[21](\mathbf{x})_5 - \cdots - q^8[4^2](\mathbf{x})_5 + \ldots \tag{5.10b}$$

It is not difficult to see that the general pattern of terms that survive are those consisting of diagrams with a core specified by either [0] or $[1^2]$ built by adding strips, or slinkies [CGR], of length $(N - 2) = 3$ to this core in all possible ways such that each slinky starts in the first row and their successive addition yields a standard diagram at each stage. Thus our two terms obtained above can be viewed as arising in the following way

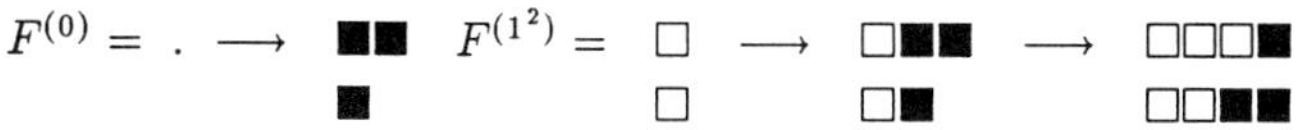

$F^{(0)} =$

```
.  ⟶  ■■
       ■
```

$F^{(1^2)} =$

```
□  ⟶  □■■  ⟶  □□□■
□      □■      □□■■
```

This leads to a reformulation of the Macdonald identity (5.8a) in terms of standard $so(N)$ characters:

COROLLARY 5.3.

$$\prod_{k=1}^{\infty} A_{q^k}(\mathbf{x})_N = \sum_{\substack{\alpha \in A \\ l(\alpha) \leq [N/2]}} \sum_{s=0}^{\infty} \sum_{\substack{\lambda \\ \lambda \equiv \alpha \bmod (N-2)}} (-1)^{|\alpha|/2+c} q^{|\alpha|/2+r} [\lambda](\mathbf{x})_N \tag{5.11}$$

where $c = c_1 + c_2 + \cdots + c_s$, $r = r_1 + r_2 + \cdots + r_s$ and λ is any partition such that F^λ is formed from F^α through the addition of s slinkies each of length $(N - 2)$ with the ith slinky starting at position $(1, r_i)$ and extending over c_i columns. □

This form, (5.11), is the optimum form of any expansion of type (5.6) in the sense that all the resulting characters are standard, requiring no further modification. To make contact with the original expansions of type (5.1) it is necessary to examine closely the lattice M and to perform the Weyl reflections bringing each vector m to the dominant chamber. Such an examination lends support to the validity of (5.8a) and consequently to (5.11), but does not offer a tractable way to prove these results. Instead we may proceed by generalising Lemma 4.3 as follows.

LEMMA 5.4. *Let the notation for quotients of series be such that*

$$s_{A_p/C_r}(\mathbf{x}) = \sum_{\alpha \in A} \sum_{\gamma \in C} (-p)^{|\alpha|/2}(-r)^{|\gamma|/2} s_{\alpha/\gamma'}(\mathbf{x}) \ .$$

Then

$$s_{A_p/C_r}(\mathbf{x}) = s_{A_p}(\mathbf{x})\ s_{A_{rp^2}/C_{p^{-1}}}(\mathbf{x}) \tag{5.12}$$

Proof. Replacing D by A_p in the derivation of Lemma 3.2 leads to the result

$$s_{A_p/\mu}(\mathbf{x}) = s_{A_p}(\mathbf{x}) \sum_{\alpha \in A} (-p)^{|\mu|-|\alpha|/2} s_{\mu'/\alpha'}(\mathbf{x}) \tag{5.13}$$

where particular care has to be taken with conjugates and powers of p. It then follows that

$$\begin{aligned} s_{A_p/C_r}(\mathbf{x}) &= s_{A_p}(\mathbf{x}) \sum_{\alpha \in A} \sum_{\gamma \in C} (-p)^{|\gamma|-|\alpha|/2}(-r)^{|\gamma|/2} s_{\gamma'/\alpha'}(\mathbf{x}) \\ &= s_{A_p}(\mathbf{x}) \sum_{\alpha \in A} \sum_{\gamma \in C} (-rp^2)^{|\gamma|/2}(-p^{-1})^{|\alpha|/2} s_{\gamma'/\alpha'}(\mathbf{x}) \\ &= s_{A_p}(\mathbf{x})\ s_{A_{rp^2}/C_{p^{-1}}}(\mathbf{x}) \end{aligned}$$

where the final step involves recognising that the series C is conjugate to A. □

Now we are in a position to prove (5.8a) of Theorem 5.1. It follows from Lemma 5.4 that

$$\begin{aligned} s_{A_q/C}(\mathbf{x}) &= s_{A_q}(\mathbf{x})\ s_{A_{q^2}/C_{q^{-1}}}(\mathbf{x}) \\ &= s_{A_q}(\mathbf{x}) s_{A_{q^2}}(\mathbf{x}) s_{A_{q^3}/C_{q^{-2}}}(\mathbf{x}) = \dots \\ &= \left\{ \prod_{k=1}^{n} s_{A_{q^k}}(\mathbf{x}) \right\} s_{A_{q^{n+1}}/C_{q^{-n}}}(\mathbf{x}) \end{aligned} \tag{5.14}$$

However

$$\begin{aligned} s_{A_{q^{n+1}}/C_{q^{-n}}}(\mathbf{x}) &= \sum_{\alpha \in A} \sum_{\gamma \in C} (-q^{n+1})^{|\alpha|/2}\ (-q^{-n})^{|\gamma|/2}\ s_{\alpha/\gamma}(\mathbf{x}) \\ &= 1 + \sum_{\substack{\alpha \in A \\ \alpha \neq 0}} \sum_{\substack{\gamma \in C \\ |\alpha| > |\gamma|}} (-q)^{|\alpha|/2}\ (-q^n)^{(|\alpha|-|\gamma|)/2}\ s_{\alpha/\gamma}(\mathbf{x}) \end{aligned} \tag{5.15}$$

where the restriction $|\alpha| > |\gamma|$ is only valid because the definitions of the partitions in the sets A and C are such that $s_{\alpha/\gamma}(\mathbf{x}) = 0$ not only for $|\alpha| < |\gamma|$ but also for $|\alpha| = |\gamma| \neq 0$. It follows that for $|q|$ sufficiently small

$$\operatorname*{Lim}_{n \to \infty} s_{A_{q^{n+1}}/C_{q^{-n}}} = 1 \ , \tag{5.16}$$

so that

$$s_{A_q/C}(\mathbf{x}) = \prod_{k=1}^{\infty} s_{A_{qk}}(\mathbf{x}) \ . \tag{5.17}$$

Thanks to the universal character identity (3.7a) of Theorem 3.1, under the specialisations (3.6b) and (3.6d) this result (5.17) is precisely the required Macdonald identity (5.8a) of Theorem 5.1. The remaining identities of (5.8) may be proved either by taking conjugates, or by specialising (5.8a) through letting one indeterminate x_i take on the fixed value 1 or -1, or by proceeding as before via an identity of the same type as that of Lemma 5.4. □

Just as (5.8a) has as its corollary (5.11) so the remaining identities (5.8b)-(5.8j) each have a corollary of the same type. Rather than enunciate these we move to a final application of these techniques, namely the proof of Theorem 5.2. The first step is to prove yet another Lemma

LEMMA 5.5. *Let*

$$s_{I_p/I_r}(\mathbf{x};\mathbf{y}) = \sum_{\zeta,\omega} (-p)^{|\zeta|}(-r)^{|\omega|} s_{\zeta/\omega}(\mathbf{x}) \ s_{\zeta'/\omega'}(\mathbf{y}) \ .$$

Then

$$s_{I_p/I_r}(\mathbf{x};\mathbf{y}) = s_{I_p}(\mathbf{x};\mathbf{y}) s_{I_{rp^2}/I_{p^{-1}}}(\mathbf{x};\mathbf{y}) \ . \tag{5.18}$$

Proof. Just as $J(\mathbf{x},\mathbf{u};\mathbf{y},\mathbf{v})$ was expanded in two ways in the proof of Lemma 4.7 we expand $I_p(\mathbf{x},\mathbf{u};\mathbf{y},\mathbf{v})$ in two ways. Comparing coefficients of the S-functions depending upon $\mathbf{u}$ and $\mathbf{v}$ then gives

$$\begin{aligned}
&\sum_{\zeta} \sum_{\vartheta\phi} (-p)^{|\zeta|} \ c^{\zeta}_{\vartheta\psi} \ c^{\zeta'}_{\phi'\omega'} \ s_\vartheta(\mathbf{x}) s'_\phi(\mathbf{y}) \\
&= \sum_{\gamma\delta\sigma\tau} \sum_{\vartheta\phi} (-p)^{|\gamma|+|\delta|+|\sigma|+|\tau|} c^{\vartheta}_{\gamma\sigma} \ c^{\phi'}_{\gamma'\tau'} \ c^{\psi}_{\delta\tau} \ c^{\omega'}_{\delta'\sigma'} s_\vartheta(\mathbf{x}) \ s_{\phi'}(\mathbf{y}) \ .
\end{aligned}$$

This can be rewritten in the form

$$\begin{aligned}
&\sum_{\zeta} (-p)^{|\zeta|} \ s_{\zeta/\psi}(\mathbf{x}) \ s_{\zeta'/\omega'}(\mathbf{y}) \\
&= \sum_{\gamma,\delta} (-p)^{|\gamma|+|\psi|+|\omega|-|\delta|} s_\gamma(\mathbf{x}) \ s_{\gamma'}(\mathbf{y}) \ s_{\omega/\delta}(\mathbf{x}) \ s_{\psi'/\delta'}(\mathbf{y}) \ .
\end{aligned}$$

Finally setting $\psi = \omega$, multiplying by $(-r)^{|\omega|}$ and summing over w gives

$$\begin{aligned}
&\sum_{\zeta,\omega} (-p)^{|\zeta|}(-r)^{|\omega|} \ s_{\zeta/\omega}(\mathbf{x}) \ s_{\zeta'/\omega'}(\mathbf{y}) \\
&= \sum_{\gamma} (-p)^{|\gamma|} s_\gamma(\mathbf{x}) \ s_{\gamma'}(\mathbf{y}) \sum_{\delta,\omega} (-rp^2)^{|\omega|}(-p^{-1})^{|\delta|} \ s_{\omega/\delta}(\mathbf{x}) \ s_{\omega'/\delta'}(\mathbf{y}),
\end{aligned}$$

which is nothing other than the required result (5.18). □

Now we may prove Theorem 5.2 by iterating as follows:

$$s_{I_q/I}(\mathbf{x};\mathbf{y}) = s_{I_q}(\mathbf{x};\mathbf{y})(1-q)^{-1} s_{I_{q^2}/I_{q^{-1}}}(\mathbf{x};\mathbf{y})(1-q)$$
$$= s_{I_q}(\mathbf{x};\mathbf{y})\,(1-q)^{-1}\, s_{I_{q^2}}(\mathbf{x};\mathbf{y})(1-q^2)^{-1} s_{I_{q^3}/I_{q^{-2}}}(\mathbf{x};\mathbf{y})\,(1-q)\,(1-q^2)$$
$$(5.19) \qquad = \cdots = \prod_{k=1}^{n} \left\{ s_{I_{q^k}}(\mathbf{x};\mathbf{y})\,(1-q^k)^{-1} \right\} s_{I_{q^{n+1}}/I_{q^{n+1}}}(\mathbf{x};\mathbf{y}) \prod_{k=1}^{n}(1-q^k)$$

However

$$s_{I_{q^{n+1}}/I_{q^{-n}}}(\mathbf{x};\mathbf{y}) = \sum_{\zeta,\omega} (-q^{n+1})^{|\zeta|}(-q^{-n})^{|\omega|} s_{\zeta/\omega}(\mathbf{x})\, s_{\zeta'/\omega'}(\mathbf{y})$$
$$= \sum_{\zeta} q^{|\zeta|} + \sum_{\substack{\zeta,\omega \\ |\zeta|>|\omega|}} (-q)^{|\zeta|}(-q^n)^{|\zeta|-|\omega|}\, s_{\zeta/\omega}(\mathbf{x})\, s_{\zeta'/\omega'}(\mathbf{y})\,.$$

Remembering that in the limit as $n \to \infty$ the last factor of (5.19) is just the inverse of the familiar partition generating function, it follows that for sufficiently small $|q|$

$$\lim_{n\to\infty} \left\{ s_{I_{q^{n+1}}/I_{q^{-n}}}(\mathbf{x};\mathbf{y}) \prod_{k=1}^{n}(1-q^k) \right\} = 1\,.$$

Hence

$$s_{I_q/I}(\mathbf{x};\mathbf{y}) = \prod_{k=1}^{\infty} \left\{ s_{I_{q^k}}(\mathbf{x};\mathbf{y})\,(1-q^k)^{-1} \right\}.$$

Finally setting $\mathbf{y} = \overline{\mathbf{x}}$ gives

$$\prod_{k=1}^{\infty} s_{K_{q^k}}(\mathbf{x}) = \prod_{k=1}^{\infty} \left\{ s_{I_{q^k}}(\mathbf{x};\overline{\mathbf{x}})\,(1-q^k)^{-1} \right\} = s_{I_q/I}(\mathbf{x};\overline{\mathbf{x}})$$
$$= \sum_{\zeta,\omega} (-q)^{|\zeta|}(-1)^{|\omega|}\, s_{\zeta/\omega}(\mathbf{x})\, s_{\zeta'/\omega'}(\overline{x}) = \sum_{\zeta} (-q)^{|\zeta|}\, s_{\overline{\zeta'};\zeta}(\mathbf{x})\,.$$

from (2.13)

Adopting the specialisation (3.6e) of these universal characters then completes the proof of Theorem 5.2. □

Once again it has to be emphasised that for any particular value of N the modification rule (2.12) must be applied. Doing this leads for example, in the case $N = 3$ to

$$\prod_{k=1}^{\infty} K_{q^k}(\mathbf{x})_3 = 1 - q\{\overline{1};1\}(\mathbf{x})_3 + q^2\,\{\overline{2};1^2\}(\mathbf{x})_3 + q^2\,\{\overline{1}^2;2\}(\mathbf{x})_3$$
$$(5.20) \qquad + \cdots + q^{20}\,\{\overline{8642};4^2 3^2 2^2 1^2\}(\mathbf{x})_3 + \ldots$$

The first four terms, and only these four, are standard. All the others require modification in accordance with (2.12). For example the non-standard term exhibited in (5.19) modifies to give, after two applications of (2.12) with $N = 3$, the standard term

$$+q^{20}\ \{\overline{8}; 4^2\}(\mathbf{x})_3$$

This is illustrated by the diagram modifications

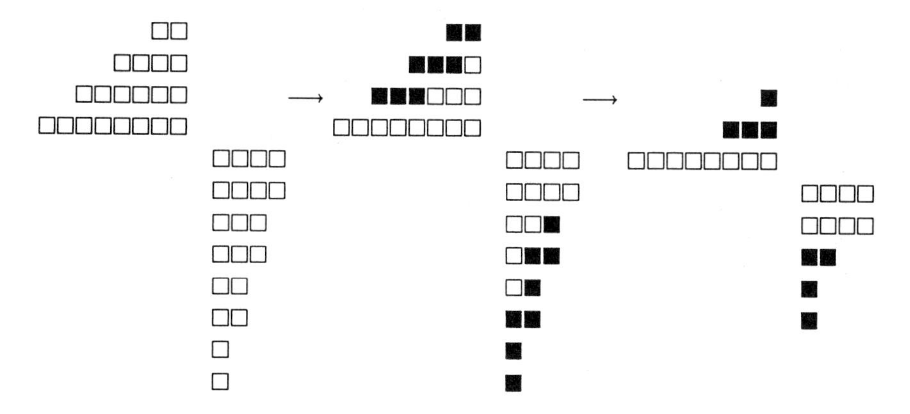

corresponding to the successive removal of a pair of strips of length 8, and then a second pair of length 4. It is not difficult to see that what remains is a standard composite diagram which can be built from a standard core by the addition of two successive pairs of slinkies each of length 3, as shown below

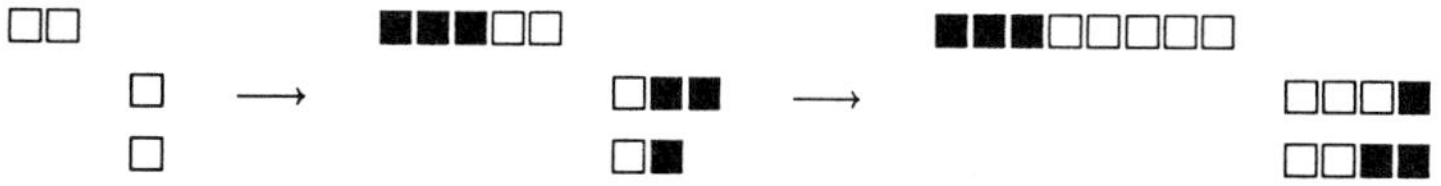

The significance of this is that it allows one to arrive at the Corollary 5.6

(5.21)

$$\prod_{k=1}^{\infty} K_{q^k}(\mathbf{x})_N = \sum_{\substack{\zeta \\ l(\zeta)+l(\zeta')\leq N}} \sum_{s=0}^{\infty} \sum_{\substack{\mu,\nu \\ \mu\equiv\zeta\mathrm{mod}(N) \\ \nu\equiv\zeta\mathrm{mod}(N)}} (-1)^{|\zeta|+c+\overline{c}}\ q^{|\zeta|+r+\overline{r}-s}\ \{\overline{\nu};\mu\}(\mathbf{x})_N$$

where $F^{\overline{\nu};\mu}$ is formed from $F^{\overline{\zeta}';\zeta}$ by adding s pairs of slinkies each of length N. The ith slinky added to F^{ζ} starts at positions $(1, r_i)$ and covers c_i columns, whilst the ith slinky added to $F^{\overline{\zeta}'}$ starts at the position $(1, \overline{r}_i)$ and covers $\overline{c}_i$ columns. The remaining parameters appearing in (5.21) are then given by $r = r_1 + r_2 + \cdots + r_s$, $c = c_1 + c_2 + \cdots + c_s$, $r = \overline{r}_1 + \overline{r}_2 + \cdots + \overline{r}_s$ and $\overline{c} = \overline{c}_1 + \overline{c}_2 + \cdots + \overline{c}_s$.

It should be pointed out that a quite distinct proof of Theorem 5.2 is implicit in the work of Stembridge [Ste1,Ste2]. The connection can be seen by noting that the transformation from $F^{\overline{\nu};\mu}$ to F^{λ} associated with (2.3) leads in the case of the core terms $(\nu;\mu) = (\overline{\zeta}';\zeta)$ of (5.21) to a partition λ whose vertical sequence taken mod(N) is, as required, a permutation of $0, 1, \ldots, N-1$. This observation was made by Stembridge [Ste1]. The fact that the use of universal characters guarantees the stability of this result as $N \to \infty$ amounts to a second proof of the form of the

Macdonald identity given in Theorem 5.2 . The new aspect of the proof presented here is that it stresses the roles of universal characters in expressing the result in a form valid for all N, and of modification rules in arriving at explicit results valid for any particular value of N.

6. Characters of superalgebras and S-functions. Much of the material of the previous sections may be extended to the domain of superalgebras. The starting point is the work of Kac [Ka1,Ka2] who not only classified simple Lie superalgebras but initiated the systematic study of their finite-dimensional irreps. The Lie superalgebra $G = G_{\overline{0}} + G_{\overline{1}}$ over C is said to be a basic classical Lie superalgebra if G is simple, if the even subalgebra $G_{\overline{0}}$ is a reductive Lie algebra and if there exists a non-degenerate invariant bilinear form on G [Ka2]. Let H be a Cartan subalgebra of $G_{\overline{0}}$. Each finite-dimensional irrep of G may be specified by means of its highest weight vector $\Lambda \in H^*$. The root systems are graded so that the set of positive roots $\Delta^+ = \Delta_0^+ \cup \Delta_1^+$ is the union of the sets of positive even and positive odd roots. It is convenient to define $\rho = \rho_0 - \rho_1$, where ρ_0 and ρ_1 are half the sum of the positive even and odd roots, respectively. The positive roots, Δ^+, of the classical Lie superalgebras considered here are given in the standard basis by

$$
\begin{array}{lllll}
(6.1a) & sl(m/n) & \Delta_0^+ = \{\epsilon_i - \epsilon_j,\ \delta_a - \delta_b\} & \Delta_1^+ = \{\epsilon_i - \delta_a\} \\
(6.1b) & osp(2m+1/2n) & \Delta_0^+ = \{\epsilon_i \pm \epsilon_j,\ \delta_a \pm \delta_b,\ 2\delta_a\} & \Delta_1^+ = \{\epsilon_i \pm \delta_a,\ \delta_a\} \\
(6.1c) & osp(2/2n) & \Delta_0^+ = \{\delta_a \pm \delta_b,\ 2\delta_a\} & \Delta_1^+ = \{\epsilon_1 \pm \delta_a\} \\
(6.1d) & osp(2m/2n) & \Delta_0^+ = \{\epsilon_i \pm \epsilon_j,\ \delta_a \pm \delta_b,\ 2\delta_a\} & \Delta_1^+ = \{\epsilon_i \pm \delta_a\} \\
(6.1e) & gl(m/n) & \Delta_0^+ = \{\epsilon_i - \epsilon_j,\ \delta_a - \delta_b\} & \Delta_1^+ = \{\epsilon_i - \delta_a\}
\end{array}
$$

where the range of the various combinations of indices are given by $1 \leq i < j \leq m$, $1 \leq a < b \leq n$, $1 \leq i \leq m$ and $1 \leq a \leq n$. It should be noted that the inner product on H^* is such that

$$(6.2) \qquad (\epsilon_i|\epsilon_j) = \delta_{ij} \qquad (\epsilon_i|\delta_a) = 0 \qquad (\delta_a|\delta_b) = -\delta_{ab} \ .$$

In the case of $sl(m/n)$, but not $gl(m/n)$, there exists the additional constraint

$$(6.3) \qquad \epsilon_1 + \epsilon_2 + \cdots + \epsilon_m - \delta_1 - \delta_2 - \cdots - \delta_n = 0 \ .$$

For an arbitrary weight vector $\mathbf{M} \in H^*$ with $\mathbf{M} = m_1\epsilon_1 + m_2\epsilon_2 + \cdots + m_m\epsilon_m + n_1\delta_1 + n_2\delta_2 + \cdots + n_n\delta_n$ it is convenient to write $\mathbf{M} = (m_1, m_2, \ldots, m_m|n_1, n_2, \ldots, n_n)$ and to express the formal exponential $e^{\mathbf{M}}$ in terms of indeterminates $x_1 = e^{\epsilon_i}$ and $y_a = e^{\delta_a}$ by writing $e^{\mathbf{M}} = x_1^{m_1} x_2^{m_2} \ldots x_m^{m_m} y_1^{n_1} y_2^{n_2} \ldots y_n^{n_n}$. The indeterminates themselves are conveniently denoted by $(\mathbf{x}/\mathbf{y}) = (x_1, x_2, \ldots, x_m/y_1, y_2, \ldots, y_n)$. The relevant Weyl group, W, is the Weyl group of the even subalgebra, $G_{\overline{0}}$. This contains $S_m \times S_n$ as a subgroup, so that with respect to the ordering we have adopted each highest weight vector Λ is such that $\Lambda_1 \geq \Lambda_2 \geq \cdots \geq \Lambda_m$ and $\Lambda_{m+1} \geq \Lambda_{m+2} \geq \cdots \geq \Lambda_{m+n}$.

So far so good, but unfortunately there does not exist an all embracing generalisation to the superalgebra case of Weyl's character formula (1.1). Nonetheless it is possible to make some progress. Kac [Ka2] distinguished between two types of irrep: typical and atypical. An irrep of highest weight Λ is said to be atypical if and only if there exists an odd root $\beta \in \Delta_{\bar{1}}^{+}$ such that $(\Lambda+\rho,\beta)=0$ Otherwise it is said to be typical. In the case of typical irreps Kac [Ka2] derived the following character formula

$$ch(\Lambda) = \frac{\prod_{\beta\epsilon\Delta_{\bar{1}}^{+}} (e^{\beta/2} - e^{-\beta/2})}{\prod_{\alpha\epsilon\Delta_{\bar{0}}^{+}} (e^{\alpha/2} - e^{-\alpha/2})} \sum_{w\in W} \epsilon(w)\, e^{w(\Lambda+\rho)}$$

$$(6.4) \qquad = \sum_{w\in W} \left\{ \epsilon(w)\, e^{w(\Lambda+\rho_0)} \prod_{\beta\epsilon\Delta_{\bar{1}}^{+}} (1-e^{-w\beta}) \right\} \Big/ \sum_{w\in W} \epsilon(w)\, e^{w\rho_0}.$$

where the second form is readily derived from the first once it is realised that $\Delta_{\bar{1}}^{+}$ and hence ρ_1 are invariant under the action of the Weyl group W. The second form makes it apparent that Kac's character formula is a natural generalisation of Weyl's formula (1.1). Moreover it is a form from which it is very easy to obtain the branching of the irrep of the superalgebra G on restriction to the even subalgrebra $G_{\bar{0}}$.

Given the validity of Kac's formula (6.4) in dealing with characters of typical irreps it is tempting to postulate the existence of a more general result covering all irreps:

Conjecture 6.1. Let Λ be the highest weight of an arbitrary finite-dimensional irrep of a basic classical Lie superalgebra. Then the character of this irrep can be written in the form

$$(6.5) \qquad ch(\Lambda) = \sum_{w\in W} \left\{ \epsilon(w)\, e^{w(\Lambda+\rho_0)} \prod_{\beta\epsilon\Delta_{\Lambda}^{+}} (1-e^{-w\beta}) \right\} \Big/ \sum_{w\epsilon W} \epsilon(w)\, e^{w\rho_0},$$

for some subset Δ_{Λ}^{+} of $\Delta_{\bar{1}}^{+}$ depending upon Λ.

Kac's formula (6.4) for the characters of typical irreps is consistent with this Conjecture in the sense that it is covered by (6.5) with

$$(6.6) \qquad \Delta_{\Lambda}^{+} = \Delta_{\bar{1}}^{+} \quad \text{if} \quad |\{\ \beta \mid \text{and } (\Lambda+\rho,\beta)=0\ \}| = 0.$$

In the case of atypical irreps several independent approaches [BL,Le,VdJ,HK,SS] to this problem have resulted in formulae of precisely this type. However it has not been possible to obtain any definition of Δ_{Λ}^{+} which covers more than a limited class of atypical irreps. For example an attractive possibility due originally to Bernstein and Leites [BL,Le], and rederived under certain conditions on Λ by Van der Jeugt [VdJ], is that

$$(6.7) \qquad \Delta_{\Lambda}^{+} = \{\ \beta \mid \beta\epsilon\Delta_{\bar{1}}^{+} \quad \text{and } (\Lambda+\rho,\beta)\neq 0\}\ .$$

An irrep with highest weight Λ is said to be singly atypical if $|\{ \beta \mid \beta \in \Delta_{\bar{1}}^+$ and $(\Lambda+\rho, \beta) = 0 \}| = 1$ and multiply atypical if $|\{ \beta \mid \beta \in \Delta_{\bar{1}}^+$ and $(\Lambda+\rho, \beta) = 0 \}| > 1$. In the case of singly atypical irreps of both $sl(m/n)$ and $osp(2/2n)$ it appears, as a result of checking many individual cases, that (6.7) is valid. However it certainly fails for some multiply atypical irreps.

Fortunately in the case of $gl(M/N)$ there is a class of irreps, including typical, singly atypical and multiply atypical irreps, for which the character can be determined using S-function methods. These irreps are each specified by a partition λ with $\lambda_{M+1} \leq N$. The corresponding highest weight vector is given by $\Lambda = (\lambda_1, \lambda_2, \ldots, \lambda_M | \sigma'_1, \sigma'_2, \ldots, \sigma'_N)$, where $\sigma'_j = \max(0, \lambda'_j - M)$ for $j = 1, 2, \ldots, N$ and the character is denoted by

$$ch(\Lambda) = ch(\lambda)_{M/N} = \{\lambda\}_{M/N} \ . \tag{6.8}$$

These characters have been obtained by a variety of Young diagram and S-function methods [DJ,BB1,BR,Ki5]. One method [DJ] leads to the analogue of (1.17) which takes the form

$$\{\lambda\}_{M/N} = \sum_{\mu,\nu} (-1)^{|\nu|} \, c^{\lambda}_{\mu\nu} \{\mu\}_M \ \{\nu'\}_N \ . \tag{6.9}$$

Another [BB1] makes use of a determinantal expansion similar to (2.7), namely

$$\{\lambda\}_{M/N} = | \ \{\lambda_i - i + j\}_{M/N} \ | \ , \tag{6.10}$$

together with the special case of (6.9) given by

$$\{k\}_{M/N} = \sum_{n=0}^{k} (-1)^n \ \{k-n\}_M \ \{1^n\}_N \ . \tag{6.11}$$

An alternative combinatorial approach to the evaluation of the character is provided by a generalisation of the standard Young tableaux T^λ defined in Section 1. A semistandard Young tableau S^λ [BR,Ki5] is defined to be a numbering of the boxes of F^λ in which

(i) the entries are taken from $\mathsf{E} \cup \mathsf{E}'$ where $\mathsf{E} = \{1, 2, \ldots, M\}$ and $\mathsf{E}' = \{1', 2', \ldots, N'\}$ and these sets are ordered so that $1 < 2 < \cdots < M$ and $1' < 2' < \cdots < N'$,

(ii) the entries from E form a standard Young tableau T^μ for some μ such that F^μ is a subdiagram of F^λ,

(iii) the entries from E' form the conjugate of a standard Young tableau $T^{\lambda'/\mu'}$ of shape $F^{\lambda/\mu}$.

For example a semistandard Young tableau is illustrated in the case $\lambda = (732^21^2)$, $M = 3$ and $N = 5$ by

$$S^\lambda = \begin{array}{l} 1\,1\,2\,2\,1'4'5' \\ 2\,1'2' \\ 3\,3' \\ 1'3' \\ 2' \end{array}$$

To each such semistandard Young tableau S^λ there corresponds a weight vector $\mathbf{M} = (m_1, m_2, \ldots, m_M | n_1, n_2, \ldots, n_N)$, where m_j is the number of entries j in the tableau, and n_k is the number of entries k' in the tableau. With this terminology one can extend the definition of an S-function in such a way that

$$s_\lambda(\mathbf{x}/\mathbf{y})_{M/N} = \sum_{S^\lambda} (-1)^{n_1+n_2+\cdots+n_N} x_1^{m_1} x_2^{m_2} \ldots x_M^{m_M} y_1^{n_1} y_2^{n_2} \ldots y_N^{n_N}. \tag{6.12}$$

It may then be shown [BR,Ki5] that the character of such an irrep of gl(M/N) is precisely this S-function, that is:

$$ch(\lambda)_{M/N} = \{\lambda\}(\mathbf{x}/\mathbf{y})_{M/N} = s_\lambda(\mathbf{x}/\mathbf{y})_{M/N} \tag{6.13}$$

where our previous notation has been generalised in an obvious way. It follows from the definition of semistandard Young tableaux that

$$\{\lambda\}(\mathbf{x}/\mathbf{y})_{M/N} = s_\lambda(\mathbf{x}/\mathbf{y})_{M/N} = 0 \qquad \text{if} \quad \lambda_{M+1} > N\ . \tag{6.14}$$

This modification rule is a direct generalisation of (1.8) which can be recovered from it by setting $N = 0$. As before the restriction from $gl(M/N)$ to $sl(M/N)$ raises no problems. Each irrep remains irreducible and its character is unchanged save for the fact that the condition $x_1 x_2 \ldots x_M \overline{y}_1 \overline{y}_2 \ldots \overline{y}_N = 1$ now applies by virtue of (6.3), where $\overline{y}_a = y_a^{-1}$.

Exactly as before universal characters may be defined by taking this time a double inverse limit in M and N

$$\{\lambda\}(\mathbf{x}/\mathbf{y}) = s_\lambda(\mathbf{x}/\mathbf{y}) = \lim_{\leftarrow M} \ \lim_{\leftarrow N} \{\lambda\}(\mathbf{x}/\mathbf{y})_{M/N}\ . \tag{6.15}$$

The characters of $gl(M/N)$ for any finite M and N are then recovered by means of the specialisation

$$\{\lambda\}(\mathbf{x}/\mathbf{y})_{M/N} = \{\lambda\}(x_1, x_2, \ldots, x_M, 0, 0, \ldots, 0 | y_1, y_2, \ldots, y_N, 0, 0, \ldots, 0). \tag{6.16}$$

The algebra of these universal characters coincides with the usual algebra of Schur functions in that products are once more given by

$$s_\mu(\mathbf{x}/\mathbf{y})\ s_\nu(\mathbf{x}/\mathbf{y}) = \sum_\lambda c_{\mu\nu}^\lambda\ s_\lambda(\mathbf{x}/\mathbf{y})\ , \tag{6.17}$$

where the coefficients are yet again the Littlewood-Richardson coefficients. It is for this reason that the determinantal formula (6.10) can be taken over from the $gl(N)$ case to that of $gl(M/N)$. Moreover it signifies that the tensor product formula appropriate to $gl(M/N)$ takes the universal form

$$\{\mu\}_{M/N} \cdot \{\nu\}_{M/N} = \sum_\lambda c_{\mu\nu}^\lambda\ \{\lambda\}_{M/N}\ , \tag{6.18}$$

so that the application of the modification rule (6.14) gives the final result

$$ch(\mu)_{M/N}\, ch(\nu)_{M/N} = \sum_{\lambda,\lambda_{M+1}\leq N} c^{\lambda}_{\mu\nu}\, ch(\lambda)_{M/N} \,. \tag{6.19}$$

It is very straightforward to deal with the branching rule appropriate to the restriction from $gl(M/N)$ to its even subalgebra $gl(M)+gl(N)$. The definition in terms of semistandard tableaux gives at once

$$\begin{aligned} s_{\lambda}(\mathbf{x}/\mathbf{y})_{M/N} &= \sum_{\nu}(-1)^{|\nu|}\, s_{\lambda/\nu}(\mathbf{x})_M\, s_{\nu'}(\mathbf{y})_N \\ &= \sum_{\mu,\nu}(-1)^{|\nu|}\, c^{\lambda}_{\mu\nu}\, s_{\mu}(\mathbf{x})_M\, s_{\nu'}(\mathbf{y})_N \,. \end{aligned} \tag{6.20}$$

This confirms the validity of (6.9) and the application of the modification rule (6.14) then gives

$$ch(\lambda)_{M/N} = \sum_{\mu,l(\mu)\leq M}\ \sum_{\nu,l(v)\leq N} (-1)^{|v|}\, c^{\lambda}_{\mu\nu}\, ch(\mu)_M\, ch(\nu')_N \,. \tag{6.21}$$

This character formula is very different in appearance from that of Kac and the Conjecture (6.5). However the examination of very many cases led to the speculation that the required definition of Δ^{+}_{λ} in (6.5) is given in the case of the $gl(M/N)$ irrep having $\Lambda = (\lambda_1,\lambda_2,\ldots,\lambda_M|\sigma'_1,\sigma'_2,\ldots,\sigma'_N)$, where $\sigma'_j = \max(0,\lambda'_j - M)$, by

$$\Delta^{+}_{\Lambda} = \left\{\beta = \epsilon_i - \delta_a \,\middle|\, (i,a)\in F^{\kappa} = \left(F^{\lambda}\cap F^{N^M}\right)\right\}. \tag{6.22}$$

This implies that the odd roots required to generate the character are those corresponding to the boxes of F^{κ}. These are precisely those boxes of F^{λ} lying in an M by N rectangle positioned so that the top left corner of the rectangle coincides with that of F^{λ}. For example in the case $\lambda = (732^2 1^2)$, $M=3$ and $N=5$ we have

$F^{\lambda} =$
□□□□□□□
□□□
□□
□□
□
□

$F^{\kappa} =$
□□□□□
□□□
□□

We may distinguish between the odd roots of Δ^{+}_{1} which are in the generating set Δ^{+}_{Λ} and those which are not by marking the former with the symbol ■ and the latter with □ in the M by N rectangle F^{N^M}. Then in our example Δ^{+}_{λ} is represented pictorially by

and consequently $\Delta_\Lambda^+ = \Delta_1^+ \setminus \{\epsilon_2 - \delta_4,\ \epsilon_2 - \delta_5,\ \epsilon_3 - \delta_3,\ \epsilon_3 - \delta_4,\ \epsilon_3 - \delta_5\}$. However $\Lambda = (732|31000)$ and the set of atypical roots is just $\{\epsilon_2 - \delta_5, \epsilon_3 - \delta_4\}$. Thus (6.22) differs from (6.7). Moreover in contrast to the use of (6.7), (6.22) gives the character correctly when used in the formula (6.5). That this is the case quite generally has been established by Sergeev [P] who proved that the use of (6.22) in (6.5) coincides with the S-function result (6.21) for all irreps of $gl(M/N)$ specified by a partition λ.

However there exist other irreps. In particular there exist irreps of $gl(M/N)$ associated with composite Young diagrams $F^{\overline{\nu};\mu}$ [DJ,BB2,CK2]. Following the pattern established earlier for characters of $gl(N)$ it might seem reasonable to hope that their characters might be given by

$$(6.23)\qquad s_{\overline{\nu};\mu}(\mathbf{x}/\mathbf{y})_{M/N} = \sum_{\zeta} (-1)^{|\zeta|} s_{\nu/\zeta}(\overline{\mathbf{x}/\mathbf{y}})_{M/M}\ s_{\mu/\zeta'}(\mathbf{x}/\mathbf{y})_{M/N}.$$

This is after all a direct generalisation of (2.13). Furthermore it may be shown to follow [CK2] from the determinantal definition of characters suggested by Balantekin and Bars [BB2] in the form of an identity like (2.8).

Unfortunately whilst (6.23) does give the character of the corresponding irrep in the typical case, it does not do so, in general, in the atypical case [DJ,BMR]. Instead it gives a linear combination of characters of irreps. To illustrate the difficulties, consider the adjoint irrep of $gl(M/N)$ with highest weight $\Lambda = (10\ldots0|0\ldots0-1)$. This is associated with the composite Young diagram $F^{\overline{1};1}$. The use of (6.20) and (6.23) gives the branching rule to $gl(M) + gl(N)$ in the form

$$(6.24)\qquad \begin{aligned}\{\overline{1};1\}_{M/N} = \{1\}_M \times \{\overline{1}\}_N - \{0\}_M \times \{\overline{1};1\}_N - \{\overline{1};1\}_M \times \{0\}_N \\ - \{0\}_M \times \{0\}_N + \{\overline{1}\}_M \times \{1\}_N\ .\end{aligned}$$

This is correct in the cases for which $M \neq N$ but in the case $M = N$ the correct result is

$$(6.25)\qquad \{\overline{1};1\}_{M/M} = \{1\}_M\times\{\overline{1}\}_M-\{0\}_M\times\{\overline{1};1\}_M-\{\overline{1};1\}_M\times\{0\}_M+\{\overline{1}\}_M\times\{1\}_M$$

The reason for the discrepancy can be traced to an extra degree of atypicality in the $M = N$ case.

Just as this example affords an illustration of the difficulties of an S-function approach, it also illustrates some of the difficulties arising in trying to generalise (6.22). The most promising generalisation of (6.22) arrived at independently by several people [SS,HKTV] can be cast in the form

$$(6.26)\qquad \Delta_\Lambda^+ = \left\{\beta = \epsilon_i - \delta_a \,\middle|\, (i,a) \in F^\sigma \cup F^\tau = \left(F^\mu \cap F^{N^M}\right) \cup \left(F^\nu \cap F^{N^M}\right)\right\}$$

where F^μ and F^ν are positioned at opposite corners of the M by N rectangle. For example in the special case of the adjoint of $gl(M/M)$ the relevant picture is

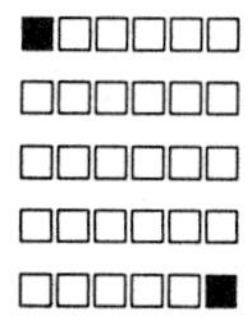

and one would expect $\Delta_\Lambda^+ = \{\ \epsilon_1 - \delta_1,\ \epsilon_M - \delta_N\ \}$. The use of this in (6.5) gives (6.25) as required. However in the case of the adjoint of $gl(M/N)$ with $M \neq N$ this set is not correct and the picture must be altered to

in order to correspond to the generating set $\Delta_\Lambda^+ = \{\ \epsilon_1 - \delta_1,\ \epsilon_1 - \delta_N, \epsilon_M - \delta_N\ \}$. This then yields (6.24) correctly.

Thus the S-function formula (6.23) is correct in the $M \neq N$ case, but fails if $M = N$, whilst (6.26) gives the correct result in the $M = N$ case but has to be altered if $M \neq N$. This strongly suggests that (6.26) is itself only a special case. In our example the necessity for deleting the root $\epsilon_1 - \delta_N$ from the generating set if $M = N$ can be ascribed to the fact that in this particular case this root is atypical in the sense that $(\Lambda + \rho, \epsilon_1 - \delta_N) = 0$.

Various attempts [HK,HKTV] have been made to look for and derive an alternative specification of the generating set. However a more radical change is required as can be seen from our final example: the irrep of $gl(3/4)$ having highest weight $\Lambda = (210|00-1-1)$. This irrep is associated with the composite Young diagram $F^{\overline{2};21}$ since the S-function $s_{\overline{2};21}(\mathbf{x}/\mathbf{y})_{3/4}$ defined by (6.23) has the same highest weight. Moreover the root $\epsilon_1 - \delta_4$ is atypical. It follows that the picture of the generating set of odd roots could be expected to be specified by one or other of the pictures

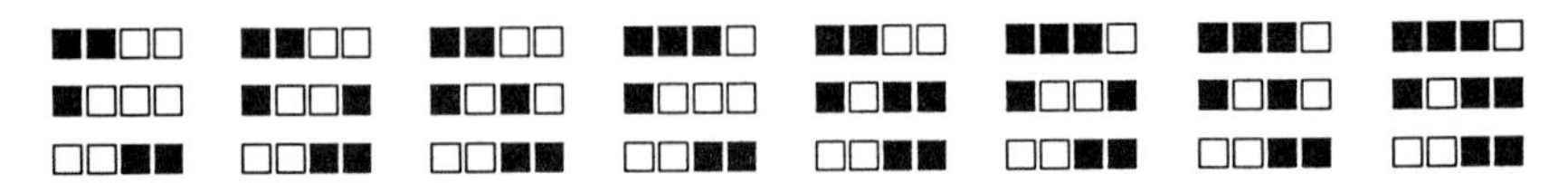

It turns out that none of these is correct. Worse still it can be shown that there exists no generating set Δ_Λ^+ which gives correctly the character of this irrep of $gl(3/4)$ when substituted into the formula of Conjecture 6.5.

Thus even for $gl(M/N)$ this conjecture must be abandoned, and there currently exists no character formula for all the irreps of even these Lie superalgebras. Even so a variety of ways forward present themselves. In particular all the required characters of irreps can certainly be expressed as linear combinations of the characters obtained from (6.5) by choosing the generating set Δ_Λ^+ to be any one of the sets (6.6), (6.7) or (6.26), or even those defined by the S-function expansion (6.23). Some progress has been made towards determining the correct linear combinations [HKTV]. At the same time parabolic subsuperalgebras of $gl(M/N)$ have been used to derive new character formulae appropriate to a class of irreps said to be non-degenerate, as well as to some others that are degenerate [PS]. In addition the use of Young diagrams in the study of indecomposable representations of $gl(M/N)$ has shed some light on the characters of irreps [DG1,DG2,H].

As far as $osp(M/N)$ is concerned progress is also limited. Young diagrams have been used in a number of quite different approaches [FJ2,MSS,G1,G2,CK1]. It is possible to define characters of $osp(M/N)$ by means of a generating function [CK1] analogous to (3.4) and (3.5), which were used for $o(N)$ and $sp(N)$, respectively. These characters are then related to characters of $gl(M/N)$ by formulae involving quotients with respect to S-function series exactly as in Theorems 3.1 and 3.2. Moreover their tensor products are given once again by the Newell-Littlewood Theorem 4.1, subject however to a rather complicated modification rule [CK1]. Unfortunately these characters are not in general the characters of irreps, although they coincide with the Kac characters if the irreps are typical. Apart from the case of low values of M or N [FJ1,FJ2,G1,G2] much remains to be done since the general algorithms suggested to date [T-M,BSS,H] all leave something to be desired in as much as they do not lead unambiguously to the characters of all atypical irreps.

Acknowledgement. The author would like to thank the organisers of the Workshop on Invariant Theory and Tableaux for giving him the opportunity to indulge himself by presenting some conjectures at the Workshop and then the time to complete proofs of some of them for the Proceedings. Without the stimulus of the meeting and the interest shown in them by some of the participants the results would probably not have seen the light of day.

Finally, heartfelt thanks go to Patricia V. Brick for her patience and skill in coping with a difficult manuscript.

REFERENCES

[AK] J. Abramsky and R. C. King, *Formation and decay of negative-parity baryon resonances in a broken $U_{6,6}$ model*, Il Nuovo Cimento, 67A (1979) 153–216.

[BB1] A. B. Balantekin and I. Bars, *Dimension and character formulas for Lie supergroups*, J. Math. Phys. 22 (1981) 1149–1162.

[BB2] A. B. Balantekin and I. Bars, *Representations of supergroups*, J. Math. Phys. 22 (1981) 1810–1818.

[BMR] I. Bars, B. Morel and H Ruegg, *Kac-Dynkin diagrams and supertableaux*, J. Math. Phys. 24 (1983) 2253–2262.

[BR] A. Berele and A. Regev, *Hook Young diagrams with applications to combinatorics and to representations of Lie superalgebras*, Adv. Math. 64 (1987) 118–175.

[BL] I. N. Bernstein and D. A. Leites, *A formula for the characters of the irreducible finite-dimensional representations of Lie superalgebras of series gl and sl*, (in Russian) C. R. Acad. Bulg. Sci. 33 (1980) 1049–1051.

[CGR] Y. M. Chen, A. M. Garsia and J. Remmel, *Algorithms for plethysm*, Contemporary Math. 34 (1984) 109–153.

[C] C. J. Cummins, *Applications of S-function techniques to representation theory of Lie superalgebras and symmetry breaking*, Ph D thesis, University of Southampton, (1986).

[CK1] C. J. Cummins and R. C. King, *Young diagrams, supercharacters of $OSp(M/N)$ and modification rules*, J. Phys. A20 (1987) 3103–3120.

[CK2] C. J. Cummins and R. C. King, *Composite Young diagrams, supercharacters of $U(M/N)$ and modification rules*, J. Phys. A20 (1987) 3121–3133.

[CK3] C. J. Cummins and R. C. King, *Some noteworthy S-function identities*, preprint, Universite de Montreal, (1987).

[DG1] F. Delduc and M. Gourdin, *Mixed supertableaux of the superunitary groups.I. $SU(n|1)$*, J. Math. Phys. 25 (1984) 1651–1661.

[DG2] F. Delduc and M. Gourdin, *Mixed supertableaux of the superunitary groups.II. $SU(n|m)$*, J. Math. Phys. 25 (1984) 1651–1661.

[DJ] P. H. DONDI AND P. D. JARVIS, *Diagram and superfield techniques in the classical superalgebras*, J. Phys A14 (1981) 547–563.

[FJ1] R. J. FARMER AND P. D. JARVIS, *Representations of low-rank orthosymplectic superalgebras by superfield techniques*, J. Phys A16 (1983) 473–487.

[FJ2] R. J. FARMER AND P. D. JARVIS, *Representations of orthosymplectic superalgebras: II. Young diagrams and weight space techniques*, J. Phys A17 (1984) 2365–2387.

[F] H. O. FOULKES, *Differential operators associated with S-functions*, J. Lond. Math. Soc. 24 (1949) 136–143.

[G1] M. GOURDIN, *The generalized atypical supertableaux of the orthosymplectic groups $OSP(2|2p)$*, J. Math. Phys. 27 (1986) 2832–2841.

[G2] M. GOURDIN, *Classification and interpretation of the supertableaux of the orthosymplectic groups $OSP(m|4)$*, J. Math. Phys. 28 (1987) 2007–2017.

[HK] J. W. B . HUGHES AND R. C. KING, *A conjectured character formula for atypical irreducible modules of the Lie superalgebra $sl(m/n)$*, J. Phys. A20 (1987) L1047–1052.

[HKTV] J. W. B. HUGHES, R. C. KING, J. THIERRY-MIEG AND J. R. VAN DER JEUGT, unpublished (1988).

[H] J. P. HURNI, *Young supertableaux of the basic Lie superalgebras*, J. Phys. A20 (1987) 5755–5821.

[Ka1] V. G. KAC, *Lie superalgebras*, Advances in Math. 26 (1977) 8–96.

[Ka2] V. KAC, *Representations of classical Lie superalgebras*, Lecture Notes in Maths. 676 (1978) 596–626.

[Ki1] R. C. KING, *Generalized Young tableaux and the general linear group*, J. Math. Phys. 11 (1970) 280–293.

[Ki2] R. C. KING, *Modification rules and products of irreducible representations of the unitary, orthogonal and symplectic groups*, J. Math. Phys. 12 (1971) 1588–1598.

[Ki3] R. C. KING, *Branching rules for classical Lie groups using tensor and spinor methods*, J. Phys. A8 (1975) 429–449.

[Ki4] R. C. KING, *Weight multiplicities of the classical groups*, Lecture Notes in Physics 50 (1976) 490–499.

[Ki5] R. C. KING, *Supersymmetric functions and the Lie supergroup $U(m/n)$*, Ars. Comb. 16A (1983) 269–287.

[KElS] R. C. KING AND N. G. I. EL-SHARKAWAY, *Standard Young tableaux and weight multiplicities of the classical Lie groups*, J. Phys. A16 (1983) 3153-3177.

[KT] K. KOIKE AND I. TERADA, *Young diagrammatic methods for the representation theory of the classical groups of type B_n, C_n and D_n*, J Algebra 107 (1987) 466–511.

[Le] D. A. LEITES, *A formula for the characters of the irreducible finite-dimensional representations of Lie superalgebras of series C*, (in Russian) C. R. Acad. Bulg. Sci. 33 (1980) 1053–1055.

[Li1] D. E. LITTLEWOOD, *The theory of group characters*, Oxford Univ. Press, Oxford, (1940).

[Li2] D. E. LITTLEWOOD, *Products and plethysms of characters with orthogonal, symplectic and symmetric groups*, Can. J. Math. 10 (1958) 17–32.

[LR] D. E. LITTLEWOOD AND A. R. RICHARDSON, *Group characters and algebra*, Phil. Trans. Roy. Soc. A233 (1934) 99–141.

[Ma1] I. G. MACDONALD, *Affine root systems and Dedekind's η-function*, Invent. Math. 15 (1972) 91-143.

[Ma2] I. G. MACDONALD, *Symmetric functions and Hall polynomials*, Oxford Univ. Press, Oxford, (1979).

[MSS] B. MOREL, A. SCIARRINO AND P. SORBA, *Representations of $OSp(M/2n)$ and Young supertableaux*, J. Phys. A18 (1985) 1597–1613.

[Mu] F. D. MURNAGHAN, *The theory of group representations*, The Johns Hopkins Press, Baltimore, (1938).

[N] M. J. NEWELL, *Modification rules for the orthogonal and symplectic groups*, Proc. Roy. Soc. Irish Acad. 54 (1951) 153-156.

[PS] I. Penkov and V. Serganova, *Character formulas for some classes of atypical $gl(m+ne)$ and $p(m)$ modules*, preprint Inst. Math., Bulg. Acad. Sci. (1988).

[P] P. Pragacz, private communication (1988).

[SS] A. N. Sergeev and V. Serganova, unpublished (1988).

[Sta] R. P. Stanley, *Theory and applications of plane partitions I*, Stud. App. Math. 50 (1971) 167–188.

[Ste1] J. R. Stembridge, *Combinatorial decompositions of characters of $SL(n,C)$*, Ph D thesis, MIT, (1985).

[Ste2] J. R. Stembridge, *First layer formulas for characters of $SL(n,C)$*, Trans. Amer. Math. Soc. 299 (1987) 319–350.

[Ste3] J. R. Stembridge, *Rational tableaux and the tensor algebra of gl_n*, J. Comb. Theory, A46 (1987) 79–120.

[Su1] S. Sundaram, *On the combinatorics of representations of $Sp(2n,C)$*, Ph D thesis, MIT, (1986).

[Su2] S. Sundaram, *Tableaux in the representation theory of the classical Lie groups*, these Proceedings, (1988).

[T-M] J. Thierry-Mieg, *Irreducible representations of the basic classical Lie superalgebras $SU(m/n)$ $SU(n/n)/U(1), OSp(m/2n), D(2/1,a), G(3)$ and $F(4)$*, Lecture Notes in Physics 201 (1984) 94–98.

[T] T. Tokuyama, *On a relation between the representations of a general linear group and a symplectic group*, Algebraic and Topological Theories, Kinokuniya Press, Tokyo, (1986) 344–368.

[VdJ] J. R. Van der Jeugt, *Structure of atypical representations of the Lie superalgebras $sl(m/n)$*, J. Phys. A20 (1987) 809–824.

[W1] H. Weyl, *Theorie der Darstellungen kontinuierlicher halb-einfacher Gruppen durch lineare Transfomationen III*, Math. Z. 24 (1926) 377–395.

[W2] H. Weyl, *The classical groups, their invariants and representations*, Princeton Univ. Press, Princeton, (1939).

THE UBIQUITOUS YOUNG TABLEAU

BRUCE E. SAGAN†

Abstract. Young tableaux have found extensive application in combinatorics [45], group representations [14], invariant theory [2,3], symmetric functions [20], and the theory of algorithms [18, pages 48-72]. This paper is an expository treatment of some of the highlights of tableaux theory. These include the hook and determinantal formulae for enumeration of both standard and generalized tableaux, their connection with irreducible representations of matrix groups, and the Robinson-Schensted-Knuth algorithm.

Key words. Young tableaux, hook and determinantal formulae, group representations, tableau algorithms

AMS(MOS) subject classifications. 05A15, 05A17, 20C30, 68C05

1. Three families of tableaux. Young tableaux were first introduced in 1901 by the Reverend Alfred Young [47, page 133] as a tool for invariant theory. Subsequently, he showed that they can give information about representations of symmetric groups. Since then, tableaux have played an important rôle in many areas of mathematics from enumerative combinatorics to algebraic geometry. This paper is a survey of some of these applications.

In recent years the number of tableaux of various types has been increasing at an impressive rate. To limit this paper to a reasonable length, our discussion will be restricted to three fundamental families of tableaux: ordinary, shifted and oscillating.

The rest of this section will be devoted to the definitions and notation need to describe these arrays. In Section 2 we present the hook and determinantal formulae for enumeration of standard tableaux. The third section examines the connection with representations of the symmetric group. The Robinson-Schensted algorithm appears in Section 4 as a combinatorial way of explaining the decomposition of the regular representation. The next four sections rework the material from the first four using generalized tableaux (those with repeated entries), representations of general linear and symplectic groups, and the theory of symmetric functions. Section 9 is a brief exposition of some open problems.

1.1. Ordinary tableaux. In what follows, $\mathbf{N}$ and $\mathbf{P}$ stand for the non-negative and positive integers respectively. A *partition* λ *of* $n \in \mathbf{N}$, written $\lambda \vdash n$, is a sequence of positive integers $\lambda = (\lambda_1, \lambda_2, \cdots, \lambda_l)$ in weakly decreasing order such that $\sum_{i=1}^{l} \lambda_i = n$. The λ_i are called the *parts of* λ. The unique partition of 0 is $\lambda = \phi$. The *shape of* λ is an array of boxes (or dots or cells) with l left-justified rows and λ_i boxes in row i. We will use λ to represent both the partition and its shape, while (i, j) will denote the cell in row i and column j. By way of illustration,

†Department of Mathematics, Michigan State University, East Lansing, MI 48824-1027

the following figure shows the shape of the partition $\lambda = (2,2,1) \vdash 5$ with cell (3,1) displayed as a diamond[1].

$$\begin{array}{cc} \square & \square \\ \square & \square \\ \diamond & \end{array}$$

A *standard Young tableau (SYT) of shape* λ, denoted P, is obtained by filling the cells of $\lambda \vdash n$ with the integers from 1 to n so that

(1) each integer is used exactly once, and

(2) the rows and columns increase.

We let $p_{i,j}$ denote the element of P in cell (i,j). There are 5 SYT of shape $(2,2,1)$:

$$\begin{array}{cc} 1 & 2 \\ 3 & 4 \\ 5 & \end{array}\ ,\quad \begin{array}{cc} 1 & 2 \\ 3 & 5 \\ 4 & \end{array}\ ,\quad \begin{array}{cc} 1 & 3 \\ 2 & 4 \\ 5 & \end{array}\ ,\quad \begin{array}{cc} 1 & 3 \\ 2 & 5 \\ 4 & \end{array}\ ,\quad \begin{array}{cc} 1 & 4 \\ 2 & 5 \\ 3 & \end{array}$$

and the first tableau has $p_{3,1} = 5$. Letting f^λ be the number of SYT of shape λ, we see that $f^{(2,2,1)} = 5$.

1.2. Shifted tableaux. A partition $\lambda^* = (\lambda_1, \lambda_2, \cdots, \lambda_l)$ of n is *strict*, $\lambda^* \models n$, if $\lambda_1 > \lambda_2 > \cdots > \lambda_l$. The *shifted shape of* λ^* is like the ordinary shape except that row i starts with its leftmost box in position (i,i). The fact that the parts of λ^* strictly decrease assures that shifting the rows in this manner does not cause any cells to stick out from the right-hand boundary. As an example, the shifted shape of $\lambda^* = (4,2,1) \models 7$ is:

$$\begin{array}{cccc} \square & \square & \square & \square \\ & \square & \square & \\ & & \square & \end{array}.$$

A *standard shifted Young tableau (SST) of shape* λ^*, P^* , is a filling of the shifted shape λ^* satisfying the same two conditions as for an SYT. The notation $p^*_{i,j}$ should be self-explanatory. The number of SST of shape λ^* is denoted g^λ. The list below displays all SST of shifted shape $(4,2,1)$, demonstrating that $g^{(4,2,1)} = 7$.

$$\begin{array}{cccc} 1 & 2 & 3 & 4 \\ & 5 & 6 & \\ & & 7 & \end{array}\ ,\quad \begin{array}{cccc} 1 & 2 & 3 & 5 \\ & 4 & 6 & \\ & & 7 & \end{array}\ ,\quad \begin{array}{cccc} 1 & 2 & 3 & 6 \\ & 4 & 5 & \\ & & 7 & \end{array}\ ,\quad \begin{array}{cccc} 1 & 2 & 3 & 7 \\ & 4 & 5 & \\ & & 6 & \end{array}\ ,$$

$$\begin{array}{cccc} 1 & 2 & 4 & 5 \\ & 3 & 6 & \\ & & 7 & \end{array}\ ,\quad \begin{array}{cccc} 1 & 2 & 4 & 6 \\ & 3 & 5 & \\ & & 7 & \end{array}\ ,\quad \begin{array}{cccc} 1 & 2 & 4 & 7 \\ & 3 & 5 & \\ & & 6 & \end{array}.$$

[1] In deference to Alfred Young's nationality, we have chosen to draw partition shapes in the English style, i.e., as if they were part of a matrix. The reader should be aware that some mathematicians (notably the French) prefer to use the conventions of coordinate geometry where λ_1 cells are placed along the x-axis, λ_2 cells are placed along the line $y = 1$, etc. To them and to René Descartes, we abjectly apologize.

1.3. Oscillating tableaux. To motivate the definition of an oscillating tableau, we must first look again at the definition of an SYT. An SYT P of shape λ gives rise to a sequence of shapes $\phi = \lambda^0 \subset \lambda^1 \subset \lambda^2 \subset \cdots \subset \lambda^n = \lambda$ where λ^m is the shape containing the numbers $1, 2, \cdots, m$ in P, i. e. , λ^m is obtained from λ^{m-1} by adding the cell (i,j) such that $p_{i,j} = m$. For example if

$$P = \begin{matrix} 1 & 3 \\ 2 & 5 \\ 4 & \end{matrix}$$

then the corresponding sequence is

$$\phi \subset \square \subset \begin{matrix}\square\\ \square\end{matrix} \subset \begin{matrix}\square & \square\\ \square & \end{matrix} \subset \begin{matrix}\square & \square\\ \square & \\ \square & \end{matrix} \subset \begin{matrix}\square & \square\\ \square & \square\\ \square & \end{matrix} = \lambda.$$

Conversely it should be clear that any sequence of shapes starting with ϕ and adding a box at each stage corresponds to a SYT. The definition of oscillating tableau generalizes this concept.

An *oscillating Young tableau (OYT) of shape λ and length k*, $\tilde{P}_k^\lambda$, is a sequence of shapes $(\phi = \lambda^0, \lambda^1, \cdots, \lambda^k = \lambda)$ such that λ^m is obtained from λ^{m-1} by adding or subtracting a cell. Oscillating tableaux are also called *up-down* or *alternating* tableaux. We let $\tilde{f}_k^\lambda$ denote the number of OYT of shape λ and length k. If $\lambda = (1)$ and $k = 3$ then a complete list of the corresponding OYT is:

$$(\phi, \square, \phi, \square)\,;\,(\phi, \square, \square\,\square, \square)\,;\,\left(\phi, \square, \begin{matrix}\square\\ \square\end{matrix}, \square\right).$$

It follows that $\tilde{f}_3^{(1)} = 3$.

2. Enumeration of tableaux. It would be useful to have an expression for the number of tableaux of a given shape, since enumeration by hand (as in the previous section) rapidly becomes unwieldy as n increases. There are two principle formulae of this type, one involving products (the hook formula) and one involving determinants.

2.1. Ordinary tableaux. If (i,j) is a cell in the shape of λ then it has *hook*

$$H_{i,j} = \{(i,j)\} \cup \{(i,j') \mid j' > j\} \cup \{(i',j) \mid i' > i\}$$

with corresponding *hooklength* $h_{i,j} = |H_{i,j}|$ (where $|\cdot|$ stands for cardinality). The sets $\{(i,j')|j' > j\}$ and $\{(i',j)|i' > i\}$ are called the *arm* and *leg* of the hook respectively. If $\lambda = (6,5,3,2)$ then the diamonds in

$$\begin{matrix} \square & \square & \diamond & \diamond & \diamond & \diamond \\ \square & \square & \diamond & \square & \square & \\ \square & \square & \diamond & & & \\ \square & \square & & & & \end{matrix}$$

represent the hook $H_{1,3}$ with hooklength $h_{1,3} = 6$. The famous *hook formula* expresses the number of SYT in terms of hooklengths.

THEOREM 2.1.1 (FRAME-ROBINSON-THRALL [4]). *If* $\lambda \vdash n$ *then*

$$f^\lambda = \frac{n!}{\prod_{(i,j)\in\lambda} h_{i,j}}.$$

Before discussing various proofs of this theorem, let us look at an example to see how easy it is to apply the hook formula. If $\lambda = (2,2,1) \vdash 5$ then the hooklengths of λ are given in the diagram

$$\begin{matrix} 4 & 2 \\ 3 & 1 \\ 1 & \end{matrix}$$

where $h_{i,j}$ is placed in cell (i,j). Thus $f^{(2,2,1)} = 5!/4 \cdot 3 \cdot 2 \cdot 1^2 = 5$ which agrees with our previous computation in Section 1.1. There are many different proofs of the hook formula; we outline a few of them next.

Proof sketches.

(1) (inductive) It is easy to prove the hook formula by induction on n. Unfortunately this type of proof gives no inkling of *why* the hooklengths should play a role.

(2) (probabilistic) Greene, Nijenhuis, and Wilf [11] have given a beautiful probabilistic proof where the hooks do enter in a very strong way. The general idea is this. Fix a shape $\lambda \vdash n$. If we can find an algorithm that produces any SYT P with probability $prob(P) = \prod h_{i,j}/n!$, then we will be done since the distribution is uniform. In what follows a *corner cell* is $(i,j) \in \lambda$ such that $(i+1,j), (i,j+1) \notin \lambda$.

(a) Pick $(i,j) \in \lambda$ with probability $1/n$.

(b) While (i,j) is not a corner cell do **begin**

(i) pick a cell $c \in H_{i,j} - \{(i,j)\}$ with probability $1/(h_{i,j}-1)$;

(ii) $(i,j) := c$, i.e., c becomes the new value for (i,j) **end**.

(c) Give the label n to the corner cell (i',j') you've reached.

(d) Go back to step (a) with $\lambda := \lambda - \{(i',j')\}$ and $n := n-1$.

Repeat this outer loop until all cells of λ are labeled.

It should be clear that this procedure gives a standard labeling of λ. It is less obvious (though not hard to prove) that all labelings are equally likely and of the right probability. The interested reader can consult [11] for the details.

(3) (combinatorial) Franzblau and Zeilberger [5] were the first to come up with a combinatorial proof of the hook formula. Rewriting the equation as $n! = f^\lambda \cdot \prod h_{i,j}$, we see that it suffices to find a bijection $S \longleftrightarrow (P,H)$ where S is an arbitrary filling of λ with $1, 2, \cdots, n$ (rows and columns need not increase), P is a SYT of shape λ, and H is a *pointer tableau* of shape λ, i.e., a placement of (computer science-type) pointers in λ such that the pointer in cell (i,j) points to some cell of $H_{i,j}$. Roughly, given a scrambled tableau S we wish to rearrange its entries to form a SYT P with the pointer tableau H keeping track of the unscrambling process. While this idea is simple, the actual bijection is long and difficult. Subsequently Zeilberger [48] found a way to turn the Greene-Nijenhuis-Wilf proof into a bijection, but the details are still not as pleasant as one would like. □

It is unfortunate that a simple combinatorial statement like the hook formula has no simple combinatorial proof. (It seems as if all the simple proofs are not combinatorial and all the combinatorial proofs are not simple!) To find such a proof is one of the tantalizing open problems in this area.

While the hook formula is relatively recent, the determinantal formula goes back to Frobenius and Young. In what follows, $1/r! = 0$ if $r < 0$.

THEOREM 2.1.2. *If* $\lambda = (\lambda_1, \lambda_2, \cdots, \lambda_l) \vdash n$ *then*

$$f^\lambda = n! \cdot det[1/(\lambda_i - i + j)!]$$

where the determinant is $l \times l$.

The simplest way to remember the denominators in the determinant is to note that the parts of λ are found along the main diagonal. The other entries in a given row are computed by increasing or decreasing the number (inside the factorial) by 1 for each step taken to the right or left respectively. If we apply this formula to our running example where $\lambda = (2, 2, 1)$ we find

$$5! \cdot \begin{vmatrix} 1/2! & 1/3! & 1/4! \\ 1/1! & 1/2! & 1/3! \\ 0 & 1/0! & 1/1! \end{vmatrix} = 5$$

which has not changed since our last computation.

Proof sketches (of the determinantal formula).

(1) (inductive) If $\lambda = (\lambda_1, \lambda_2, \cdots, \lambda_l)$ then it is easy to see directly from the definitions that $\lambda_i - i = h_{i,1} - l$. Hence it is enough to show that

$$f^\lambda = n! \cdot det[1/(h_{i,1} - l + j)!].$$

This can be done using induction and the hook formula.

(2) (combinatorial) A SYT can be represented as a family of non-intersecting lattice paths in the plane. Such families are also counted by determinants, as shown by Gessel [unpublished manuscript]. Remmel [25] combined these ideas with the Garsia-Milne involution principle [7] to give a bijective proof of both the hook and determinantal formulae. Gessel and Viennot [8,9] have extended this idea to a multitude of interesting applications. $\square$

2.2. Shifted tableaux. A shifted hook is like a hook with an extra appendage. The ***shifted hook*** of $(i, j) \in \lambda^*$ is

$$H^*_{i,j} = \{(i,j)\} \cup \{(i,j') \mid j' > j\} \cup \{(i',j) \mid i' > i\} \cup \{(j+1,j') | j' > j\}$$

with ***hooklength*** $h^*_{i,j} = |H^*_{i,j}|$. The diamonds below outline the hook $H^*_{2,3}$ of the partition $\lambda^* = (6, 5, 3, 2, 1)$ which has hooklength $h^*_{2,3} = 7$:

$$\begin{array}{cccccc} \square & \square & \square & \square & \square & \square \\ & \square & \diamond & \diamond & \diamond & \diamond \\ & & \diamond & \square & \square & . \\ & & & \diamond & \diamond & \\ & & & & \square & \end{array}$$

One way to motivate this definition is to note that if λ^* is a shifted shape then one can paste together λ^* and $\lambda^{*\,t}$ (where t denotes the transpose) to form a left justified shape λ. If $(i, j) \in \lambda^*$ then $H^*_{i,j} \subseteq \lambda^*$ is the same as $H_{i,j} \subseteq \lambda$ except that the bottom part of its leg has been twisted. For example, if we use $\lambda^* = (6, 5, 3, 2, 1)$

as before and represent $\lambda^{*\,t}$ using circles, then the shifted hook above corresponds to the normal hook in

$$\begin{array}{ccccccc}
\circ & \square & \square & \square & \square & \square & \square \\
\circ & \circ & \square & \diamond & \diamond & \diamond & \diamond \\
\circ & \circ & \circ & \diamond & \square & \square & \\
\circ & \circ & \circ & \diamond & \square & \square & \\
\circ & \circ & \circ & \diamond & \circ & \square & \\
\circ & \circ & & & & &
\end{array}$$

The shifted analog of the hook formula is

THEOREM 2.2.1 ([43]). *If $\lambda^* \models n$ then*

$$g^\lambda = \frac{n!}{\prod_{(i,j)\in\lambda^*} h^*_{i,j}}. \quad \square$$

As an example, $\lambda^* = (4,2,1) \models 7$ has shifted hooklengths

$$\begin{array}{cccc}
6 & 5 & 4 & 1 \\
 & 3 & 2 & \\
 & & 1 &
\end{array}$$

and so $g^{(4,2,1)} = 7!/6 \cdot 5 \cdot 4 \cdot 3 \cdot 2 \cdot 1^2 = 7$ as noted in Section 1.2.

It is easy to give an inductive proof of this result. A probabilistic analog of the Greene-Nijenhuis-Wilf proof was given by Sagan [28]. Surprisingly, although the algorithm for producing a tableau at random is identical (merely use shifted hooks in place of normal hooks), the proof that every tableau of a given shape is equally likely is much more complicated. It would be nice to find a simple proof that the shifted algorithm works. It is also an open problem to find analogs of the combinatorial proofs of [5] and [48].

The reader may be wondering why Thrall's [43] paper with the shifted hook formula appeared two years earlier than his article with Frame and Robinson [4] containing the unshifted version. In fact the 1952 paper contains an expression for g^λ which is mid-way between the shifted versions of the hook and determinantal formulae (although hooks are never mentioned explicitly) and from which either can be derived by simple manipulations. This brings us to the determinantal version.

THEOREM 2.2.2 ([43]). *If $\lambda^* = (\lambda_1, \lambda_2, \cdots, \lambda_l) \models n$ then*

$$g^\lambda = \frac{n!}{\prod_{i<j}(\lambda_i + \lambda_j)} \cdot det[1/(\lambda_i - l + j)!]$$

where the determinant is $l \times l$. $\square$

Note that the parts of λ^* are now found in the last column of the determinant. Also the extra product in the denominator is precisely the set of hooklengths for the cells $(i,j) \in \lambda^*$ such that $i < j$. Thrall proved this theorem by induction. It seems probable that the techniques of Remmel [25] and Gessel-Viennot [8] can also be used.

2.3. Oscillating tableaux. If $\lambda \vdash n$ and $\tilde{f}_k^\lambda \neq 0$ then we must take at least k steps to reach λ and so $k \geq n$. Furthermore, the number of extra additions and subtractions of cells must cancel out, so $k \equiv n \pmod 2$ and thus $k - n = 2d$ for some $d \in \mathbf{N}$. With these preliminaries we can state a formula of Sundaram for the number of OYT of given length and shape.

THEOREM 2.3.1 ([41]). *If $\lambda \vdash n$ and $k - n = 2d$ for some $d \in \mathbf{N}$, then*

$$\tilde{f}_k^\lambda = \binom{k}{n}(2d)!!\ f^\lambda$$

where $\binom{k}{n}$ is a binomial coefficient and $(2d)!! = 1 \cdot 3 \cdot 5 \cdot 7 \cdots (2d-1)$, i.e., $(2d)!!$ is the number of fixed-point free involutions in the symmetric group S_{2d}.

Proof sketch. It suffices to find a bijection

$$\tilde{P} \longleftrightarrow (P, \pi)$$

where $\tilde{P}$ is an OYT of shape λ and length k, P is an 'SYT' of shape λ , and π is a fixed-point free involution in 'S_{2d}'. The reason for the quotes is that the tableau need not contain the integers from 1 to n and the involution need not be a permutation of 1 to $2d$. Rather, the integers from 1 to k are to be partitioned into 2 subsets of size n and $k - n = 2d$ (accounting for the binomial coefficient), with the elements of P taken from the first subset and those of π from the second.

The general idea is that P keeps track of those steps in the construction of $\tilde{P}$ where a box of λ is added for the last time, while π stores information about deletions. See [41] for details. □

3. Representations of groups. Since one of Young's original applications for his tableaux came from group representation theory, it behooves us to look at the connection. In what follows, let G be a group and let V be a finite dimensional vector space over the complex numbers, $\mathbf{C}$.

3.1. Ordinary representations. A *representation of* G is a homomorphism $\rho : G \to GL(V)$ where $GL(V)$ is the general linear group of V, i.e., the group of all invertible linear transformations from V to itself. Alternatively, a representation may be viewed as a vector space V together with an action of G on V by invertible linear transformations. The space V is called a *G-module* and if $g \in G$, $v \in V$ then the action $\rho(g)v$ is abbreviated to $g \cdot v$ or just gv. We call dim V the *degree* of the representation.

As an example, consider any group G and let $V = \mathbf{C}$. Then the map that sends every $g \in G$ to the identity linear transformation (i.e., $gv = v$ for all $g \in G$ and $v \in V$) is a representation called the *trivial* representation. The trivial representation has degree 1.

For a more substantive example, let G be the symmetric group S_n and let V be the set of all formal linear combinations

$$V = \{c_1\vec{1} + c_2\vec{2} + \cdots + c_n\vec{n} \mid c_k \in \mathbf{C} \text{ for all } k\}$$

which is a vector space over $\mathbf{C}$ with basis $\mathcal{B} = \{\vec{1}, \vec{2}, \cdots, \vec{n}\}$. If $\pi \in S_n$ then we define the action of π on a basis vector by letting

$$\pi(\vec{k}) = \overrightarrow{\pi(k)}.$$

Thus the matrix of π in the canonical basis is just the usual permutation matrix associated with π , e.g., if $\pi = (1,2)(3) \in S_3$ then $\pi(\vec{1}) = \vec{2}$, $\pi(\vec{2}) = \vec{1}$ and $\pi(\vec{3}) = \vec{3}$ so that

$$\rho(\pi) = \begin{bmatrix} 0 & 1 & 0 \\ 1 & 0 & 0 \\ 0 & 0 & 1 \end{bmatrix}$$

in the basis $\mathcal{B}$. This representation is called the *natural* or *defining* representation of S_n and is of degree n.

A G-module V is called *irreducible* if there is no proper subspace W of V which is invariant under the set of linear transformations $\rho(G) = \{\rho(g) \mid g \in G\}$. Equivalently, V is irreducible if it has no basis $\mathcal{B}$ that simultaneously brings all the matrices to block form:

$$\rho(g) = \begin{bmatrix} A_g & B_g \\ 0 & C_g \end{bmatrix}$$

for all $g \in G$.

Obviously the trivial representation is irreducible, as are all degree one representations. On the other hand, the natural representation of S_n is not irreducible for $n \geq 2$ since the one-dimensional subspace generated by $\vec{1}+\vec{2}+\cdots+\vec{n}$ is invariant.

The irreducible representations of a group are important because they are the building blocks of all other representations under certain conditions. A G-module is said to be *completely reducible* if it is a direct sum of irreducible G-modules. Mashke's theorem gives us a large supply of completely reducible modules.

THEOREM 3.1.1. *If G is a finite group then every G-module V is completely reducible.* □

Although we do not have room here to prove the results that we will need from representation theory, the reader is encouraged to consult the excellent text of Ledermann [19] or the up-coming book of Sagan [31].

The next question to ask is: given G, how many irreducible G-modules are there? First, however, we must know when two modules are the same. We say that G-modules V and W are *equivalent*, written $V \cong W$, if there is a vector space isomorphism $\phi : V \to W$ that preserves the action of G, i.e.,

$$\phi(gv) = g\phi(v) \text{ for all } v \in V, g \in G.$$

THEOREM 3.1.2. *If G is finite then the number of inequivalent irreducible G-modules is equal to the number of conjugacy classes of G.* □

If $G = S_n$ then a conjugacy class consists of all permutations of a given cycle-type. But a cycle-type is just a partition of n.

COROLLARY 3.1.3. *The number of inequivalent irreducible S_n-modules is the number of partitions of n.* $\square$

So to find the number of irreducible S_3-modules we merely list all partitions of 3:

$$(3) = \square\ \square\ \square\ , \quad (2,1) = \begin{matrix} \square & \square \\ \square & \end{matrix}\ , \quad (1,1,1) = \begin{matrix} \square \\ \square \\ \square \end{matrix}.$$

Since there are 3 partitions, there are 3 irreducible modules for S_3.

The irreducible S_n-module indexed by $\lambda \vdash n$ is usually denoted S^λ and called the *Specht module corresponding to* λ . It would be nice to know their dimensions.

THEOREM 3.1.4. *If $\lambda \vdash n$ then $\dim S^\lambda = f^\lambda$.* $\square$

Returning to S_3, we can compute the dimensions of each irreducible by listing all the SYT of the appropriate shape (or by using the hook or determinantal formulae):

$$(3) : \quad 1\ 2\ 3\ ;$$

$$(2,1) : \quad \begin{matrix} 1\ 2 \\ 3 \end{matrix}, \qquad \begin{matrix} 1\ 3 \\ 2 \end{matrix};$$

$$(1,1,1) : \quad \begin{matrix} 1 \\ 2 \\ 3 \end{matrix};$$

so $\dim S^{(3)} = 1$, $\dim S^{(2,1)} = 2$, and $\dim S^{(1,1,1)} = 1$. Now for any n we have $\dim S^{(n)} = \dim S^{(1^n)} = 1$ where $(1^n) = (\overbrace{1, \cdots, 1}^{n})$. It turns out that $S^{(n)}$ corresponds to the trivial representation and $S^{(1^n)}$ corresponds to the one-dimensional *sign* representation that sends every $\pi \in S_n$ to the matrix $[sgn(\pi)]$.

For any finite group $G = \{g_1, g_2, \cdots, g_m\}$, the *group algebra* is the vector space

$$\mathbf{C}(G) = \{c_1\vec{g}_1 + c_2\vec{g}_2 + \cdots + c_m\vec{g}_m \mid c_k \in \mathbf{C} \text{ for all } k\}.$$

Clearly $\mathbf{C}(G)$ is a G-module under the action $h \cdot \vec{g} = \overrightarrow{hg}$ for all $g, h \in G$. Hence we can ask how $\mathbf{C}(G)$ decomposes into irreducibles.

THEOREM 3.1.5. *Let G be finite and let $S^1, S^2, \cdots, S^c$ be a complete list of inequivalent irreducible G-modules. Then*

$$\mathbf{C}(G) \cong \bigoplus_{i=1}^{c} m_i S^i$$

where $m_i = \dim S^i$, i.e., every irreducible module appears in $\mathbf{C}(G)$ with multiplicity equal to its dimension. $\square$

Taking dimensions on both sides of the previous equation we obtain:

COROLLARY 3.1.5.

$$|G| = \sum_{i=1}^{c} m_i^2. \quad \square$$

Finally, specializing to the symmetric group yields:

COROLLARY 3.1.6.

$$n! = \sum_{\lambda \vdash n} (f^\lambda)^2. \quad \square$$

3.2. Projective representations. A *projective representation* of a group G is a homomorphism $\rho : G \to PGL(V)$ where $PGL(V)$ is the projective general linear group, i.e., $GL(V)$ modulo the scalar multiples of the identity transformation. Below we will list the projective analogs of the results from the previous section. For a more complete discussion, see the articles of Stembridge [40] and Józefiak [15].

The irreducible projective representations of S_n are indexed by strict partitions λ^* . Unfortunately the indexing is no longer 1-to-1 as in the ordinary case. Define the *length of* $\lambda = (\lambda_1, \lambda_2, \cdots, \lambda_l)$, denoted $l = l(\lambda)$, to be the number of parts of λ. This should not be confused with the length of an OYT. Now if $\lambda^* \models n$ has length l, then when $n - l$ is even there is a single irreducible projective S_n-module S_0^λ corresponding to λ^*, but when $n - l$ is odd there are two such: S_1^λ and S_{-1}^λ.

By way of illustration, consider S_6. The corresponding strict partitions, λ^*, are

$$\begin{matrix}\square&\square&\square&\square&\square&\square\end{matrix}\;,\quad \begin{matrix}\square&\square&\square&\square&\square\\ &\square&&&\end{matrix}\;,\quad \begin{matrix}\square&\square&\square&\square\\ &\square&\square&\end{matrix}\;,\quad \begin{matrix}\square&\square&\square\\ &\square&\square\\ &&\square\end{matrix}$$

with $n - l$ being $6 - 1 = 5$, $6 - 2 = 4$, $6 - 2 = 4$, and $6 - 3 = 3$ respectively. Thus the number of irreducible projective representations of S_6 is $2 + 1 + 1 + 2 = 6$. (It is an accident that in all the examples we have seen, the number of irreducible representations of the S_n in question is always n.)

As in the ordinary case, we can also compute the dimensions of the irreducibles using tableaux. In the following theorem, $\lfloor \cdot \rfloor$ is the greatest integer function (also called the floor or round-down function). The power of 2 enters because the Schur multiplier for the symmetric group has order 2.

THEOREM 3.2.1. *If $\lambda^* \models n$ has length l then*

$$\dim S_i^\lambda = 2^{\lfloor \frac{n-l}{2} \rfloor} \cdot g^\lambda$$

where $i = 0$ if $n - l$ is even and $i = \pm 1$ if $n - l$ is odd. $\square$

To finish our computation for S_6, we list the shifted tableaux with shapes given by the shapes above:

$$
\begin{array}{llll}
(6): & 1\,2\,3\,4\,5\,6\,; & & \\[2ex]
(5,1): & \begin{array}{l}1\,2\,3\,4\,5\,,\\ \;6\end{array} & \begin{array}{l}1\,2\,3\,4\,6\,,\\ \;5\end{array} & \\[2ex]
 & \begin{array}{l}1\,2\,3\,5\,6\,,\\ \;4\end{array} & \begin{array}{l}1\,2\,4\,5\,6\,;\\ \;3\end{array} & \\[2ex]
(4,2): & \begin{array}{l}1\,2\,3\,4\,,\\ \;5\,6\end{array} & \begin{array}{l}1\,2\,3\,5\,,\\ \;4\,6\end{array} & \begin{array}{l}1\,2\,3\,6\,,\\ \;4\,5\end{array}\\[2ex]
 & \begin{array}{l}1\,2\,4\,5\,,\\ \;3\,6\end{array} & \begin{array}{l}1\,2\,4\,6\,;\\ \;3\,5\end{array} & \\[2ex]
(3,2,1): & \begin{array}{l}1\,2\,3\,,\\ \;4\,5\\ \;\;6\end{array} & \begin{array}{l}1\,2\,4\,.\\ \;3\,5\\ \;\;6\end{array} &
\end{array}
$$

Hence $\dim S_1^{(6)} = \dim S_{-1}^{(6)} = 2^{\lfloor 5/2 \rfloor} \cdot 1 = 4$, $\dim S_0^{(5,1)} = 2^{\lfloor 4/2 \rfloor} \cdot 4 = 16$, $\dim S_0^{(4,2)} = 2^{\lfloor 4/2 \rfloor} \cdot 5 = 20$, and $\dim S_1^{(3,2,1)} = \dim S_{-1}^{(3,2,1)} = 2^{\lfloor 3/2 \rfloor} \cdot 1 = 4$.

Because Corollary 3.1.6 continues to hold for projective representations we have

$$
\begin{aligned}
n! &= \sum_{\substack{\lambda^* \models n \\ n-l \text{ even}}} (\dim S_0^\lambda)^2 + \sum_{\substack{\lambda^* \models n \\ n-l \text{ odd}}} (\dim S_1^\lambda)^2 + (\dim S_{-1}^\lambda)^2 \\
&= \sum_{\substack{\lambda^* \models n \\ n-l \text{ even}}} (2^{\frac{n-l}{2}} \cdot g^\lambda)^2 + \sum_{\substack{\lambda^* \models n \\ n-l \text{ odd}}} 2 \cdot (2^{\frac{n-l-1}{2}} \cdot g^\lambda)^2.
\end{aligned}
$$

Conveniently, the powers of two in both summations turn out to be the same, and so

COROLLARY 3.2.2.

$$
n! = \sum_{\lambda^* \models n} 2^{n-l} (g^\lambda)^2.
$$

To bring oscillating tableaux into the act, we need to talk about representations of the symplectic group (rather than the symmetric group). This discussion will be postponed until Section 7.

4. The Robinson-Schensted correspondence. Corollaries 3.1.7 and 3.2.2 were obtained from general theorems about group representations. However, the equations themselves can be viewed as purely combinatorial statements about tableaux. Hence it would be nice to have purely combinatorial (i.e., bijective) proofs of these results. The celebrated Robinson-Schensted correspondence [26, 32] does exactly that. Although Robinson was the first to discover this algorithm, Schensted's form of the correspondence (discovered independently) is easier to understand. For that reason we will follow the latter's presentation.

4.1. Left-justified tableaux. We restate Corollary 3.1.7 for ease of reference.

THEOREM 4.1.1.

$$n! = \sum_{\lambda \vdash n} (f^\lambda)^2.$$

Combinatorial Proof. It suffices to find a bijection

$$\pi \overset{\text{R-S}}{\longleftrightarrow} (P, Q) \tag{1}$$

between permutations $\pi \in S_n$ and pairs of SYT P, Q of the same shape $\lambda \vdash n$. We first exhibit a map which, given a permutation, produces a tableaux pair.

$\underline{\pi \overset{\text{R-S}}{\rightarrow} (P, Q)}$. Suppose π is given in two-line form as

$$\pi = \begin{matrix} 1 & 2 & \cdots & n \\ x_1 & x_2 & \cdots & x_n \end{matrix}.$$

We will construct a sequence of tableaux

$$(P_0, Q_0) = (\phi, \phi);\ (P_1, Q_1);\ (P_2, Q_2);\ \cdots;\ (P_n, Q_n) = (P, Q) \tag{2}$$

where $x_1, x_2, \cdots, x_n$ will be *inserted* into the P's and $1, 2, \cdots, n$ will be *placed* in the Q's so that P_k and Q_k will have the same shape for all k. The operations of insertion and placement can be described as follows.

Suppose P is a *partial* tableau of shape μ, i.e., a filling of μ with a subset of the integers from 1 to n so that rows and columns increase. Let x be an element not in P. To *row insert x in P*, we use the following sequence of steps.

(1) If x is bigger than every element of the first row of P, then put x at the end of that row ($p_{1,\mu_1+1} \leftarrow x$) and stop.

(2) Otherwise, find the left-most element of the first row of P such that $p_{1,j_1} > x$ and replace this element by x (after storing its value for future use). We say that *x bumps p_{1,j_1}* from the first row.

(3) Now iterate the first two steps. If p_{1,j_1} is bigger than every element in row 2 then put it at the end of the row and stop. Otherwise p_{1,j_1} replaces the left-most p_{2,j_2} larger than itself and this element is inserted into the third row, etc.

(4) Since the p_{i,j_i} form an increasing sequence, at some point the algorithm must terminate with an element coming to rest at the end of some row.

As an example, suppose $x = 3$ and

$$P = \begin{matrix} 1\,2\,5\,8 \\ 4\,7 \\ 6 \end{matrix}.$$

The insertion of x into P causes elements to be bumped as follows:

$$\begin{matrix} 1\,2\,5\,8 \leftarrow 3 \\ 4\,7 \\ 6 \end{matrix},\quad \begin{matrix} 1\,2\,3\,8 \\ 4\,7 \ \leftarrow 5 \\ 6 \end{matrix},\quad \begin{matrix} 1\,2\,3\,8 \\ 4\,5 \\ 6 \ \leftarrow 7 \end{matrix},\quad \begin{matrix} 1\,2\,3\,8 \\ 4\,5 \\ 6\,7 \end{matrix}.$$

If row insertion of x into P yields partial tableau P' then we write $R_x(P) = P'$. It is easy to verify that P' will still have increasing rows and columns.

Placement of an element in a tableaux is an easy construction. Suppose that Q is a partial tableau of shape μ and that (i,j) is an *outer corner of* μ, meaning that $(i,j) \notin \mu$ but $\mu \cup (i,j)$ is the shape of a partition. If k is an integer, then the *placement of* k *in* Q *at cell* (i,j) is the tableau obtained by merely putting k in cell (i,j), i.e., $q_{i,j} := k$.

If we let

$$Q = \begin{array}{lll} 1 & 2 & 5 \\ 4 & 7 & \\ 6 & & \\ 8 & & \end{array},$$

then placing $k = 9$ in cell $(i,j) = (2,3)$ yields

$$\begin{array}{lll} 1 & 2 & 5 \\ 4 & 7 & 9 \\ 6 & & \\ 8 & & \end{array}.$$

Clearly, if k is bigger than every element of Q then the array will remain a partial tableau.

We can finally describe how to build the sequence (2) from the permutation

$$\pi = \begin{array}{cccc} 1 & 2 & \cdots & n \\ x_1 & x_2 & \cdots & x_n \end{array}.$$

Start with a pair of empty tableaux (P_0, Q_0). Assuming that the pair (P_{k-1}, Q_{k-1}) has been constructed, define (P_k, Q_k) by

$$\begin{array}{lll} P_k & = & R_{x_k}(P_{k-1}), \text{ and} \\ Q_k & = & \text{the placement of } k \text{ into } Q_{k-1} \text{ at the cell } (i,j) \\ & & \text{where the row insertion terminates.} \end{array}$$

Note that the definition of Q_k insures that the shapes of P_k and Q_k are equal for all k. We call $P = P_n$ and $Q = Q_n$ the P-*tableau* and Q-*tableau* of π respectively.

We now give an example of the complete algorithm. If

$$\pi = \begin{array}{cccccc} 1 & 2 & 3 & 4 & 5 & 6 \\ 2 & 4 & 3 & 6 & 5 & 1 \end{array}, \tag{3}$$

then the tableaux constructed by the algorithm are:

$$P_k: \quad \phi, \; 2, \; 24, \; \begin{array}{l} 23 \\ 4 \end{array}, \; \begin{array}{l} 236 \\ 4 \end{array}, \; \begin{array}{l} 235 \\ 46 \end{array}, \; \begin{array}{l} 135 \\ 26 \\ 4 \end{array} = P$$

$$Q_k: \quad \phi, \; 1, \; 12, \; \begin{array}{l} 12 \\ 3 \end{array}, \; \begin{array}{l} 124 \\ 3 \end{array}, \; \begin{array}{l} 124 \\ 35 \end{array}, \; \begin{array}{l} 124 \\ 35 \\ 6 \end{array} = Q.$$

Thus

$$\begin{matrix} 1 & 2 & 3 & 4 & 5 & 6 \\ 2 & 4 & 3 & 6 & 5 & 1 \end{matrix} \stackrel{\text{R-S}}{\rightarrow} \left(\begin{matrix} 1\,3\,5 \\ 2\,6 \\ 4 \end{matrix} \quad , \quad \begin{matrix} 1\,2\,4 \\ 3\,5 \\ 6 \end{matrix} \right).$$

To show that this map is a bijection, we create its inverse.

$(P, Q) \stackrel{\text{R-S}}{\rightarrow} \pi$. We merely reverse the above procedure step by step. Define $(P_n, Q_n) = (P, Q)$. Assuming that the pair (P_k, Q_k) has been constructed, we obtain x_k (the kth element of π) and (P_{k-1}, Q_{k-1}) as follows.

Find the cell (i, j) containing the k in Q_k. Since this is the largest element in Q_k, $p_{i,j}$ must have been the last element to be bumped in the construction of P_k. Furthermore, the element that bumped it must be the right-most entry in row $i-1$ such that $p_{i-1,j_{i-1}} < p_{i,j}$. So replace $p_{i-1,j_{i-1}}$ by $p_{i,j}$ and find the entry of row $i-2$ that displaced $p_{i-1,j_{i-1}}$, etc. Working back up the rows in this manner, we will finally remove an element p_{1,j_1} from the first row. Thus $x_k = p_{1,j_1}$, P_{k-1} is P_k after the deletion process described above is complete, and Q_{k-1} is Q_k with the k erased. Continuing in this way, we will eventually recover all the elements of π in reverse order. □

The Robinson-Schensted algorithm has many beautiful and surprising properties. The literature on this subject is so vast that we can only present a sampling of results here. The interested reader can consult the extensive bibliography in [45] for other sources.

4.1.1. Column insertion. One can obviously define *column insertion* of x into P by reversing the roles of rows and columns in the definition of insertion: x displaces the highest element of the first column of P larger than x, this element is bumped into the second column, etc. If the result of column inserting x into P is P' we write $C_x(P) = P'$. It turns out that the row and column insertion operators commute (operators should be read right to left).

PROPOSITION 4.1.2 ([32]). *For any partial tableau P and positive integers $x, y \notin P$*

$$C_y R_x(P) = R_x C_y(P).$$

Proof. This proposition follows from the definitions of the two operators by an easy case-by-case argument. □

In the next result, π^r stands for the *reversal* of π, i.e., if $\pi = x_1 x_2 \cdots x_n$ then $\pi^r = x_n x_{n-1} \cdots x_1$.

COROLLARY 4.1.3. *If $\pi \stackrel{\text{R-S}}{\rightarrow} (P, Q)$. then $\pi^r \stackrel{\text{R-S}}{\rightarrow} (P^t, \cdot)$ where t denotes transpose.*

Proof. By definition, the P-tableau of π^r is

$$\begin{aligned} R_{x_1} \cdots R_{x_{n-1}} R_{x_n}(\phi) &= R_{x_1} \cdots R_{x_{n-1}} C_{x_n}(\phi) && \text{(initial tableau empty)} \\ &= C_{x_n} R_{x_1} \cdots R_{x_{n-1}}(\phi) && \text{(commutivity)} \\ &\ \ \vdots \\ &= C_{x_n} C_{x_{n-1}} \cdots C_{x_1}(\phi) && \text{(induction)} \\ &= P^t && \text{(def. of col. insertion).} \ \square \end{aligned}$$

4.1.2. The jeu de taquin. It would be nice to describe the Q-tableau for π^r. In order to do so, we must introduce a powerful operation of Schützenberger [35].

Suppose λ and μ are shapes such that $\lambda \subseteq \mu$. Then they form the *skew partition* $\lambda/\mu \stackrel{\text{def}}{=} \{(i,j) \in \lambda \mid (i,j) \notin \mu\}$. For example if

$$\lambda = \begin{array}{llll} \square & \square & \square & \square \\ \square & \square & \square & \\ \square & \square & \square & \end{array}$$

and

$$\mu = \begin{array}{ll} \square & \square \\ \square & \end{array}$$

then

$$\lambda/\mu = \begin{array}{llll} \bullet & \bullet & \square & \square \\ \bullet & \square & \square & \\ \square & \square & \square & \end{array}$$

where the missing boxes have been replaced by dots (black holes). *Skew tableaux* of both the standard and partial varieties are defined in the obvious way, filling the skew shape with an appropriate subset of the integers so that the rows and columns increase.

Now let Q be a skew tableau of shape λ/μ and let (i,j) be a corner cell of μ (called an *inner corner* of the skew shape). An (i,j)-*slide* is accomplished by performing the following sequence of operations.

(1) If neither $(i,j+1)$ nor $(i+1,j)$ are in λ/μ then (i,j) is eliminated from the shape of Q and the algorithm terminates.
(2) Otherwise, let $q_{i_1,j_1} = \min\{q_{i,j+1}, q_{i+1,j}\}$. (If one of the arguments of the min doesn't exist, then define it's value to be the element of Q which does appear.) Slide q_{i_1,j_1} into cell (i,j), creating a hole in position (i_1,j_1).
(3) Now repeat the first two steps with (i,j) and (i_1,j_1) replaced by (i_1,j_1) and (i_2,j_2) respectively, etc. After a finite number of iterations, the hole will slide to the boundary of Q and be eliminated, at which point we stop.

If applying an (i,j)-slide to Q yields Q' then we write $\Delta_{i,j}Q = Q'$

To illustrate this procedure, let

$$Q = \begin{array}{llll} \bullet & \bullet & 3 & 7 \\ \bullet & 1 & 4 & \\ 2 & 5 & 6 & \end{array} .$$

We now apply a (1,2)-slide:

$$\begin{array}{llll} \bullet & \bullet & 3 & 7 \\ \bullet & 1 & 4 & \\ 2 & 5 & 6 & \end{array}, \quad \begin{array}{llll} \bullet & 1 & 3 & 7 \\ \bullet & \bullet & 4 & \\ 2 & 5 & 6 & \end{array}, \quad \begin{array}{llll} \bullet & 1 & 3 & 7 \\ \bullet & 4 & \bullet & \\ 2 & 5 & 6 & \end{array}, \quad \begin{array}{llll} \bullet & 1 & 3 & 7 \\ \bullet & 4 & 6 & \\ 2 & 5 & & \end{array} = \Delta_{1,2}Q.$$

Given an SYT Q we will build another SYT $S(Q)$ in the following manner. First,a sequence of partial tableaux is constructed $Q = Q_n, Q_{n-1}, \cdots, Q_0 = \phi$.

To get Q_{k-1} from Q_k, we first erase the element in cell $(1,1)$ of Q_k to form a tableau of shape $\mu/(1)$ for some μ. We let Q_{k-1} be the tableau obtained by applying $\Delta_{1,1}$ to this skew tableau. Finally, we put a k in cell (i,j) of $S(Q)$ if that was the box eliminated from the boundary when passing from Q_k to Q_{k-1}.

Applying this algorithm to the Q-tableaux of the Robinson-Schensted example above, we obtain

$$
\begin{array}{llllllll}
Q_k: & \begin{matrix}1\,2\,4\\3\,5\\6\end{matrix}, & \begin{matrix}2\,4\\3\,5\\6\end{matrix}, & \begin{matrix}3\,4\\5\\6\end{matrix}, & \begin{matrix}4\\5\\6\end{matrix}, & \begin{matrix}5\\6\end{matrix}, & 6, & \phi \\
\\
S(Q): & \begin{matrix}\bullet\bullet\bullet\\\bullet\bullet\\\bullet\end{matrix}, & \begin{matrix}\bullet\bullet\,6\\\bullet\bullet\\\bullet\end{matrix}, & \begin{matrix}\bullet\bullet\,6\\\bullet\,5\\\bullet\end{matrix}, & \begin{matrix}\bullet\,4\,6\\\bullet\,5\\\bullet\end{matrix}, & \begin{matrix}\bullet\,4\,6\\\bullet\,5\\3\end{matrix}, & \begin{matrix}\bullet\,4\,6\\2\,5\\3\end{matrix}, & \begin{matrix}1\,4\,6\\2\,5\\3\end{matrix}.
\end{array}
$$

Furthermore, it can be verified that the Q-tableaux of $\pi^r = 1\ 5\ 6\ 3\ 4\ 2$ is just the transpose of $S(Q)$. This is not an accident.

THEOREM 4.1.4 ([35]). *If* $\pi \overset{\text{R-S}}{\rightarrow} (P,Q)$ *then* $\pi^r \overset{\text{R-S}}{\rightarrow} (P^t, S(Q)^t)$. □

Slides can also be used to prove another theorem of Schützenberger.

THEOREM 4.1.5 ([35]). *If* $\pi \overset{\text{R-S}}{\rightarrow} (P,Q)$ *then* $\pi^{-1} \overset{\text{R-S}}{\rightarrow} (Q,P)$.

Proof sketch. This result can also be demonstrated using a geometric form of the Robinson-Schensted correspondence due to Viennot [44]. Imagine π represented as a permutation matrix in the plane, i.e., the kth element of π is represented by a point with cartesian coordinates (k, x_k). Suppose that the plane is illuminated from the origin so that each point of π casts a shadow whose boundaries are half-lines parallel to the coordinate axes. By reading this diagram from left to right, one obtains a picture of the Robinson-Schensted algorithm as if on a time line (the kth insertion takes place as we pass the line $x = k$). One can read off the entries of the P- and Q-tableaux as certain coordinates on the y- and x-axes respectively. Once this is established, the theorem is immediate since passing from π to π^{-1} merely interchanges the two axes. □

Two more definitions are needed before we will be able to define the 'jeu de taquin' (or 'teasing game'). An *anti-diagonal strip* is the skew shape consisting of the cells $(n+1,1)$; $(n,2)$; $\cdots$; $(1,n+1)$. If $\pi = x_1x_2\cdots x_n$ then the corresponding *anti-diagonal strip tableau* has x_j in column j. For example, $\pi = 1\ 4\ 3\ 2$ corresponds to the tableau

$$
\begin{matrix}
\bullet & \bullet & \bullet & 2\\
\bullet & \bullet & 3 & \\
\bullet & 4 & & \\
1 & & &
\end{matrix}
.
$$

Now given an anti-diagonal strip tableaux, we can play *jeu de taquin*. Start by choosing any inner corner (i,j) and applying $\Delta_{i,j}$. Now choose any inner corner

(i',j') of the new skew shape and apply $\Delta_{i',j'}$, etc., until we get a left-justified (non-skew) SYT. One possible game that could be played on the tableau above is

$$\begin{array}{llll} \bullet & \bullet & \bullet & 2 \\ \bullet & \bullet & 3 & \\ \bullet & 4 & & \\ 1 & & & \end{array} \quad \overset{\Delta_{3,1}}{\longrightarrow} \quad \begin{array}{llll} \bullet & \bullet & \bullet & 2 \\ \bullet & \bullet & 3 & \\ 1 & 4 & & \end{array} \quad \overset{\Delta_{1,3}}{\longrightarrow} \quad \begin{array}{lll} \bullet & \bullet & 2 \\ \bullet & \bullet & 3 \\ 1 & 4 & \end{array} \quad \overset{\Delta_{2,2}}{\longrightarrow}$$

$$\begin{array}{lll} \bullet & \bullet & 2 \\ \bullet & 3 & \\ 1 & 4 & \end{array} \quad \overset{\Delta_{2,1}}{\longrightarrow} \quad \begin{array}{lll} \bullet & \bullet & 2 \\ 1 & 3 & \\ 4 & & \end{array} \quad \overset{\Delta_{1,2}}{\longrightarrow} \quad \begin{array}{ll} \bullet & 2 \\ 1 & 3 \\ 4 & \end{array} \quad \overset{\Delta_{1,1}}{\longrightarrow} \quad \begin{array}{ll} 1 & 2 \\ 3 & \\ 4 & \end{array} .$$

If applying jeu de taquin to the anti-diagonal strip of π yields a SYT P we write $J(\pi) = P$. It is not clear that the operation J is well-defined. However,

THEOREM 4.1.6 ([36]). *The tableau $J(\pi)$ is independent of the choice of inner corners made while playing the game. Furthermore, if P is the P-tableau of π then $J(\pi) = P$.*

Proof sketch. This proof is due to Thomas [42] .Showing that $J(\pi)$ does not depend on the order in which the inner corners are filled is a delicate case-by-case argument. Once this is established, a beautiful connection between the jeu de taquin and the Robinson-Schensted map appears. Let's choose to fill all the corners in a given row from right to left, starting with the lowest row and working up. After $k-1$ rows have been filled, the portion of the array in these rows will be the partial tableau P_k of the sequence (2). From this, it is easy to see that filling the kth row from the bottom is equivalent to the row insertion of x_{k+1} into P_k. Hence the lower portion of the array must now contain P_{k+1} and induction yields the fact that $J(\pi) = P$. □

4.1.3. Increasing and decreasing subsequences. One of Schensted's original motivations for constructing his map was to investigate lengths of increasing and decreasing subsequences . An *increasing subsequence of* $\pi = x_1 x_2 \cdots x_n$ is a subsequence $x_{i_1} x_{i_2} \cdots x_{i_k}$ such that $x_{i_1} < x_{i_2} < \cdots < x_{i_k}$. *Decreasing subsequences* are similarly defined. For example, 2 3 6 and 4 3 1 are respectively increasing and decreasing subsequences of the permutation listed as equation (3).

THEOREM 4.1.7 ([32]). *The length of a longest increasing subsequence of π is the length of the first row of it's P-tableau. The length of a longest decreasing subsequence of π is the length of the first column.*

Proof sketch. For increasing subsequences, one can inductively prove a stronger result, viz., if k enters P_{k-1} in column j then the length of the longest increasing subsequence ending in x_k must be j. The statement for decreasing subsequences now follows from Corollary 4.1.3. □

The reader should note that while the length of P's first row is the length of a longest increasing subsequence of π, the elements themselves do not form an increasing subsequence. This can be verified using our running example. It is possible, with a little more care, to recover an increasing subsequence of maximum length from the Robinson-Schensted algorithm.

Greene [10] has generalized this theorem to other types of subsequences. A subsequence of π is called *k-increasing* (respectively *k-decreasing*) if it is the union of k increasing (respectively decreasing) subsequences. Thus a 1-increasing subsequence is merely an increasing one. The subsequence $2\ 4\ 3\ 6\ 5 = 2\ 3\ 6 \cup 4\ 5$ of the permutation in (3) is 2-increasing but not 1-increasing, while π itself is 3-increasing.

THEOREM 4.1.8 ([10]). *The length of a longest k-increasing (respectively k-decreasing) subsequence of π is the sum of the lengths of the first k rows (respectively columns) of it's P-tableau.*

Proof sketch. Given a tableau P, we define the *row word* of P to be the permutation obtained by reading of the rows of P from left to right, starting with the last row and moving up. Our running example has row word $4\ 2\ 6\ 1\ 3\ 5$. It is easy to see that if P has row word π then

(1) the P-tableau of π is P itself, and
(2) the first k rows of P form a k-increasing subsequence of π of maximum length.

Hence the theorem is true for permutations that are row words. To show that it holds in general, Greene proves that any permutation with P-tableau P can be transformed into P's row word by a sequence of adjacent transpositions (the so-called *Knuth transpositions* [17]) which leave both the P-tableau and the maximum length of a k-increasing subsequence invariant. □

4.2. Shifted tableaux. Suppose λ^* is a strict partition of n having length l. The *main diagonal* of the shape of λ^* consists of the cells (i,i) where $1 \leq i \leq l$. All other cells are *off-diagonal*, so the number of off-diagonal cells is $n - l$. This fact will be useful in the combinatorial proof of Corollary 3.2.2 which we restate here.

THEOREM 4.2.1.

$$n! = \sum_{\lambda^* \models n} 2^{n-l}(g^{\lambda})^2.$$

Combinatorial proof. It suffices to find a bijection

$$\pi \longleftrightarrow (P^*, Q^*) \tag{4}$$

between permutations $\pi \in S_n$ and pairs of SST P^*, Q^* of the same shape $\lambda^* \models n$ where Q^* has a subset of it's off-diagonal elements distinguished in some manner. We will distinguish an element k by writing k'. Sagan [27] was the first to find such a map, but his correspondence did not have many of the properties of the original Robinson-Schensted algorithm. Later Sagan [30] and Worley [46] independently found a better bijection that does enjoy most of these properties. It is this version that we present.

To construct the map from permutations to tableau pairs, we create a sequence of shifted partial tableau pairs analogous to the sequence (2). Thus we need only discuss the analogs of insertion and placement for shifted tableaux.

To insert x_k into a partial tableau P^*_{k-1}, we start row inserting x_k as usual. If a diagonal element is never displaced, then insertion stops with an element coming to rest at the end of some row. This is called a *Schensted* insertion. If, on the other hand, some $p^*_{i,i}$ is bumped, then insert it in *column* $i+1$. Continue column inserting until an element comes to rest at the end of some column. This type of insertion is called *non-Schensted.* In either case, let (i,j) be the cell filled by the last bump.

Q^*_k is obtained by placing an element in cell (i,j) of Q^*_{k-1}. This element will be either a k if the insertion was Schensted, or a k' if the insertion was non-Schensted. Since a non-Schensted insertion can never terminate on the main diagonal, all primed elements will be off-diagonal.

It is an easy matter to construct the inverse map, since the distinguished elements in Q^* indicate whether to start the deletion process by rows or columns. The details are left to the reader. □

If we compute the shifted tableaux associated with the permutation $\pi = 2\,1\,6\,5\,4\,3$, we get the sequence

$$
\begin{array}{llllllll}
P^*_k: & \phi, & 2, & 1\,2, & 1\,2\,6, & \begin{array}{l}1\,2\,5,\\ \ \ \ 6\end{array} & \begin{array}{l}1\,2\,4,\\ \ \ \ 5\,6\end{array} & \begin{array}{l}1\,2\,3\,6\\ \ \ \ 4\,5\end{array} & = P^*
\end{array}
$$

$$
\begin{array}{llllllll}
Q^*_k: & \phi, & 1, & 1\,2', & 1\,2'\,3, & \begin{array}{l}1\,2'\,3,\\ \ \ \ 4\end{array} & \begin{array}{l}1\,2'\,3,\\ \ \ \ 4\,5'\end{array} & \begin{array}{l}1\,2'\,3\,6'\\ \ \ \ 4\,5'\end{array} & = Q^*.
\end{array}
$$

We should mention that MacLarnan [21] has found a way to construct other bijections between permutations and tableaux pairs using recursions satisfied by f^λ g^λ, and $\tilde{f}^\lambda_k$. All of these maps have the property that inverting the permutation interchanges the tableaux, which is *not* true for the algorithm above. However, Haiman [12] has developed a procedure called *mixed insertion* which does interchange the outputs of the Sagan-Worley algorithm when applied to π^{-1}.

5. Tableaux with repetitions. Many of the results of the first four sections can be generalized to tableaux with repeated entries. First we must define such arrays precisely.

5.1. Generalized Young tableaux. A *generalized Young tableau* (GYT), T, of shape λ is a filling of the shape with positive integers such that the rows weakly increase and the columns strictly increase. These arrays are also called *semi-standard tableaux* or *column strict reverse plane partitions* (the term 'reverse' is an historical accident coming from the fact that partitions were usually listed with parts in *decreasing* order). One possible GYT of shape $(4,4,1)$ is

$$
\begin{array}{llll}
1 & 1 & 2 & 2\\
2 & 3 & 4 & 4\\
4
\end{array}
$$

Note that if $\mu = (\mu_1, \mu_2, \cdots, \mu_l)$ is a partition of n then we can also write $\mu = (1^{m_1}, 2^{m_2}, \cdots, n^{m_n})$ where m_k is the number of parts of μ equal to k. The same notation applies if the parts of μ are arranged in a GYT, T. In this case, μ is called the *content of* T. For example, the tableau above has content $(1^2, 2^3, 3^1, 4^3)$. Using the set of variables $\mathbf{x} = \{x_1, x_2, \cdots, x_n\}$ we can associate with T a *monomial* $m(T) = x_1^{m_1} x_2^{m_2} \cdots x_n^{m_n}$. The monomial of the tableau above is $m(T) = x_1^2 x_2^3 x_3^1 x_4^3$.

Finally, if λ is a partition of length at most n, we define the corresponding *Schur function* to be

$$s_\lambda(\mathbf{x}) = \sum_{T \in \mathcal{T}_\lambda(n)} m(T)$$

where $\mathcal{T}_\lambda(n)$ is the set of all GYT of shape λ and entries of size at most n. If we let $\lambda = (2,1)$ and $n = 3$, then the tableaux in $\mathcal{T}_{(2,1)}(3)$ are

$$\begin{matrix}1\,1\\2\end{matrix}, \quad \begin{matrix}1\,2\\2\end{matrix}, \quad \begin{matrix}1\,1\\3\end{matrix}, \quad \begin{matrix}1\,3\\3\end{matrix}, \quad \begin{matrix}2\,2\\3\end{matrix}, \quad \begin{matrix}2\,3\\3\end{matrix}, \quad \begin{matrix}1\,2\\3\end{matrix}, \quad \begin{matrix}1\,3\\2\end{matrix}$$

and so

$$s_{(2,1)}(x_1, x_2, x_3) = x_1^2 x_2 + x_1 x_2^2 + x_1^2 x_3 + x_1 x_3^2 + x_2^2 x_3 + x_2 x_3^2 + 2x_1 x_2 x_3 .$$

Clearly if λ is a partition of n, then the coefficient of the monomial $x_1 x_2 \cdots x_n$ in $s_\lambda(x_1, x_2, \cdots, x_n)$ is just f^λ. It is also true (although it is not obvious from our definition) that $s_\lambda(\mathbf{x})$ is a *symmetric function*, i.e., permuting the variables in $\mathbf{x}$ does not change the polynomial s_λ. We will have more to say about symmetric functions shortly.

5.2. Generalized shifted tableaux. Let λ^* be a strict partition. A *generalized shifted tableau* (GST), T^* , of shape λ^* is obtained by filling the shifted shape with elements from the totally ordered alphabet $\mathcal{A}' = \{1' < 1 < 2' < 2 < 3' < 3 < \cdots\}$ so that

(1) T^* is weakly increasing along rows and columns and strictly increasing along diagonals, and

(2) for every integer k, there is at most one k' in each row and at most one k in each column of T^* .

An example of such a tableau is

$$T^* = \begin{matrix} 1' & 1 & 1 & 2' & 3' & 3 & 3 & 4' \\ & 2 & 2 & 2 & 3' & 4' & 4 & 4 \\ & & 3' & 3 & 3 & & & \\ & & & 4 & 4 & & & \end{matrix} .$$

Note that conditions 1 and 2 imply that, for fixed k, the cells occupied by all the elements of the form k or k' form a union of skew hooks. A *skew hook* is a skew shape that is connected (i.e., one can travel from one cell to any other by passing through cells adjacent by an edge) and contains at most one cell on every diagonal.

The elements 4 and $4'$ in T^* above lie in the union of two skew hooks while all other fixed integers and their primes lie in only one. Furthermore, condition 2 determines the character (primed or not) of each element in a skew hook, except at the lower left-hand end. This observation will be important when we discuss the analog of Knuth's algorithm for generalized shifted tableaux.

If T^* is a GST with $t^*_{i,j} \le n$ for all $(i,j) \in \lambda^*$ then it's associated monomial is $m(T^*) = x_1^{m_1} x_2^{m_2} \cdots x_n^{m_n}$ where m_k is the number of entries of T^* equal to k or k'. Our example tableau has monomial $m(T^*) = x_1^3 x_2^4 x_3^7 x_4^6$. Associated with each strict partition λ^* , $l(\lambda^*) \le n$, is a *Schur Q-function* defined by

$$Q_\lambda(\mathbf{x}) = \sum_{T^* \in \mathcal{T}^*_\lambda(n)} m(T^*),$$

$\mathcal{T}^*_\lambda(n)$ being the set of all GST with shape λ^* and entries of size at most n. Like $s_\lambda(\mathbf{x})$, $Q_\lambda(\mathbf{x})$ is a symmetric function. However, because of the presence of primed elements, the coefficient of $x_1 x_2 \cdots x_n$ in $Q_\lambda(\mathbf{x})$ for $\lambda^* \models n$ turns out to be $2^n g^\lambda$.

5.3. Symplectic tableaux. Consider the alphabet $\bar{\mathcal{A}} = \{1 < \bar{1} < 2 < \bar{2} < 3 < \bar{3} < \cdots\}$. A *symplectic tableau* (SPT), $\tilde{T}$, of shape $\lambda \vdash n$ is a GYT with entries from $\bar{\mathcal{A}}$ satisfying the extra constraint

(5) for all $i \le l(\lambda)$, the elements in row i are all of size at least i.

Equation (5) is called the *symplectic condition.* An example of such an array is

$$\tilde{T} = \begin{matrix} 1 & \bar{1} & 2 & 2 & 2 & \bar{3} \\ 2 & 2 & \bar{2} & \bar{2} & \bar{2} & \\ \bar{3} & \bar{3} & & & & \end{matrix} \quad .$$

For SPT we use the variables $\mathbf{x}^{\pm 1} = \{x_1, x_1^{-1}, x_2, x_2^{-1}, \cdots, x_n, x_n^{-1}\}$. The monomial of $\tilde{T}$ is given by

$$m(\tilde{T}) = x_1^{m_1} x_1^{-\bar{m}_1} x_2^{m_2} x_2^{-\bar{m}_2} \cdots x_n^{m_n} x_n^{-\bar{m}_n}$$

where m_k (respectively $\bar{m}_k$) is the number of k's (respectively $\bar{k}$'s) in $\tilde{T}$. The SPT above has $m(\tilde{T}) = x_1^1 x_1^{-1} x_2^5 x_2^{-3} x_3^{-3}$. The *symplectic Schur function* associated with a partition λ of length at most n is

$$sp_\lambda(\mathbf{x}^{\pm 1}) = \sum_{\tilde{T} \in \tilde{\mathcal{T}}_\lambda(n)} m(\tilde{T})$$

where the reader will already have guessed that $\tilde{\mathcal{T}}_\lambda(n)$ is the set of all SPT of shape λ with entries of size at most n. The symplectic Schur function is symmetric in the variables $\mathbf{x}^{\pm 1}$.

6. Enumeration of generalized tableaux. We will now present analogs of the hook and determinantal formulae for generalized tableaux.

6.1. The ordinary case. Let T be a GYT of shape λ , then we say *T partitions m* if $\sum_{(i,j)\in\lambda} t_{i,j} = m$. Letting $p_\lambda(m)$ be the number of such tableaux, we have the following generating function analog of the hook formula.

THEOREM 6.1.1.

$$\sum_{m\geq 0} p_\lambda(m)x^m = x^{N(\lambda)} \prod_{(i,j)\in\lambda} \frac{1}{1-x^{h_{i,j}}}$$

where $N(\lambda) = \sum_{i\geq 1} i\lambda_i$.

Proof. Stanley [38] was the first to prove this using his theory of poset partitions. We will present a beautiful bijective proof of Hillman and Grassl [13].

A *reverse plane partition* is like a GYT except that the columns need only weakly increase and 0 is allowed as an array entry. Let $r_\lambda(m)$ be the number of reverse plane partitions of m. There is a simple bijection between GYT and reverse plane partitions of the same shape λ . Merely take the GYT and subtract 1 from every element of the first row, 2 from every element of the second, etc. Since this takes away a total of $N(\lambda)$ from each GYT, it suffices to prove that

$$\sum_{m\geq 0} r_\lambda(m)x^m = \prod_{(i,j)\in\lambda} \frac{1}{1-x^{h_{i,j}}}. \tag{6}$$

The right hand side of this equation counts (linear) partitions ν all of whose parts come from the multiset (i.e., a set with repeated elements) $\{h_{i,j} | (i,j) \in \lambda\}$. Thus we need a bijection

$$T \longleftrightarrow \nu$$

between GYT's T and partitions $\nu = (\nu_1, \cdots, \nu_l)$ whose parts are all of the form $h_{i,j}$ such that $\sum_{(i,j)\in\lambda} t_{i,j} = \sum_{k\geq 1} \nu_k$.

<u>$T \rightarrow \nu$.</u> Given T, we will produce a sequence of reverse plane partitions

$$T = T_0,\ T_1,\ T_2,\ \cdots, T_f = \text{ tableau of 0's}$$

where T_k will be obtained from T_{k-1} by subtracting one from all the elements of a skew hook H of T_k such that $|H| = h_{a,c}$ for some $(a,c) \in \lambda$. The cells of the skew hook are defined recursively as follows.

Let (a,b) be the right-most highest cell of T containing a non-zero element. Then

$$(a,b) \in H \text{ and if } (i,j) \in H \text{ then } \begin{cases} (i,j-1) \in H & \text{if } t_{i,j-1} = t_{i,j} \\ (i+1,j) \in H & \text{otherwise} \end{cases}$$

i.e., move down unless forced to move left so as not to violate the weakly increasing condition along the rows (once the ones are subtracted). Continue this process until the induction rule fails. At this point we must have stopped at the end of some column, say column c. It is easy to see that after subtracting one from the elements in H, the array remains a reverse plane partition and the amount subtracted is $h_{a,c}$.

As an example, let

$$T = \begin{matrix} 1 & 2 & 2 & 2 \\ 3 & 3 & 3 & \\ 3 & & & \end{matrix} .$$

Then the skew hook H consists of the diamonds in the shape

$$\begin{matrix} \square & \diamond & \diamond & \diamond \\ \diamond & \diamond & \square & \\ \diamond & & & \end{matrix} .$$

After subtraction, we have

$$T_1 = \begin{matrix} 1 & 1 & 1 & 1 \\ 2 & 2 & 3 & \\ 2 & & & \end{matrix} .$$

To obtain the rest of the T_k, we iterate this process. The complete list for our example array, together with the corresponding $h_{i,j}$, is

$$\begin{array}{lllllll} T_k: & 1\,2\,2\,2\,, & 1\,1\,1\,1\,, & 0\,0\,0\,0\,, & 0\,0\,0\,0\,, & 0\,0\,0\,0\,, & 0\,0\,0\,0\,. \\ & 3\,3\,3 & 2\,2\,3 & 1\,2\,3 & 1\,2\,2 & 1\,1\,1 & 0\,0\,0 \\ & 3 & 2 & 1 & 1 & 1 & 0 \\ & & & & & & \\ h_{i,j}: & h_{1,1} & h_{1,1} & h_{2,3} & h_{2,2} & h_{2,1}\,. & \end{array}$$

Hence $\nu = (h_{1,1}, h_{1,1}, h_{2,3}, h_{2,2}, h_{2,1})$.

$\underline{\nu \rightarrow T.}$ Now given a partition of hooklengths ν we must rebuild T. First, however, we must know in what order the hooklengths were removed. It is easy to see that if $h_{i,j}, h_{i',j'} \in \nu$ then $h_{i,j}$ was removed before $h_{i',j'}$ if an only if $i < i'$, or $i = i'$ and $j \geq j'$. Once this is established reversing the subtraction process is straight-forward. For details the reader can consult [13] □

We can use Theorem 6.1.1 to derive the hook formula. It follows from general facts about poset partitions that if λ is a partition of n then

$$\sum_{m \geq 0} r_\lambda(m) x^m = \frac{p(x)}{\prod_{k=1}^{n}(1 - x^k)}$$

where $p(x)$ is a polynomial in x such that $p(1) = f^\lambda$. Combining this with equation (6) we obtain

$$p(x) = \frac{\prod_{k=1}^{n}(1 - x^k)}{\prod_{(i,j)\in\lambda}(1 - x^{h_{i,j}})}.$$

Now taking the limit as x approaches 1 yields $f^\lambda = n!/\prod_{(i,j)\in\lambda} h_{i,j}$.

We saw in Section 4.1.2 that the Robinson-Schensted map and the jeu de taquin are equivalent (Theorems 4.1.5 and 4.1.6). Kadell [16] has shown that the Hillman-Grassl algorithm is just another form of the jeu de taquin. Hence all three constructs are really the same. (Gansner [6] also noted this for the special case of rectangular arrays in his thesis.)

Before discussing the analog of the determinantal formula, we must talk briefly about the theory of symmetric functions. A *symmetric function* in the variables $\mathbf{x} = (x_1, x_2, \cdots, x_n)$ is a polynomial $f(\mathbf{x})$ with coefficients in $\mathbf{C}$ which is invariant under permutation of variables, i.e., for all $\pi \in S_n$ we must have $f(x_{\pi(1)}, x_{\pi(2)}, \cdots, x_{\pi(n)}) = f(x_1, x_2, \cdots, x_n)$. The set of all symmetric functions in n variables forms an algebra denoted Λ_n.

There are several well-known bases for Λ_n. The obvious one consists of the polynomials obtained by symmetrizing a given monomial. Specifically the *monomial symmetric function* corresponding to a partition $\lambda = (\lambda_1, \lambda_2, \cdots, \lambda_n)$ (where we permit λ to have parts equal to 0) is the polynomial

$$m_\lambda(\mathbf{x}) = \sum_{\pi \in S_n} x_{\pi(1)}^{\lambda_1} x_{\pi(2)}^{\lambda_2} \cdots x_{\pi(n)}^{\lambda_n}.$$

For example

$$m_{(2,1)}(x_1, x_2, x_3) = x_1^2 x_2 + x_1 x_2^2 + x_1^2 x_3 + x_1 x_3^2 + x_2^2 x_3 + x_2 x_3^2.$$

The three other important families of symmetric functions are as follows.

(1) The kth *elementary symmetric function* defined by

$$e_k(\mathbf{x}) = m_{(1^k)}(\mathbf{x}) = \sum_{i_1 < \cdots < i_k} x_{i_1} \cdots x_{i_k}.$$

(2) The kth *power sum symmetric function* defined by

$$p_k(\mathbf{x}) = m_{(k)}(\mathbf{x}) = \sum_{i \geq 1} x_i^k.$$

(3) The kth *complete homogeneous symmetric function* defined by

$$h_k(\mathbf{x}) = \sum_{\lambda \vdash k} m_\lambda(\mathbf{x}) = \sum_{i_1 \leq \cdots \leq i_k} x_{i_1} \cdots x_{i_k}.$$

By way of illustration, when $k = 2$ and $n = 3$:

$$\begin{aligned} e_2(x_1, x_2, x_3) &= x_1 x_2 + x_1 x_3 + x_2 x_3, \\ p_2(x_1, x_2, x_3) &= x_1^2 + x_2^2 + x_3^2, \\ h_2(x_1, x_2, x_3) &= x_1^2 + x_2^2 + x_3^2 + x_1 x_2 + x_1 x_3 + x_2 x_3. \end{aligned}$$

We can extend these definitions to partitions λ by letting

$$\begin{aligned} e_\lambda(\mathbf{x}) &= e_{\lambda_1}(\mathbf{x}) e_{\lambda_2}(\mathbf{x}) \cdots e_{\lambda_n}(\mathbf{x}), \\ p_\lambda(\mathbf{x}) &= p_{\lambda_1}(\mathbf{x}) p_{\lambda_2}(\mathbf{x}) \cdots p_{\lambda_n}(\mathbf{x}), \text{ and} \\ h_\lambda(\mathbf{x}) &= h_{\lambda_1}(\mathbf{x}) h_{\lambda_2}(\mathbf{x}) \cdots h_{\lambda_n}(\mathbf{x}). \end{aligned}$$

In the next result, the *length* of a partition will be the number of *non-zero* parts.

THEOREM 6.1.2. *The following sets are all bases for* Λ_n

(1) $\{m_\lambda \mid l(\lambda) \leq n\}$,

(2) $\{e_\lambda \mid l(\lambda) \leq n\}$,

(3) $\{p_\lambda \mid l(\lambda) \leq n\}$, *and*

(4) $\{h_\lambda \mid l(\lambda) \leq n\}$. □

The proof of Theorem 6.1.2 can be found in any book on symmetric functions, e.g., [20].

The Jacobi-Trudi identity is the Schur function analog of the determinantal formula.

THEOREM 6.1.3. *If* $\lambda = (\lambda_1, \lambda_2, \cdots, \lambda_n)$ *then*

$$s_\lambda(\mathbf{x}) = \det[h_{\lambda_i - i + j}(\mathbf{x})]$$

where the determinant is $n \times n$. □

Both the proofs that we gave of the determinantal formula can be generalized to prove this Theorem. In particular, weighting the lattice paths of Gessel and Viennot appropriately results in a combinatorial proof.

The Jacobi-Trudi formula also has a dual version using elementary symmetric functions.

THEOREM 6.1.4. *If* $\lambda = (\lambda_1, \lambda_2, \cdots, \lambda_n)$ *then*

$$s_{\lambda^t}(\mathbf{x}) = \det[e_{\lambda_i - i + j}(\mathbf{x})]_{n \times n}$$

where λ^t *is the transpose of the shape* λ *(also called the conjugate)* □

6.2. The shifted case. A *shifted reverse plane partition* is defined in exactly the same way as an ordinary one, only using a shifted shape. Let $q_\lambda(m)$ be the number of shifted reverse plane partitions of m having shape λ^*. Then we have the following analog of Theorem 6.1.1

THEOREM 6.2.1.

$$\sum_{m \geq 0} q_\lambda(m) x^m = \prod_{(i,j) \in \lambda^*} \frac{1}{1 - x^{h^*_{i,j}}}. \quad \square$$

Gansner [6] was the first to prove Theorem 6.2.1. He used generating function manipulations to obtain the shifted result from facts about symmetric (left-justified) reverse plane partitions. Later, Sagan [29] gave a bijective proof based on the Hillman-Grassl algorithm. He also showed that similar techniques yield many other product generating function identities.

6.3. The symplectic case. A symplectic analog of the Jacobi-Trudi identity can be derived from the Weyl character formula, a deep result in the representation theory of Lie groups. In what follows, we will use the definition

$$\tilde{h}_k(\mathbf{x}^{\pm 1}) \stackrel{\text{def}}{=} h_k(x_1, x_1^{-1}, x_2, x_2^{-1}, \cdots, x_n, x_n^{-1}).$$

THEOREM 6.3.1. *If* $\lambda = (\lambda_1, \lambda_2, \cdots, \lambda_n)$ *then*

$$sp_\lambda(\mathbf{x}^{\pm 1}) = \frac{1}{2}\det[\tilde{h}_{\lambda_i - i + j}(\mathbf{x}^{\pm 1}) + \tilde{h}_{\lambda_i - i - j + 2}(\mathbf{x}^{\pm 1})]. \quad \square$$

7. Characters of representations.

7.1. Ordinary characters. Let V be an n-dimensional G-module. By picking a basis for V we can view the corresponding representation ρ as a homomorphism from G to GL_n, the group of all $n \times n$ matrices over $\mathbf{C}$. This viewpoint will be useful in our discussion of group characters.

If $\rho : G \to GL_n$ is a representation then its *character* is the map $\chi : G \to \mathbf{C}$ defined by $\chi(g) = tr\ \rho(g)$ for all $g \in G$. If V is a G-module for ρ we say that V *affords* χ. Since the trace function is invariant under change of basis, $\chi(g)$ is well-defined. Furthermore, if g_1 and g_2 are conjugate in G, then $g_1 = hg_2h^{-1}$ for some $h \in G$ and so $\rho(g_1) = \rho(h)\rho(g_2)\rho(h)^{-1}$. Thus $\chi(g_1) = \chi(g_2)$ since similar matrices have the same trace. This means that χ is a *class function*, i.e., a function constant on conjugacy classes of G.

Let us look at a some examples. If $\rho : S_n \to GL_n$ is the defining representation, then $\rho(\pi)$ is the permutation matrix of $\pi \in S_n$. Thus $\chi(\pi)$ is just the number of fixed-points of π. Now let $G = \{g_1, g_2, \cdots, g_n\}$ be any group and let ρ be the regular representation which has the group algebra $\mathbf{C}(G)$ as module. Thus for any $g \in G$, $\chi(g)$ is the number of fixed-points of g acting on the basis $\vec{g}_1, \vec{g}_2, \cdots, \vec{g}_n$. It follows that

$$\chi(g) = \begin{cases} n & \text{if } g = e \\ 0 & \text{if } g \neq e \end{cases}$$

where e is the identity element of G. Finally note that for any G-module V, $\rho(e)$ is the identity matrix so $\chi(e) = \dim V$.

Recall that the irreducible representations of S_n are given by the Specht modules S^λ for $\lambda \vdash n$. If χ^λ is the corresponding character, then we know that $\chi^\lambda(e) = f^\lambda$. To describe the rest of the character values, we will use the notation χ^λ_μ for the value of χ^λ on the class of permutations of cycle-type $\mu \vdash n$.

THEOREM 7.1.1. *If* $\lambda \vdash n$ *then*

$$s_\lambda(\mathbf{x}) = \frac{1}{n!}\sum_{\mu \vdash n} c_\mu\ \chi^\lambda_\mu\ p_\lambda(\mathbf{x})$$

where c_μ *is the number of elements of* S_n *in the class* μ. $\square$

Hence the Schur function $s_\lambda(\mathbf{x})$ is just the cycle index generating function (in the sense of Pólya theory) for the character of the corresponding Specht module.

Now let us consider representations of matrix groups. If G is a group of matrices, then $\rho : G \rightarrow GL_n$ is called a *polynomial representation* if, for every $X \in G$, the entries of the matrix $\rho(X)$ are polynomials in the entries of X. As examples, the trivial representation is clearly polynomial. The identity map $id : GL_n \rightarrow GL_n$ is a polynomial representation, called the *defining representation.* Also the determinant $\det : GL_n \rightarrow GL_1$ is a representation which is polynomial. It follows from the work of Schur [33] that polynomial representations are very nice.

THEOREM 7.1.2. *Polynomial representations of GL_n are completely reducible.* □

Lest the reader get the idea that every representation is completely reducible, consider the representation of GL_n defined by

$$\rho(X) = \begin{bmatrix} 1 & \log|\det X| \\ 0 & 1 \end{bmatrix}$$

for all $X \in GL_n$. The x-axis is an invariant subspace, so if ρ were completely reducible it would have two decompose as the direct sum of two invariant one-dimensional subspaces. But this would mean that there would exist a fixed matrix Y such that

$$Y\rho(X)Y^{-1} = \begin{bmatrix} 1 & 0 \\ 0 & 1 \end{bmatrix}$$

for every $X \in GL_n$ (since both eigenvalues of $\rho(X)$ equal 1) which is absurd.

Now let $\rho : GL_n \rightarrow GL(V)$ be a polynomial representation with character χ. Let $X \in GL_n$ be diagonalizable with eigenvalues $\mathbf{x} = \{x_1, x_2, \cdots, x_n\}$ and consider $\mathrm{diag}(x_1, x_2, \cdots, x_n)$, the corresponding diagonal matrix . As χ is a class function, $\chi(X) = \chi(\mathrm{diag}(x_1, x_2, \cdots, x_n))$ which is a polynomial in $x_1, x_2, \cdots, x_n$. Since the diagonalizable matrices are dense in GL_n and ρ is continuous (being polynomial) it follows that $\chi(X)$ is a polynomial in the eigenvalues of X for any $X \in GL_n$. Furthermore this polynomial must be a symmetric function of $x_1, x_2, \cdots, x_n$ (since permuting the elements in a diagonal matrix leaves one in the same conjugacy class). As an example, note that if ρ is the defining representation then its character is $\chi(X) = x_1 + x_2 + \cdots + x_n$.

It is natural to ask which symmetric functions give the characters of the irreducible polynomial GL_n-modules. Again, the Schur functions play a role.

THEOREM 7.1.3 ([33]). *The irreducible polynomial representations of GL_n are indexed by partitions λ of length at most n. If λ is such a partition with corresponding module V^λ then the character afforded by V^λ is*

$$\phi^\lambda(X) = s_\lambda(\mathbf{x})$$

where $\mathbf{x} = \{x_1, x_2, \cdots, x_n\}$ is the set of eigenvalues of X. □

If V is any GL_n-module, then so is the kth tensor power $V^{\otimes k}$ since we have the natural action

$$X(\vec{v}_1 \otimes \vec{v}_2 \otimes \cdots \otimes \vec{v}_k) \stackrel{\text{def}}{=} X\vec{v}_1 \otimes X\vec{v}_2 \otimes \cdots \otimes X\vec{v}_k.$$

If V affords the character χ and we denote the character of $V^{\otimes k}$ by $\chi^{\otimes k}$, then it is easy to see that $\chi^{\otimes k}(X) = (\chi(X))^k$ for all $X \in GL_n$. In particular, if V is the module for the defining representation then $\chi^{\otimes k}(X) = (x_1 + x_2 + \cdots + x_n)^k$ where the x_i are the eigenvalues of X. Clearly if V corresponds to a polynomial representation then so does $V^{\otimes k}$.

Suppose that V is a module for the defining representation of GL_n. Decomposing $V^{\otimes k}$ into irreducibles produces the following beautiful theorem.

THEOREM 7.1.4 ([33]). *If V is the defining module for GL_n then*

$$V^{\otimes k} \cong \bigoplus_{\substack{\lambda \vdash k \\ l(\lambda) \leq n}} m_\lambda V^\lambda$$

where $m_\lambda = f^\lambda$. □

Taking characters on both sides of Theorem 7.1.4, we immediately obtain

COROLLARY 7.1.5.

$$(x_1 + x_2 + \cdots + x_n)^k = \sum_{\substack{\lambda \vdash k \\ l(\lambda) \leq n}} f^\lambda s_\lambda(\mathbf{x}). \quad \square$$

7.2. Projective characters. The Schur Q-functions give information about the characters of projective representations of S_n. Recall that given a strict partition $\lambda^* \models n$ of length l, there is a single irreducible projective S_n-module S_0^λ when $n - l$ is even and two, S_1^λ and S_{-1}^λ, when $n - l$ is odd. Let ζ_i^λ for $i = 0, \pm 1$ be the corresponding characters. It turns out that these characters are only non-zero on two families of partitions $\mu = (\mu_1, \mu_2, \cdots, \mu_m)$: those where the μ_i are all odd, and those where μ is a strict partition with an odd number of even parts.

The Schur Q-functions only give information about the values of ζ_i^λ on partitions from the first family, but there is an explicit formula, rather than a generating function, for their values on the second (see [Mor 77] for details). If μ has only odd parts, the values of all three characters are the same and we will denote this common value by ζ_μ^λ.

THEOREM 7.2.1 ([34]). *If $\lambda^* \models n$ then*

$$Q_\lambda(\mathbf{x}) = \frac{1}{n!} \sum_{\mu \vdash n} 2^{\lceil \frac{l(\lambda)+l(\mu)}{2} \rceil} c_\mu \, \zeta_\mu^\lambda \, p_\mu(\mathbf{x})$$

where the sum is over all partitions μ with only odd parts, and $\lceil \cdot \rceil$ is the round-up or ceiling function. □

Corollary 7.1.5 also has a projective analog.

THEOREM 7.2.2.

$$(x_1 + x_2 + \cdots x_n)^k = \sum_{\substack{\lambda^* \models k \\ l(\lambda^*) \le n}} 2^{-l(\lambda^*)} \, g^\lambda \, Q_\lambda(\mathbf{x}). \quad \square$$

7.3. Symplectic characters. Let V be a $2n$-dimensional vector space over $\mathbf{C}$ equipped with a non-degenerate skew-symmetric bilinear form $\langle \cdot, \cdot \rangle$. The *symplectic group*, $Sp_{2n} = Sp(V)$, is the subgroup of GL_{2n} that preserves the bilinear form, i.e.,

$$Sp_{2n} = \{X \in GL(V) \mid \langle X\vec{v}, X\vec{w} \rangle = \langle \vec{v}, \vec{w} \rangle \text{ for all } \vec{v}, \vec{w} \in V\}.$$

Polynomial representations and characters are defined as for GL_n. Furthermore, all polynomial representations of Sp_{2n} are completely reducible. Since $X \in Sp_{2n}$ stabilizes a skew-symmetric form, its set of eigenvalues must be of the form $\mathbf{x}^{\pm 1} = \{x_1, x_1^{-1}, x_2, x_2^{-1}, \cdots, x_n, x_n^{-1}\}$. This motivates the symplectic analog of Theorem 7.1.3.

THEOREM 7.3.1. *The irreducible polynomial representations of Sp_{2n} are indexed by partitions λ of length at most n. If λ is such a partition with corresponding module $\tilde{V}^\lambda$ then the character afforded by $\tilde{V}^\lambda$ is*

$$\tilde{\phi}^\lambda(X) = sp_\lambda(\mathbf{x}^{\pm 1}) \,,$$

where $\mathbf{x}^{\pm 1} = \{x_1, x_1^{-1}, x_2, x_2^{-1}, \cdots, x_n, x_n^{-1}\}$ is the set of eigenvalues of X. $\square$

Now consider $Sp(V)$ and it's defining module V. We can take tensor powers as before and study the decomposition into irreducibles. In what follows, $\tilde{f}_k^\lambda(n)$ is the number of OYT $(\phi = \lambda^0, \lambda^1, \cdots, \lambda^k)$ such that $l(\lambda^i) \le n$ for all $i = 1, 2, \cdots, k$.

THEOREM 7.3.2. *If V is the defining module for Sp_{2n} then*

$$V^{\otimes k} \cong \bigoplus_{\substack{\lambda \vdash k \\ l(\lambda) \le n}} m_\lambda \tilde{V}^\lambda$$

where $m_\lambda = \tilde{f}_k^\lambda(n)$. $\square$

Taking characters on both sides above, we obtain:

COROLLARY 7.3.3.

$$(x_1 + x_1^{-1} + x_2 + x_2^{-1} + \cdots + x_n + x_n^{-1})^k = \sum_{\substack{\lambda \vdash k \\ l(\lambda) \le n}} \tilde{f}_k^\lambda(n) sp_\lambda(\mathbf{x}^{\pm 1}). \quad \square$$

Our discussion of symplectic representations has been rather cursory. For more details, see the paper of Sundaram, *Tableaux in representation theory of the classical Lie groups*, elsewhere in this volume.

8. The Knuth correspondence.

We now generalize the Robinson-Schensted map to give combinatorial proofs of Corollary 7.1.5, Theorem 7.2.2, and Corollary 7.3.3. These proofs are due to Knuth [17], Sagan-Worley [30,46] and Berele [1] respectively.

8.1. Left-justified tableaux.

The Robinson-Schensted correspondence is a map between permutations and SYT. To obtain the analog for GYT (which have repeated entries), we will have to introduce permutations with repetitions. A *generalized permutation* is a two line array

$$\pi = \begin{matrix} k_1 & k_2 & \cdots & k_m \\ l_1 & l_2 & \cdots & l_m \end{matrix}$$

which is in lexicographic order where the top line takes precedence. For example,

$$\pi = \begin{matrix} 1 & 1 & 1 & 2 & 2 & 3 \\ 2 & 3 & 3 & 1 & 2 & 1 \end{matrix}.$$

We say that π is a *generalized permutation of n* if every element in π is less than or equal to n. We let GP_n stand for the set of all generalized permutations of n. Associated with each $\pi \in GP_n$ is a pair of contents $(1^{b_1}, 2^{b_2}, \cdots, n^{b_n})$ and $(1^{t_1}, 2^{t_2}, \cdots, n^{t_n})$ where b_i (respectively t_i) is the number of occurrences of i in the bottom (respectively top) line of π. The example above has contents $(1^2, 2^2, 3^2)$ and $(1^3, 2^2, 3^1)$. Introducing a new set of variables $\mathbf{y} = \{y_1, y_2, \cdots, y_n\}$ we can define the *monomial of π* to be

$$m(\pi) = x_1^{b_1} x_2^{b_2} \cdots x_n^{b_n} y_1^{t_1} y_2^{t_2} \cdots y_n^{t_n}.$$

Our example has monomial $m(\pi) = x_1^2 x_2^2 x_3^2 y_1^3 y_2^2 y_3^1$.

We claim that the generating function for generalized permutations of n is

$$\sum_{\pi \in GP_n} m(\pi) = \prod_{i,j=1}^{n} \frac{1}{1 - x_i y_j}.$$

To see this, note that

$$\frac{1}{1 - x_i y_j} = 1 + x_i y_j + x_i^2 y_j^2 + \cdots + x_i^k y_j^k + \cdots.$$

Thus the term $x_i^k y_j^k$ corresponds to having the column $\binom{i}{j}$ repeated k times in π.

We should note that there is a bijection between generalized permutations of n and $n \times n$ matrices with non-negative integral entries. This is because having $\binom{i}{j}$ repeated k times in π is equivalent to having the (i, j) entry of the matrix equal to k. Knuth's original proof of Theorem 8.1.1 below was stated in terms of matrices and generalized Young tableaux.

In Schensted's paper [32], he gave what amounts to a combinatorial proof of Corollary 7.1.5 (although Schur functions were never mentioned explicitly). If remained for Knuth [17] to give a combinatorial proof of Cauchy's identity, which is a generalization of this corollary, and to make the connection with $s_\lambda(\mathbf{x})$.

THEOREM 8.1.1.

$$\prod_{i,j=1}^{n} \frac{1}{1-x_i y_j} = \sum_{l(\lambda)\leq n} s_\lambda(\mathbf{x}) s_\lambda(\mathbf{y}) \ .$$

Proof. We wish to find a bijection

$$\pi \longleftrightarrow (T, U)$$

between generalized permutations $\pi \in GP_n$ and pairs of GYT T, U of the same shape such that the content of T (respectively U) equals the content of the lower (respectively upper) line of π. For the forward direction we form, as before, a sequence of tableaux pairs

$$(T_0, U_0) = (\phi, \phi);\ (T_1, U_1);\ (T_2, U_2);\ \cdots;\ (T_m, U_m) = (T, U)$$

where the elements of the bottom line of π are inserted into the T's and the elements of the top line are placed in the U's. Furthermore, the rules of insertion and placement are exactly the same.

Applying this algorithm to the permutation above, we obtain

$$\begin{array}{lccccccccl} & \phi, & 2, & 2\,3, & 2\,3\,3, & 1\,3\,3, & 1\,2\,3, & 1\,1\,3 & \\ T_i: & & & & & 2 & 2\,3 & 2\,2 & = T \\ & & & & & & & 3 & \end{array}$$

$$\begin{array}{lccccccccl} & \phi, & 1, & 1\,1, & 1\,1\,1, & 1\,1\,1, & 1\,1\,1, & 1\,1\,1 & \\ U_i: & & & & & 2 & 2\,2 & 2\,2 & = U. \\ & & & & & & & 3 & \end{array}$$

It is easy to verify that the insertion rules make sure that T is a GYT and that U always has weakly increasing rows. To verify the column condition for U, we must make sure that no two equal elements of the top row of π can end up in the same column. But if $k_i = k_{i+1} = k$ in the upper row, then by the lexicographic condition on π we must have $l_i \leq l_{i+1}$. This implies that the insertion path of l_{i+1} will always lie strictly to the right of the path for l_i which gives the desired result. Note that we have shown that all elements equal to k are placed in U from left to right as the algorithm proceeds.

For the inverse map, the only problem is deciding which of the maximum elements of U corresponds to the last insertion. But from the observation just made, the right-most of these maxima is the correct choice to start the deletion process. □

Many of the properties of the Robinson-Schensted algorithm also hold for Knuth's generalization. For details, the reader can consult [6].

8.2. Shifted tableaux. The analog of Cauchy's identity that corresponds to Theorem 7.2.2 is:

THEOREM 8.2.1.

$$\prod_{i,j=1}^{n} \frac{1+x_i y_j}{1-x_i y_j} = \sum_{l(\lambda^*)\leq n} 2^{-l(\lambda^*)} Q_\lambda(\mathbf{x}) Q_\lambda(\mathbf{y}) \,.$$

Proof sketch. The left-hand side of this equation counts generalized permutations π where, if a column $\binom{i}{j}$ appears in π, then it's first occurrence can be distinguished by being putting a prime on the j. (Picking the '1' or the '$x_i y_i$' in the numerator's $1+x_i y_j$ corresponds respectively to not distinguishing or distinguishing the j).

To see what the right side counts, remember that primed and unprimed versions of the same integer lie in skew hooks and that the nature of every element of a given hook is completely determined except at it's lower-left end. But $l(\lambda^*)$ skew hooks have their lower-left end on the main diagonal, so $2^{-l(\lambda^*)} Q_\lambda(\mathbf{y})$ counts GST with primes only on off-diagonal elements.

Thus it suffices to find a bijection

$$\pi^* \longleftrightarrow (T^*, U^*)$$

where π^* is a primed generalized permutation of n and T^*, U^* are GST of the same shape such that U only has off-diagonal primes and the content of the lower (respectively upper) line of π^* equals the content of T^* (respectively U^*). Primes are ignored when taking contents. Details of the bijection can be found in [30] or [46]. □

8.3. Symplectic tableaux. Berele's algorithm [1] was constructed to give a combinatorial proof of Corollary 7.3.3, which we restate here.

THEOREM 8.3.1.

$$(x_1 + x_1^{-1} + x_2 + x_2^{-1} + \cdots + x_n + x_n^{-1})^k = \sum_{\substack{\lambda \vdash k \\ l(\lambda)\leq n}} \tilde{f}_k^\lambda(n) sp_\lambda(\mathbf{x}^{\pm 1}).$$

Proof. The left-hand side counts permutations $\tilde{\pi}$ of length k over the alphabet $\{1 < \bar{1} < 2 < \bar{2} < \cdots < n < \bar{n}\}$ (possibly with repetitions in the lower line, but none in the upper). As expected, $\tilde{\pi}$ has monomial

$$m(\tilde{\pi}) = x_1^{m_1} x_1^{-\bar{m}_1} x_2^{m_2} x_2^{-\bar{m}_2} \cdots x_n^{m_n} x_n^{-\bar{m}_n}$$

with m_k (respectively $\bar{m}_k$) being the number of k's (respectively $\bar{k}$'s) in $\tilde{\pi}$. Thus we want to give a bijection

$$\tilde{\pi} \longleftrightarrow (\tilde{T}, \tilde{P}_k^\lambda)$$

where $\tilde{\pi}$ is as described above; $\tilde{T}, \tilde{P}_k^\lambda$ are an SPT and an OYT respectively having the same shape λ; and $m(\tilde{\pi}) = m(\tilde{T})$.

In the forward direction, we create a sequence

$$(\tilde{T}_0, \lambda^0) = (\phi, \phi);\ (\tilde{T}_1, \lambda^1);\ (\tilde{T}_2, \lambda^2); \cdots ; (\tilde{T}_k, \lambda^k)$$

so that at the end we can let $\tilde{T} = \tilde{T}_k$ and $\tilde{P}_k^\lambda = (\lambda^0, \lambda^1, \lambda^2, \cdots, \lambda^k)$. We will also build the pairs so that the shapes of $\tilde{T}_i$ and λ^i are the same for all $i = 0, 1, 2, \cdots, k$.

Suppose that $(\tilde{T}_{r-1}, \lambda^{r-1})$ has been constructed and $\tilde{\pi} = l_1\ l_2\ \cdots\ l_k$. We start row inserting l_r into $\tilde{T}_{r-1}$ as usual. If the symplectic condition (equation (5)) is never violated during the insertion, then we let $\tilde{T}_r = R_{l_r}(\tilde{T}_{r-1})$ and add a box to λ_{r-1} to mark the location of termination.

Suppose, on the other hand, that a violation of equation (5) is about to occur at some point of the insertion. It is easy to see that this could only happen if an i added to row i is trying to bump an $\bar{\imath}$ into row $i+1$. In this case the i and the $\bar{\imath}$ annihilate each other, creating a (black) hole in cell (i,j) for some j. But now the hole can be filled using an (i,j)-slide, resulting in the new tableau $\tilde{T}_r$ which has one less box. In this case we delete the corresponding box of λ^{r-1} to form λ^r. Note that this cancellation of an i and an $\bar{\imath}$ corresponds to the cancellation of an x_i with an x_i^{-1}.

By way of illustration, we consider the insertion of $\bar{\imath}$ into the SPT below:

$$\begin{matrix} 1 & \bar{1} & 2 & \bar{2} \\ 2 & \bar{2} & 3 & \\ 3 & 3 & \bar{3} & \end{matrix} \leftarrow \bar{1}, \quad \begin{matrix} 1 & \bar{1} & \bar{1} & \bar{2} \\ 2 & \bar{2} & 3 & \\ 3 & 3 & \bar{3} & \end{matrix} \leftarrow 2 \quad ,$$

$$\begin{matrix} 1 & \bar{1} & \bar{1} & \bar{2} \\ 2 & \bullet & 3 & \\ 3 & 3 & \bar{3} & \end{matrix}, \quad \begin{matrix} 1 & \bar{1} & \bar{1} & \bar{2} \\ 2 & 3 & 3 & \\ 3 & \bullet & \bar{3} & \end{matrix}, \quad \begin{matrix} 1 & \bar{1} & \bar{1} & \bar{2} \\ 2 & 3 & 3 & \\ 3 & \bar{3} & & \end{matrix}.$$

For a look at the whole algorithm, let's compute the image of $\tilde{\pi} = 2\bar{2}\bar{1}211$:

$$\tilde{T}_r:\ \phi\ ,\ 2\ ,\ 2\ \bar{2}\ ,\ \begin{matrix} \bar{1} & \bar{2} \\ 2 & \end{matrix}\ ,\ \begin{matrix} \bar{1} & 2 \\ 2 & \bar{2} \end{matrix}\ ,\ \begin{matrix} 2 & 2 \\ \bar{2} & \end{matrix}\ ,\ 1\ 2$$

$$\lambda^r:\ \phi\ ,\ \square\ ,\ \square\,\square\ ,\ \begin{matrix} \square & \square \\ \square & \end{matrix}\ ,\ \begin{matrix} \square & \square \\ \square & \square \end{matrix}\ ,\ \begin{matrix} \square & \square \\ \square & \end{matrix}\ ,\ \square\,\square.$$

Thus $\tilde{T} = 1\ 2$ and

$$\tilde{P}_k^\lambda = \phi\ ,\ \square\ ,\ \square\,\square\ ,\ \begin{matrix} \square & \square \\ \square & \end{matrix}\ ,\ \begin{matrix} \square & \square \\ \square & \square \end{matrix}\ ,\ \begin{matrix} \square & \square \\ \square & \end{matrix}\ ,\ \square\,\square\ .$$

We leave it as an exercise to the reader to construct the inverse map. □

9. Open questions. Now that the reader has gained some familiarity with tableaux and their relation with representations and symmetric functions, it seems appropriate to propose some outstanding problems using these ideas.

1. Shifted analogs. We have seen (Theorem 4.1.4) that reversing a permutation transposes its P-tableau. What effect does this have on the P^* tableau of the

shifted Robinson-Schensted map? More generally, what does it mean to 'transpose' a shifted tableau? There are a host of other problems concerning shifted analogs of known results in the left-justified case. The reader can consult [30] for further information.

2. Restricted partitions. There are many beautiful product generating functions for various families of tableaux. For example, we have the following theorem of MacMahon:

THEOREM 9.0.2 ([22]). *Fix positive integers k, l and m and let λ be the rectangular partition (k^l) . Then the generating function for reverse plane partitions with at most k rows, at most l columns (i.e., having shape contained in λ), and with largest part at most m is*

$$\prod_{(i,j)\in\lambda} \frac{1-x^{h_{i,j}+m}}{1-x^{h_{i,j}}}. \quad \square$$

It would be nice to have a combinatorial proof of this theorem, perhaps using a Hillman-Grassl type bijection. The paper [39] of Stanley is a good source for problems of this type. His survey article [37] is also very informative.

3. Projective modules. Since the dimension of the Specht module S^λ is just the number of SYT of shape λ, it is desirable to have a basis constructed out of these tableaux. This can be done using *tabloids* which are equivalence classes of tableaux (two tableaux are equivalent if corresponding rows contain the same set of elements; here, rows and columns need not increase). The symmetric group acts on tabloids in a natural way, and from these permutation modules one can construct the irreducibles. See [14] for details.

Only recently have the matrices for the irreducible projective representations been constructed by Nazarov [24]. Can one find a way to use an S_n action on shifted tableaux to accomplish the same task?

4. Hall-Littlewood polynomials. Both the normal Schur functions and the Schur Q-functions are special cases of the Hall-Littlewood polynomials, $Q_\lambda(\mathbf{x};t)$. These polynomials are symmetric in the variables $\mathbf{x}$ with an additional parameter, t. When $t=0$ or -1 they specialize to $s_\lambda(\mathbf{x})$ or $Q_\lambda(\mathbf{x})$ respectively. More information about these functions can be found in Macdonald's book [20] or in the survey article of Morris [23].

The $Q_\lambda(\mathbf{x};t)$ satisfy the identity

$$\prod_{i,j=1}^{n} \frac{1-tx_iy_j}{1-x_iy_j} = \sum_{l(\lambda^*)\leq n} \frac{1}{b_\lambda(t)} Q_\lambda(\mathbf{x};t)Q_\lambda(\mathbf{y};t)$$

where $b_\lambda(t)$ is a polynomial in t. This generalizes both Cauchy's identity (Theorem 8.1.1) and Theorem 8.2.1. Perhaps it is possible to find a Robinson-Schensted-Knuth type map to give a combinatorial proof of this result. Thus the left-justified and shifted correspondences would be combined into one.

REFERENCES

[1] A. BERELE, *A Schensted-type correspondence for the symplectic group*, J. Combin. Theory, Ser. A, 43 (1986), pp. 320–328.

[2] J. DÉSARMÉNIEN, J. P. S. KUNG AND G.-C. ROTA, *Invariant theory, Young bitableaux, and Combinatorics*, Advances in Math., 27 (1978), pp. 63–92.

[3] P. DOUBILET, G.-C. ROTA AND J. STEIN, *On the foundations of combinatorial theory: IX. Combinatorial methods in invariant theory*, Studies in Applied Math., 53 (1974), pp. 185–216.

[4] J. S. FRAME, G. DE B. ROBINSON AND R. M. THRALL, *The hooklengths of S_n*, Canad. J. Math., 6 (1954), pp. 316–325.

[5] D. S. FRANZBLAU AND D. ZEILBERGER, *A bijective proof of the hook-length formula*, J. of Algorithms, 3 (1982), pp. 317–343.

[6] E. GANSNER, *Matrix correspondences and the enumeration of plane partitions*, Ph.D. thesis, M.I.T., 1978.

[7] A. GARSIA AND S. MILNE, *A Rogers-Ramanujan bijection*, J. Combin. Theory, Ser. A, 31 (1981), pp. 289–339.

[8] I. GESSEL AND G. VIENNOT, *Binomial determinants, paths, and hooklength formulae*, Adv. in Math., 58 (1985), pp. 300–321.

[9] ———, *Determinants, paths, and plane partitions*, in preparation.

[10] C. GREENE, *An extension of Schensted's theorem*, Advances in Math., 14 (1974), pp. 254–265.

[11] C. GREENE, A. NIJENHUIS AND H. S. WILF, *A probabilistic proof of a formula for the number of Young tableaux of a given shape*, Adv. in Math., 31 (1979), pp. 104–109.

[12] M. D. HAIMAN, *On mixed insertion, symmetry, and shifted Young tableaux*, preprint.

[13] A. P. HILLMAN AND R. M. GRASSL, *Reverse plane partitions and tableau hook numbers*, J. Combin. Theory, Ser. A, 21 (1976), pp. 216–221.

[14] G. D. JAMES, *The representation theory of symmetric groups*, Lecture Notes in Math. No. 682, Springer-Verlag, New York, 1978.

[15] T. JÓZEFIAK, *Characters of projective representations of symmetric groups*, preprint.

[16] K. KADELL, *Schützenberger's "jeu de taquin" and plane partitions*, preprint.

[17] D. E. KNUTH, *Permutations, matrices and generalized Young tableaux*, Pacific J. Math., 34 (1970), pp. 709–727.

[18] D.E. KNUTH,, *The art of computer programming, vol. 3: Sorting and searching*, Addison–Wesley, Reading, 1973.

[19] W. LEDERMANN, *Introduction to group characters*, Cambridge University Press, Cambridge, 1977.

[20] I. G. MACDONALD, *Symmetric functions and Hall polynomials*, Oxford University Press, Oxford, 1979.

[21] T. J. MACLARNAN, *Tableau recursions and symmetric Schensted correspondences for ordinary, shifted and oscillating tableaux*, Ph.D. thesis, UCSD, 1986.

[22] P. A. MACMAHON, *Combinatorial Analysis, vols. 1 and 2,*, Cambridge University Press, Cambridge, 1915–1916; reprinted by Chelsea, New York.

[23] A. O. MORRIS, *A survey of Hall-Littlewood functions and their applications to representation theory*, in *Combinatoire et Représentation du Groupe Symétrique*, D. Foata ed., Lecture Notes in Math. No. 579, Springer-Verlag, New York, 1977, pp. 136–154.

[24] M. L. NAZAROV, *An orthogonal basis of irreducible projective representations of the symmetric group*, Functional Anal. Appl., 22 (1988), pp. 77–78.

[25] J. B. REMMEL, *Bijective proofs of formulae for the number of standard Young tableaux*, Linear and Multilin. Alg., 11 (1982), pp. 45–100.

[26] G. DE B. ROBINSON, *On the representations of the symmetric group*, Amer. J. Math., 60 (1938), pp. 745–760.

[27] B. E. SAGAN, *An analog of Schensted's algorithm for shifted Young tableaux*, J. Combin. Theory, Ser. A, 27 (1979), pp. 10–18.

[28] ———, *On selecting a random shifted Young tableau*, J. Algorithms, 1 (1980), pp. 213–234.

[29] ———, *Enumeration of partitions with hooklengths*, Europ. J. Combinatorics, 3 (1982), pp. 85–94.

[30] ———, *Shifted tableaux, Schur Q-functions and a conjecture of R. Stanley*, J. Combin. Theory, Ser. A, 45 (1987), pp. 62–103.

[31] ———, *The symmetric group: representations, combinatorics and symmetric functions*, in preparation, Wadsworth and Brooks/Cole,, Monterey.

[32] C. Schensted, *Longest increasing and decreasing subsequences*, Canad. J. Math., 13 (1961), pp. 179–191.

[33] I. Schur, *Über eine Klasse von Matrizen, die sich einer gegeben Matrix zuorden lassen,*, Dissertation, Berlin,1901, in *Gesammelte Abhandlungen*, Springer-Verlag, New York, 1973.

[34] ———, *Über die Darstellung der symmetrischen und der alterneinden Gruppe durch gebrochene lineare Substitutionen*, J. Reine Angew. Math, 139 (1911), pp. 155–250.

[35] M. P. Schützenberger, *Quelques remarques sur une construction de Schensted*, Math. Scand., 12 (1963), pp. 117–128.

[36] ———, *La correspondence de Robinson*, in *Combinatoire et Représentation du Groupe Symétrique,*, D. Foata ed., Lecture Notes in Math. No. 579, Springer-Verlag, New York, 1977, pp. 59–135.

[37] R. P. Stanley, *Theory and application of plane partitions I and II*, Stud. in Applied Math., 50 (1971), pp. 167–188 and 259–279.

[38] ———, *Ordered Structures and Partitions*, Memoirs of the A.M.S. No. 119, A.M.S., Providence, 1972.

[39] ———, *A baker's dozen of conjectures on plane partitions*, in *Combinatoire Enumerative*, G. Labelle and P. Leroux eds., Lecture Notes in Math. No. 1234,, Springer-Verlag, New York, 1986, pp. 285–296.

[40] J. R. Stembridge, *Shifted tableaux and the projective representations of symmetric groups*, preprint.

[41] S. Sundaram, *On the combinatorics of representations of $Sp(2n, \mathbf{C})$*, Ph.D. Thesis, M.I.T., 1978.

[42] G. P. Thomas, *On a construction of Schützenberger*, Discrete Math. 17 (1977), pp. 107–118.

[43] R. M. Thrall, *A combinatorial problem*, Michigan Math. J., 1 (1952), pp. 81–88.

[44] G. Viennot, *Une forme géométrique de la correspondance de Robinson-Schensted*, in *Combinatoire et Représentation du Groupe Symétrique*, D. Foata ed., Lecture Notes in Math. No. 579, Springer-Verlag, New York, 1977, pp. 29–58.

[45] ———, *Chain and antichain families grids and Young tableaux*, Annals of Discrete Math., 23 (1984), pp. 409–464.

[46] D. R. Worley, *A theory of Shifted Young tableaux*, Ph.D. thesis, M.I.T..

[47] A. Young, *Quantitative substitutional analysis I*, Proc. London Math. Soc., 33 (1901), pp. 97–146.

[48] D. Zeilberger, *A short hook-lengths bijection inspired by the Greene-Nijenhuis-Wilf proof*, Discrete Math., 51 (1984), pp. 101–108.